Understanding Advanced
Organic and Analytical Chemistry

The Learner's Approach

Revised Edition

Understanding Advanced Organic and Analytical Chemistry

The Learner's Approach

Kim Seng Chan

BSc (Hons), PhD, PGDE (Sec), MEd, MA (Ed Mgt),
MEd (G Ed), MEd (Dev Psy)

Jeanne Tan

BSc (Hons), PGDE (Sec), MEd (LST)

W⊜ WS Education

NEW JERSEY · LONDON · SINGAPORE · BEIJING · SHANGHAI · HONG KONG · TAIPEI · CHENNAI · TOKYO

Published by

WS Education, an imprint of
World Scientific Publishing Co. Pte. Ltd.
5 Toh Tuck Link, Singapore 596224
USA office: 27 Warren Street, Suite 401-402, Hackensack, NJ 07601
UK office: 57 Shelton Street, Covent Garden, London WC2H 9HE

British Library Cataloguing-in-Publication Data
A catalogue record for this book is available from the British Library.

UNDERSTANDING ADVANCED ORGANIC AND ANALYTICAL CHEMISTRY
The Learner's Approach
Revised Edition

Copyright © 2017 by World Scientific Publishing Co. Pte. Ltd.

For photocopying of material in this volume, please pay a copying fee through the Copyright Clearance Center, Inc., 222 Rosewood Drive, Danvers, MA 01923, USA. In this case permission to photocopy is not required from the publisher.

ISBN 978-981-4733-98-4 (pbk)

Typeset by Stallion Press
Email: enquiries@stallionpress.com

Printed in Singapore

Preface

The majority of the learning of organic chemistry in classrooms happens with the help of textbooks and lecture notes. Unfortunately, most textbooks are of the expository and non-refutational type, presenting facts rather than explaining them. In addition, the links between concepts are often not made explicit, presupposing that learners would be able to make the necessary integration with the physical chemistry concepts they have come across, forgetting that some learners lack the prior knowledge and metacognitive skills to do so. Hence, learners would at most be able to reproduce the information that is structured and organized by the textbook writer, but not able to construct a meaningful conceptual mental model for oneself.

This current book is a continuation of our first book, *Understanding Advanced Physical Inorganic Chemistry*, retaining the main refutational characteristics of the first book by strategically planting think-aloud questions to promote conceptual understanding, knowledge construction and discourse opportunities. It is hoped that these essential questions will make learners aware of the possible conflict between their prior knowledge, which may be counter intuitive or misleading with those presented in the text, and hence in the process, make the necessary conceptual changes. In essence, we are trying to effect metaconceptual awareness — awareness of the theoretical nature of one's thinking — while learners are trying to master organic chemistry concepts. We hope that by pointing out the differences between possible misconceptions and the actual chemistry content, we can promote metaconceptual awareness and thus assist the learner in constructing a meaningful conceptual model of understanding for organic chemistry. We want our learners to not only know what they know, but at the same time have a sense of how they know what they know, and how their new learning is interrelated within the discipline.

Lastly, the substance of this book would be both informative and challenging to the practices of teachers. This book would certainly illuminate the teaching of all chemistry teachers that strongly believe in teaching organic chemistry with a meaningful and integrative approach, from the learners' perspective. The integrated questions at the end of each chapter will certainly prove useful to students in helping them to revise fundamental concepts learnt from previous chapters and also to see the importance and relevance in the application to their current learning. Collectively, this book offers a vision of understanding organic chemistry meaningfully and fundamentally from the learners' approach and to fellow chemistry teachers, we hope that it will help you develop a greater insight into what makes you tick, explain, enthuse and develop in the course of your teaching.

This revised edition has been updated to meet the minimum requirements of the new Singapore GCE A level syllabus that would be implemented in the year 2016. Nevertheless, this book is also highly relevant to students who are studying chemistry for other examination boards. In addition, the authors have also included more Q&A to help students better understand and appreciate the chemical concepts that they are mastering.

Acknowledgements

We would like to express our sincere thanks to the staff at World Scientific Publishing Co. Pte. Ltd. for the care and attention which they have given to this book, particularly our editors Lim Sook Cheng and Sandhya Devi, our editorial assistant Chow Meng Wai and Stallion Press.

Special appreciation goes to Ms Ek Soo Ben (Principal of Victoria Junior College), Mrs Foo Chui Hoon, Mrs Toh Chin Ling and Mrs Ting Hsiao Shan for their unwavering support to Kim Seng Chan.

Special thanks go to all our students who have made our teaching of chemistry fruitful and interesting. We have learnt a lot from them just as they have learnt some good chemistry from us.

Finally, we thank our families for their wholehearted support and understanding throughout the period of writing this book. We would like to share with all the passionate learners of chemistry two important quotes from the Analects of Confucius:

學而時習之，不亦悅乎？(Isn't it a pleasure to learn and practice what is learned time and again?)

學而不思則罔，思而不學則殆 (Learning without thinking leads to confusion, thinking without learning results in wasted effort)

Contents

CHAPTER 1

Structure and Bonding

1.1 Introduction

Hearing the words "organic chemistry" may strike fear in some students, but this should not be the case at all. You may have come across complex-looking structures, alien-looking mechanisms and reactions. Exam questions about such may confuse you and heighten anxiety, but the answers to these questions need not be highly sophisticated, as these usually stem from the basic principles of organic chemistry that are covered in this book.

Organic chemistry is mainly the study of carbon-containing compounds, excluding those classified as inorganic compounds, such as carbonates and oxides.

It is commendable that the carbon element is able to form chains upon chains, resulting in millions of uniquely different compounds. This remarkable feat cannot be reproduced even by its closest group member, silicon, in the periodic table.

Although there are numerous organic compounds in existence, one can easily identify various specific combinations and arrangements of atoms that are responsible for the chemical behavior of these compounds. These atoms or group of atoms are known as *functional groups*, and we will set out to discuss their characteristic properties in this book.

In subsequent chapters, we will look into the properties of simple organic compounds grouped into homologous series according to the functional group present. Some of these are presented in Table 1.1.

Table 1.1 Common Functional Groups and Homologous Series.

Homologous Series	Functional Group Structure	Example	
Alkane	–		Ethane
Alkene	$\diagdown$C$=$C$\diagup$		Ethene
Alkyne	$-$C$\equiv$C$-$	H$-$C$\equiv$C$-$H	Ethyne
Arene			Benzene
Halogenoalkane	$-$C$-$X (X = halogen)	H$-$C$-$Br	Bromomethane
Alcohol	$-$C$-$OH	H$-$C$-$OH	Methanol
Ether	$-$C$-$O$-$C$-$	H$-$C$-$O$-$C$-$H	Dimethyl ether
Aldehyde	$-$C$-$H (with =O)	H$-$C$-$C$-$H	Ethanal
Ketone	$-$C$-$C$-$C$-$ (with =O)	H$-$C$-$C$-$C$-$H	Propanone
Carboxylic acid	$-$C$-$O$-$H (with =O)	H$-$C$-$C$-$O$-$H	Ethanoic acid

(*Continued*)

Table 1.1 (*Continued*)

Homologous Series	Functional Group Structure	Example	
Acyl chloride or acid chloride	O‖ —C—Cl	H—C—C—Cl (with H above and below left C, O above right C)	Ethanoyl chloride
Ester	O‖ —C—O—C—	H—C—C—C—O—C—H (with H atoms, O double bond)	Methyl propanoate
Acid anhydride	O‖ O‖ —C—O—C—	H—C—C—C—O—C—C—C—H (with H atoms, O double bonds)	Propanoic anhydride
Amine (Primary amine)	H —C—N—H	H H —C—N—H (H—C—N—H)	Methylamine
Amide (Primary amide)	O‖ H —C—N—H	H O H H—C—C—N—H	Ethanamide
Nitrile	—C—C≡N	H —C—C≡N (H—C—C≡N)	Ethanenitrile

Before we get into the organic chemistry proper, we need to be able to interpret what structures such as those in Table 1.1 represent. The naming of simple organic compounds will be covered under each chapter on the homologous series.

1.2 Constitutional/Structural Formulae of Organic Compounds

The molecular formula of a compound informs us of the *number* and *type* of atoms present in a molecule. But it may not give us information on how the various atoms are actually connected to one another. For instance, the molecular formula $C_4H_8O_2$ can represent more than one compound, and these include the following carboxylic acid and ester:

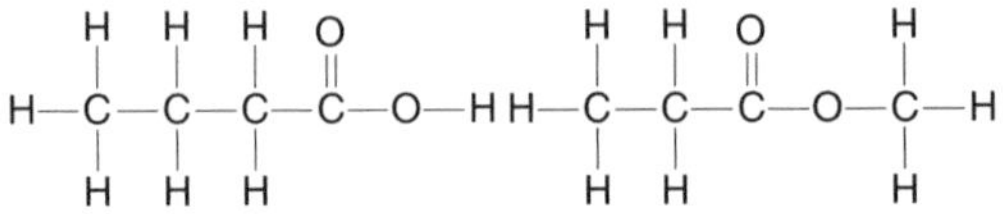

$$
\begin{array}{c}
\text{H} \quad \text{H} \quad \text{H} \quad \text{O} \\
\text{H--C--C--C--C--O--H} \\
\text{H} \quad \text{H} \quad \text{H}
\end{array}
\qquad
\begin{array}{c}
\text{H} \quad \text{H} \quad \text{O} \quad \text{H} \\
\text{H--C--C--C--O--C--H} \\
\text{H} \quad \text{H} \quad \text{H}
\end{array}
$$

Butanoic acid   Methyl propanoate

We need to rely on some visuals to depict the arrangement of atoms within a molecule or, in other words, the consitutional/structural formula of an organic compound. In a displayed formula, otherwise known as the full consitutional/structural formula, all the atoms and the bonds between them are shown. One of the displayed formulae for a carboxylic acid with the molecular formula C_3H_7COOH looks like this:

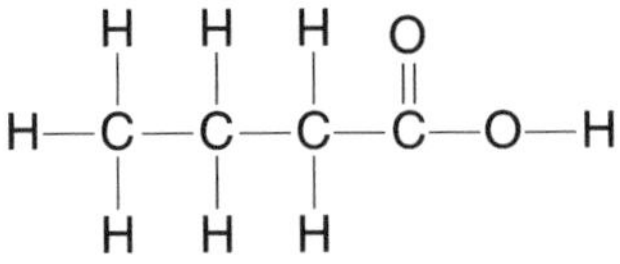

$$
\begin{array}{c}
\text{H} \quad \text{H} \quad \text{H} \quad \text{O} \\
\text{H--C--C--C--C--O--H} \\
\text{H} \quad \text{H} \quad \text{H}
\end{array}
$$

A skeletal formula is a simplified drawing in the sense that all $C-H$ bonds are not shown and each $C-C$ single bond is depicted as a line "$-$" with C atoms present, but not shown, at both ends of the line. If heteroatoms (such as O, N, Cl, etc.) are present, their elemental symbol is included to depict the bonding to the carbon skeletal structure. The skeletal formula for the same carboxylic acid, C_3H_7COOH, that was discussed above is shown as:

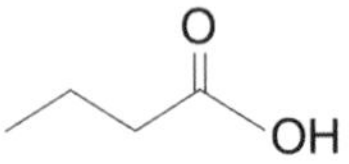

There are no hard and fast rules in drawing a consitutional/structural formula. Basically, a diagram drawn must not be ambiguous. For instance, the following condensed consitutional/structural formulae are valid in depicting the same carboxylic acid molecule:

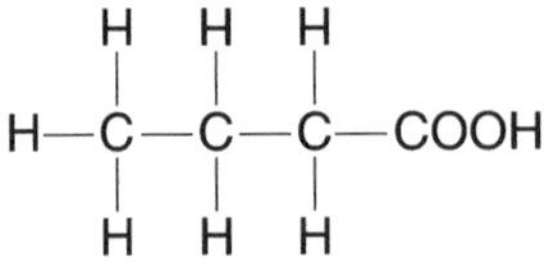

$$
\begin{array}{c}
\text{H} \quad \text{H} \quad \text{H} \\
\text{H--C--C--C--COOH} \\
\text{H} \quad \text{H} \quad \text{H}
\end{array}
\qquad
CH_3CH_2CH_2COOH
\qquad
CH_3(CH_2)_2COOH
$$

All these structures show the same sequential arrangement of atoms in the molecule when one reads from left to right. However, ambiguity arises when the formula for the acid is written as C_3H_7COOH because it can represent more than one unique compound. The ambiguity resides in the $-C_3H_7$ segment as it can be either one of the following two structures:

$$CH_3CH_2CH_2COOH \qquad\qquad (CH_3)_2CHCOOH$$

With all that said, the various types of consitutional/structural formulae mentioned so far only show the connectivity of one atom to another. They do not indicate the actual geometry or bond angles in the molecule. To depict the three-dimensional (3-D) geometry about a particular atom, we make use of the 3-D structural formula which utilizes the following conventions:

Solid line represents a bond on the plane of the paper

Dashed wedge represents a bond protruding below the plane of the paper

Solid wedge represents a bond protruding above the plane of the paper

Q: How can we tell the number of C–H bonds present in a particular carbon atom by the drawing of a skeletal formula?

A: Every carbon atom in a stable molecule fulfills the octet rule and can have a maximum of four sigma bonds or combinations of sigma(σ) and pi(π) bonds (for double or triple bonds). Based on this, we can account for the number of C–H bonds that are present for each of the carbon atoms in the following example:

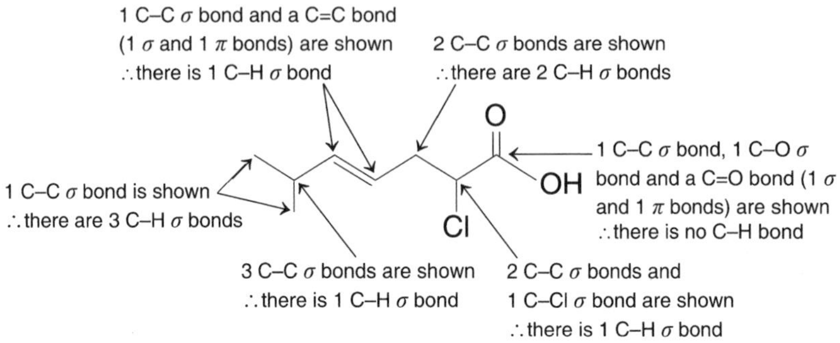
1 C–C σ bond and a C=C bond
(1 σ and 1 π bonds) are shown
∴ there is 1 C–H σ bond
2 C–C σ bonds are shown
∴ there are 2 C–H σ bonds
O
1 C–C σ bond, 1 C–O σ bond and a C=O bond (1 σ and 1 π bonds) are shown
∴ there is no C–H bond
OH
1 C–C σ bond is shown
∴ there are 3 C–H σ bonds
Cl
3 C–C σ bonds are shown
∴ there is 1 C–H σ bond
2 C–C σ bonds and
1 C–Cl σ bond are shown
∴ there is 1 C–H σ bond

1.3 Bonding and Shapes of Organic Molecules

Looking at the various molecules in Table 1.1, you will find that a carbon atom has one of the common bonding patterns as shown below:

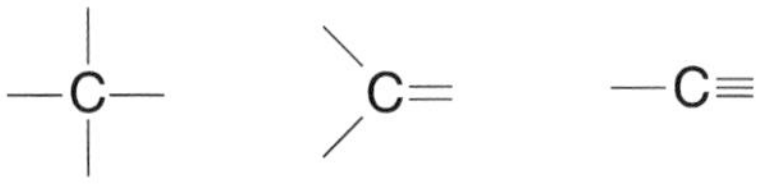

Each of these bonding patterns corresponds to a different geometry around the carbon atom:

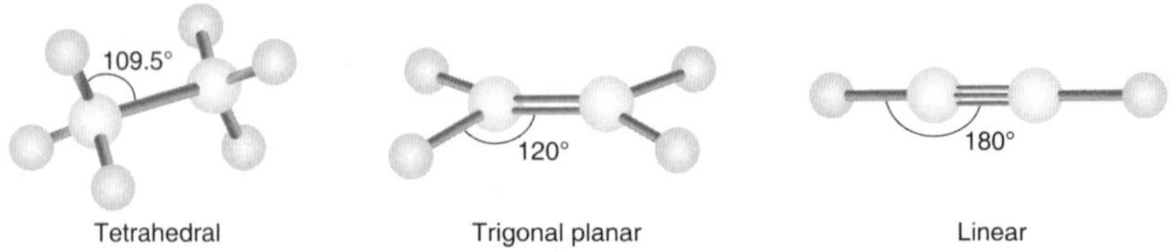
109.5°
120°
180°
Tetrahedral
Trigonal planar
Linear

But when we think about the electronic configuration of a carbon atom, $1s^2 2s^2 2p^2$, there is a lack of agreement between the observed experimental data and the idea of atomic orbital overlap during the formation of covalent bond. What do we mean by this? Consider the methane molecule, CH_4.

H

C 109.5°

H—C⁝⁝⁝⁝H

H

The carbon atom is covalently bonded to four hydrogen atoms. But there are only two $2p$ orbitals, each containing an electron that can be used to form two covalent bonds with other atoms. May be this discrepancy can be resolved by simply unpairing the pair of electrons in the $2s$ orbital and promoting one of them to the $2p$ subshell. This process does not really require too much energy as the energy gap between the $2s$ and $2p$ subshells is relatively small, thus it can be easily achieved. We now have four unpaired electrons sitting in four orbitals (the $2s$ orbital and the three $2p$ orbitals):

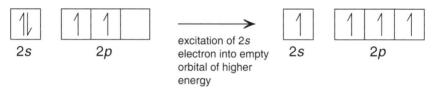

2s	2p

excitation of 2s
electron into empty
orbital of higher
energy

2s	2p

These four electrons reside in various types of orbitals, and each of these orbitals comes with a particular shape and orientation in space. If these four orbitals overlap with the $1s$ orbital of four hydrogen atoms, we are going to get four C—H bonds. However, since the three $2p$ orbitals are perpendicular to each other, we will get three C—H bonds at 90° angles to each other. Furthermore, the four C—H bonds will have different bond lengths since different types of orbitals are used by the carbon atom in bond formation.

But experimentally, it has been found that all the bond angles in a CH_4 molecule are the same at 109.5°, and all the C—H bonds are of the same strength and length! We therefore need a different model to explain what we have observed experimentally. The hybridization model was formulated mathematically for this very purpose. Hybridization is essentially a mathematical linear combination of atomic orbitals to create new orbitals (known as hybrid orbitals) that are then used for bonding. Combinations of a specific number of atomic orbitals will give rise to the same number of hybrid orbitals.

We will now look at the various types of hybridization and understand their purpose and usage, especially for organic molecules.

Q: What is a "linear combination?"

A: Linear combination refers to a resultant mathematical function that is obtained by simply adding up other mathematical functions together. But before adding these mathematical functions together, each of these mathematical functions is multiplied by a coefficient, which indicates the loading factor or the "amount of contribution" this particular mathematical function has made to the resultant function. An example would be, $w = a_1 x + a_2 y$, where both x and y are mathematical functions and a_1 and a_2 are coefficients.

1.3.1 sp^3 *Hybridization*

It can be used to account for the tetrahedral geometry about a carbon atom covalently bonded to four other atoms.

Consider the valence orbitals of the C atom:

Four atomic orbitals (one $2s$ and three $2p$ orbitals) give rise to four equivalent sp^3 hybrid orbitals which are orientated at 109.5° apart at the corners of a tetrahedron. In forming methane, each of the four sp^3 hybrid orbitals of the C atom overlaps head-on with the $1s$ orbital of the H atom to form four sigma bonds. This accounts for the tetrahedral geometry of the molecule.

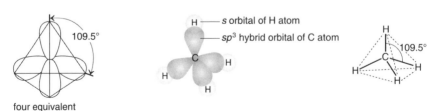

four equivalent
sp^3 hybrid orbitals

In the case of ethane, C_2H_6, three of the four sp^3 hybrid orbitals of a C atom overlap head-on with the $1s$ orbital of three H atoms to form three sigma bonds. The same goes for the other C atom. The remaining sp^3 hybrid orbitals of both C atoms overlap head-on with each other to form the C–C sigma bond. The geometry about each C atom is tetrahedral, which is in accordance with the experimental data.

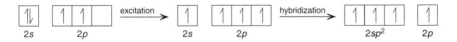

1.3.2 sp^2 Hybridization

To account for the trigonal planar geometry about a carbon atom covalently bonded to three other atoms, such as those in the ethene molecule (C_2H_4), we have to consider another type of hybridization — the sp^2 hybridization.

Consider the valence orbitals of the C atom:

This time round, we have three atomic orbitals (one $2s$ and two $2p$ orbitals) giving rise to three equivalent sp^2 hybrid orbitals which are orientated at $120°$ apart in a triangular plane. The remaining unhybridized p orbital lies perpendicular to this plane.

Q: Why is the energy of the sp^2 hybrid orbitals lower than that of the p orbitals?

A: Since the s orbital has a lower energy than the p orbital, when both s and p orbitals undergo hybridization, the resultant hybrid orbitals would have higher energy than the s orbital but lower than the unused p orbital.

Q: Does it matter which of the two $2p$ orbitals are used for hybridization?

A: No! Hybridization is a model created by scientist to explain the phenomenon that is observed regarding the bond angle and molecular shape. The atom itself does not know anything about it. Anyway all the three $2p$ orbitals are of the same energy (they are degenerate). Thus, it does not matter which two of the three $2p$ orbitals are used during hybridization.

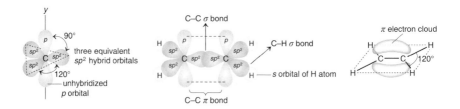

In forming ethene, two of the three sp^2 hybrid orbitals of each C atom overlap head-on with the $1s$ orbital of the H atoms forming a total of four C−H sigma bonds. The remaining sp^2 hybrid orbitals of both C atoms overlap with each other head-on to form the C−C sigma bond. When this happens, the unhybridized p orbitals are brought close to each other, and they are able to overlap side-on to form the pi bond. As such, the bonding pattern for a sp^2 hybridized carbon atom is two single bonds to two atoms and a double bond to a third atom — giving a total of eight valence electrons around the bonded carbon atom.

The geometry about each type of hybridized carbon atom is specific — the arrangement of substituents for a sp^3 hybridized carbon atom is tetrahedral and that for a sp^2 hybridized carbon atom is trigonal planar. Shown here is an example of a molecule that contains both types of hybridized carbon atoms. Can you figure out the geometry about each carbon atom?

Q: Can we talk about the hybridization of the Cl atom in the above molecule?

A: What is the purpose of hybridization? It is used to explain shape. It is pointless to describe a hybridization state for the Cl atom in the above molecule as it is only bonded to another atom and the shape is obviously linear. Similarly for the O atom in the C=O functional group. But it is alright to describe the hybridization state of the O atom in the –OH group as it would tell us the shape about this O atom. It is in fact sp^3 hybridized with two other atoms (a C and a H atom) bonded to it. In addition, there are two lone pairs of electrons and, as a result of these two lone pairs of electrons, the molecular geometry is a bent shape.

You may find that it is easier to first account for all the C−H bonds in the molecule before deciding on the hybridization and hence geometry about each carbon atom.

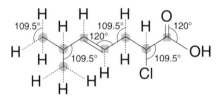

1.3.3 *sp Hybridization*

The *sp* hybridization is used to account for the linear geometry about a carbon atom covalently bonded to two other atoms, such as those in the ethyne molecule, C_2H_2.

Consider the valence orbitals of the C atom:

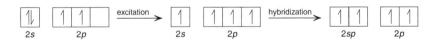

Two atomic orbitals (one $2s$ and one $2p$ orbitals) give rise to two equivalent *sp* hybrid orbitals which are orientated at $180°$ apart. Two unhybridized p orbitals lie perpendicular to the plane containing the hybrid orbitals. In forming ethyne, one of the two *sp* hybrid orbitals of each C atom overlaps head-on with the $1s$ orbital of the H atom forming a total of two C−H sigma bonds altogether. The remaining *sp* hybrid orbitals of the C atoms overlap with each other head-on to form the C−C sigma bond. Consequently, the two unhybridized *p* orbitals on each C atom overlap side-on to form two pi bonds.

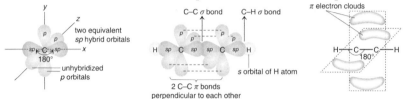

Exercise: State the type of hybridization of the carbon atoms, labeled 1 to 5, in the following molecule and hence the geometry about each of them.

Solution:
Carbon 1: sp^2 hybridized; trigonal planar geometry
Carbon 2: sp^3 hybridized; tetrahedral geometry
Carbon 3: sp^2 hybridized; trigonal planar geometry
Carbon 4: sp^2 hybridized; trigonal planar geometry
Carbon 5: sp hybridized; linear geometry

Q: What is the hybridization state of the carbon atoms in the following molecule?

A: For both C1 and C3, each of them forms two single bonds and one double bond. Hence, they are sp^2 hybridized carbon atoms. As for C2, it forms two double bonds which consists a total of two sigma bonds and two pi bonds. In order for C2 to be able to form two pi bonds, it must have two unused p orbitals not involved in hybridization. Therefore, C2 must be sp hybridized.

In short, to determine the hybridization state of an atom, we can make use of the following guidelines:

- *Four* single bonds *OR four* sigma bonds *OR* four regions of electron densities $\Rightarrow$ atom is sp^3 hybridized with tetrahedral electron-pair geometry;
- *Two* single bonds and *one* double bond *OR three* sigma bonds and *one* pi bond *OR* three regions of electron densities $\Rightarrow$ atom is sp^2 hybridized with trigonal planar electron-pair geometry;
- *One* single bond and *one* triple bond *OR two* double bonds *OR two* sigma bonds and *two* pi bonds *OR* two regions of electron densities $\Rightarrow$ atom is sp hybridized with linear electron-pair geometry.

Q: What is electron-pair geometry? Is it the same as molecular geometry or shape?

A: Electron-pair geometry gives the geometrical distribution of electron densities (pairs) around a central atom. It MAY NOT be equal to molecular geometry. The latter is the geometrical distribution of the nuclei of atoms around the central atom. For example, the electron-pair geometry can be tetrahedral (because there are 4 regions of electron densities) but the shape can be trigonal pyramidal because there is a lone pair of electrons amongst the 4 regions of electron densities. And if there are two lone pairs of electrons, then the molecular geometry becomes bent in shape. Take note that it is the massive nucleus of an atom and not its electron or electron cloud that helped us pinpoint the position of an atom. Electrons are too small for their exact position to be determined.

With these simple guidelines, we can determine the hybridization states of the nitrogen atom in the following molecules:

In the molecule of NH_3, there are three sigma bonds and a lone pair of electrons around the nitrogen atom. The lone pair can be considered equivalent to a sigma bond. Hence, the nitrogen atom is sp^3 hybridized. In the molecule of N_2H_2, there are two sigma bonds, one pi bond and a lone pair of electrons around the nitrogen atom. Here again, the lone pair can be considered equivalent to a sigma bond. Hence, the nitrogen atom is sp^2 hybridized.

Q: What happens when the lone pair of electrons on the nitrogen resides in a p orbital instead of in a sp^3 hybridized orbital for the NH_3?

A: If the lone pair of electrons resides in a p orbital, this would mean that we cannot use the sp^3 hybridization model. Instead, may be we should be using the sp^2 hybridization model. But if this is so, then you would find that both sp^2 and sp hybridization models cannot be used to account for the actual experimental geometry around the nitrogen atom of NH_3. In addition, a lone pair of electrons residing in a p orbital that is perpendicular to the nitrogen atom would induce too much inter-electronic repulsion compared to if it were to reside in a hybridized orbital, which is pointing away from the atom. Thus, the hybridization model that we used to explain the shape around an atom has to be supported by experimental observation.

1.3.4 *Delocalized Bonding/Resonance*

What we have covered thus far are examples that use the hybridization model to account for the shapes and bonding of molecules. But there are certain molecules and ions with bonding pattern that cannot be adequately explained based on this model.

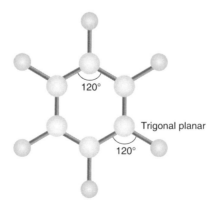

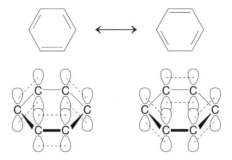

Fig. 1.1. Canonical forms of benzene: Localized structures of benzene.

Take, for instance, the benzene molecule, C_6H_6. It contains six sp^2 hybridized carbon atoms with trigonal planar geometry about each of them — as accounted for by the hybridization model. If we are to draw either one of the localized structures using Lewis structures (shown in Fig. 1.1), we would expect to have two different types of bond length — that of the $C-C$ single bond and the $C=C$ double bond.

However, experimental data shows that all the carbon–carbon bond lengths in benzene are identical and the bond length is intermediate between the $C-C$ single and $C=C$ double bonds. The only way to be in sync with the observed experimental evidence is to imagine that each p orbital overlaps simultaneously with its two neighboring p orbitals, creating two rings of delocalized electron cloud above and below the planar six-member carbon ring structure (see Fig. 1.2). To describe this phenomenon of electron delocalization through the side-way overlapping of p orbitals, the term *resonance* is used. As a result, a resonance hybrid, which is described by an "average" of the two equivalent resonance forms (or canonical structures) is created. Take note that the term resonance DOES NOT mean that the molecule is constantly flipping from one resonance structure to the other. There is only one particular structure for benzene at any one time, and for this, it is represented by the resonance hybrid structure.

1.4 Bonding and Reactivity of Organic Compounds

Chemical bonds are electrostatic forces of attraction (positive charge attracting negative charge) that bind particles together to form matter. When different types of particles interact electrostatically,

Fig. 1.2.   Resonance hybrid: Delocalized structure of benzene.

different types of chemical bonds are formed. Since the carbon atoms in an organic molecule are bonded by covalent bonds to either other carbon atoms or some other heteroatoms such as oxygen, nitrogen, halogens, etc., the reactivity of the organic compounds is mostly determined by the relative ease of breaking these covalent bonds.

1.4.1 *Nature and Strength of Covalent Bonds*

Understanding the nature of covalent bonds provides us with deeper insight as to how reactions occur and also the factors that affect the rate of reaction.

When we talk about two different species reacting with each other, there must be a reason for them to want to approach each other. For instance, an atom that is electron-rich can "attract" an attack by a complementary species that is electron-deficient and vice versa — a scenario that stems from the all-too-familiar notion of "opposites attract."

As transformation occurs in producing new compounds, a preliminary step consists of the cleaving of bonds within the reactant molecules. Different types of chemical bonds have different strengths. Consequently, energy that is needed for bond cleavage varies, and thus the rates of reactions are affected differently. More details on the different types of organic reactions are given in Chapter 3.

An electron-rich or electron-deficient atom in a molecule results from the unequal sharing of electrons between the two bonding atoms due to a difference in their electronegativity. Electronegativity denotes the ability of an atom in a molecule to attract electrons to itself. When the two bonding atoms have the same electronegativity, the bonding electrons are shared equally between the two nuclei, forming a pure covalent bond. This normally happens when the bonding atoms consist of identical atoms. A difference in electronegativity creates a permanent separation of charges (dipole) in the bond. The more electronegative atom has a stronger hold onto the bonding electrons. It is thus "electron rich" and gains a partial negative charge ($\delta-$). The other bonding atom partially loses its hold on the bonding electrons and acquires a partial positive charge ($\delta+$), an indication of it being "electron-deficient." Such covalent bonds with unequal sharing of electrons are termed "polar bonds."

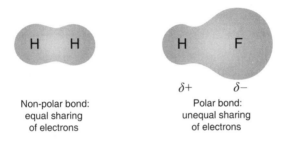

Non-polar bond:
equal sharing
of electrons

Polar bond:
unequal sharing
of electrons

Bond energy (BE) is a good indicator of the strength of covalent bonds. It is the average amount of energy required to break one mole of a particular bond in any compound in the gas phase. A greater value of bond energy implies a stronger covalent bond.

A covalent bond is basically the electrostatic attraction between the localized shared electrons and the two positively charged nuclei. A greater electron density in the inter-nuclei region contributes to a stronger bond. Since an electron is perceived as residing in a region of space known as the orbital, a covalent bond is formed when two partially filled valence orbitals overlap with the same type of orbital from the carbon atom. Hence, the size of the orbital used in bond formation affects bond strength.

Example 1.1: Account for the trend of the C−X bond energies whereby X = Cl, Br, and I.
BE(C−Cl): $340 \, \text{kJ mol}^{-1}$; BE(C−Br): $280 \, \text{kJ mol}^{-1}$; BE(C−I): $240 \, \text{kJ mol}^{-1}$.

Solution: The orbital used for covalent bonding increases in size from Cl to Br to I. The larger orbitals are more diffuse, and when these orbitals overlap with the same type of orbital from the carbon atom, the accumulation of electron density within the inter-nuclei region is lower. Thus, the orbital overlap is less effective, and this accounts for the weaker bond strength as reflected in the bond energies.

Another factor affecting bond strength is the number of bonds between the bonding atoms, as shown in Example 1.2.

Example 1.2: Account for the trend of bond energies for the different carbon–carbon bonds:
BE(C–C): $350\,\text{kJ}\,\text{mol}^{-1}$;   BE(C=C): $610\,\text{kJ}\,\text{mol}^{-1}$;   BE(C≡C): $840\,\text{kJ}\,\text{mol}^{-1}$.

Solution: As the number of shared electrons increases within the inter-nuclei region for multiple bonds, the attractive force for this greater number of electrons increases. Thus, the carbon–carbon covalent bond strength increases from single to double to triple bonds.

Q: Why aren't the bond energies as follows: $\text{BE(C–C)} = 1/2 \times \text{BE(C=C)} = 1/3 \times \text{BE(C≡C)}$?

A: The C=C double bond consists of a sigma bond and a pi bond whereas the C≡C triple bond is made up of one sigma and two pi bonds. The bond energies of the C=C and C≡C bonds are not exact multiples of the C–C single bond as the pi bond is weaker than a sigma bond.

Q: Does it mean that all the sigma bonds in each of the carbon–carbon bonds, C–C, C=C and C≡C, are not of the same strength?

A: Yes, they are indeed not of the same strength. The presence of a pi bond in the C=C bond would weaken the sigma bond within the double bond itself. Further weakening effect of the sigma bond would be observed in the C≡C bond due to the presence of two pi bonds. Similarly, do not expect the strength of the pi bond in the C=C bond to be of the same strength as each of the two pi bonds in the C≡C bond. In fact, the pi bond in the C=C bond should be stronger than each of the two pi bonds in the C≡C bond.

Q: So, are the two pi bonds in the C≡C bond of the same strength?

A: Yes, they are of the same strength. Since the two pi bonds cannot be really differentiated from one another, and each of the pi bonds would affect the other, on the average, they would have the same strength.

The electron density accumulated in the inter-nuclei region of a sigma bond is greater than that in the pi bond. A sigma bond is formed when two orbitals overlap head-on, whereas a pi bond usually occurs when two p orbitals overlap side-on. Therefore, there is *more effective overlap* of orbitals in forming the sigma bond as compared to the pi bond.

The type and strength of covalent bonds influence the chemical properties of organic compounds. These bonds are strong, and some are even comparable in strength to the ionic bonds present in ionic compounds such as sodium chloride. But how does this similarity in bond strength between covalent and ionic compounds account for the difference in the physical properties of these compounds? Why do covalent compounds, organic compounds included, have lower melting and boiling points than ionic compounds?

Melting and boiling are phase changes for a substance. When a simple molecular organic compound undergoes a phase change, it still retains its molecular entity. Ethanol, be it in the solid, liquid or vapor phase, still consists of CH_3CH_2OH molecules. No intramolecular bonds are broken in transiting from one physical phase to another. It is the extent of interactions an ethanol molecule has with other ethanol molecules and the degree of freedom of motion of these molecules that causes the lower melting and boiling points for these compounds. So what kinds of interactions are we talking about that exist between ethanol molecules?

1.4.2   *Types of Intermolecular Attractive Forces*

Depending on the type of molecules, one of the following intermolecular forces of attractions, or combinations of these, could exist between molecules:

- van der Waals forces of attraction
  - Instantaneous dipole–induced dipole interactions (id–id)
  - Permanent dipole–permanent dipole interactions (pd–pd)
- Hydrogen bonding

1.4.2.1   *Instantaneous dipole–induced dipole (id–id) interactions*

Id–id interactions exist for all types of molecules and it may be used to account for the observed physical properties of matter.

As electrons are always in random motion, there is an uneven distribution of electrons in the molecule at any point in time. This separation of charges creates an instantaneous dipole in the molecule. The instantaneous dipole in one molecule can induce the formation of dipoles in nearby unpolarized molecules. As a result, a weak electrostatic attraction forms between these dipoles.

Instantaneous   Induced
    dipole   dipole

Due to the very fact that electrons are constantly moving, these dipoles are short-lived. However, new dipoles are soon formed again, and the process repeats itself so that on average, there are "permanent" forces of attraction between these molecules that give rise to their observed physical properties such as melting point, boiling point, non-ideal gas behavior and so forth.

The strength of id–id interactions depends on:

- The number of electrons in the molecule: The greater the number of electrons (i.e. the bigger the electron cloud), the more polarizable would be the electron cloud, and the stronger the id–id interactions.
- The surface area for contact of the molecule: The greater the surface area of contact that is possible between the molecules, the greater would be the extent of id–id interactions.

Q: Why would a greater number of electrons lead to a more polarizable electron cloud?

A: A greater number of electrons would imply that more electrons would be further away from the nucleus, hence subjected to weaker attractive force by the nucleus. Thus, the electron cloud would be more polarizable.

Q: But a greater number of electrons would also imply a greater number of protons in the nucleus. So, shouldn't the greater number of protons lead to stronger attractive force by the nucleus?

A: Yes, you are right that the greater number of protons would lead to a greater attractive force by the nucleus. Electrostatic force can be perceived as an inverse square law, $(F \propto \frac{1}{r^2})$, depending on the separation between the electron and nucleus. The effect of an increase in distance leads to a more drastic decrease in the electrostatic force of attraction than an increase in the quantity of charge. In addition, with an increase in the number of electrons, the inter-electronic repulsion would also increase significantly. These two factors coupled together would result in a weaker net attractive force for the electrons in the outermost principal quantum shell. Hence, ionization energy decreases and polarizability increases.

1.4.2.2 *Permanent dipole–permanent dipole (pd–pd) interactions*

Pd–pd interactions exist for polar molecules only. As there is an uneven distribution of electrons in polar bonds, permanent separation of charges (dipole) is found within polar molecules. The permanent dipoles in neighboring polar molecules attract each other. As a result, a weak electrostatic attraction forms between these dipoles, known as pd–pd interactions.

The strength of pd–pd interactions depends on the magnitude of the molecule's net dipole moment. The greater the magnitude of the dipole moment of a polar bond, the more polar is the bond. The magnitude of the dipole, in turn, depends on the magnitude of the electronegativity difference between the bonding atoms. The greater the difference in electronegativity, the greater the dipole moment and the more polar the covalent bond. Dipole moment is a vector quantity that has both magnitude and direction. It is represented by the symbol $\longmapsto$, which depicts the $\delta+$ end pointing towards the $\delta-$ end of the dipole.

But one needs to take note that the net polarity of a molecule is also dependent on the geometrical shape of the molecule itself. A molecule may contain polar covalent bonds, but if the dipole moments of these polar bonds cancel each other out due to its geometrical shape, overall this molecule is still a non-polar molecule. Hence, there would not be any pd–pd interaction due to the absence of a net dipole moment.

1.4.2.3 *Hydrogen bonding*

Hydrogen bonding is present between molecules that have one highly electronegative atom $(X) - F$, O, or N – covalently bonded to an H atom.

X, being more electronegative than H, attracts the bonding electrons in the $H-X$ bond closer towards itself. As a result, it is more "electron rich" and gains a partial negative charge $(\delta -)$. The H atom, on the other hand, acquires a partial positive charge $(\delta +)$.

Being "electron deficient," this H atom is strongly attracted to the lone pair of electrons on another highly electronegative atom (electron-rich region) in other molecules. This electrostatic attraction between the H atom and the lone pair of electrons on the highly electronegative atom, F, O or N, is known as hydrogen bonding.

Hydrogen bonding can be regarded as a more specific type of pd–pd interaction that is applicable to certain types of polar molecules only. So, take note that the terms "hydrogen bonding" and "pd–pd interactions" are non-interchangeable. Examples of compounds that exhibit hydrogen bonding include H_2O, HF, NH_3 and organic compounds such as alcohols, carboxylic acids, amines and amides.

The strength of a hydrogen bond depends on:

- Dipole moment of the $H-X$ bond, where X is O, F, or N:
  $F-H- - -F-H > O-H- - -O-H > N-H- - -N-H$

- Ease of donation of a lone pair on Y, where Y is O, F, N:
  N−H- - -N−H > O−H- - -O−H > F−H- - -F−H

Overall, hydrogen bond strength is in the following order:
F−H- - -F−H > O−H- - -O−H > N−H- - -N−H
This indicates that the dipole moment factor predominates over the ease of donation factor.

1.4.3 Bonding and Physical Properties of Organic Compounds

1.4.3.1 Melting and boiling points

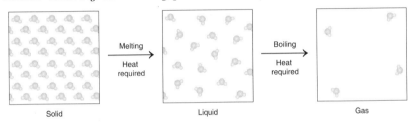

For melting to occur, the crystal lattice has to be broken down so that the discrete molecules are free to move about. Therefore, energy (in the form of heat) is required to overcome the intermolecular forces of attraction among the molecules. Likewise, for boiling to occur, energy is needed to overcome the intermolecular attractions between the mobile molecules so that they can break away from one another and have greater freedom of movement.

In trying to relate the type of structure, bonding or intermolecular forces of attraction to the physical properties of substances, it is useful (but <u>NOT</u> necessary always true) to start with the general trend:

- The order of decreasing strength of intermolecular forces of attraction:

Hydrogen bonding >> pd–pd interactions > id–id interactions
(strongest)   (weakest)

Example 1.3: Account for the lower boiling point of ethanal as compared to ethanol.

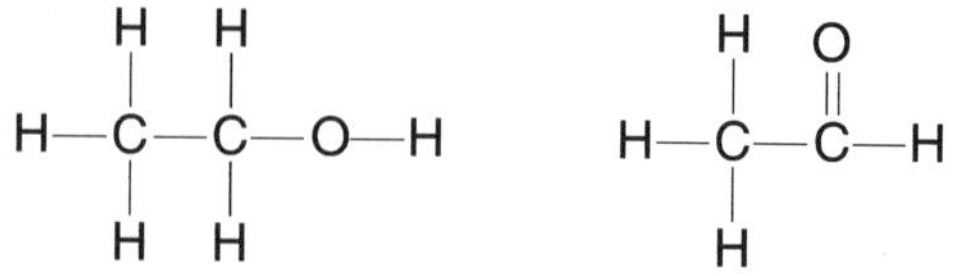

Ethanol
(b.p. 78.4 °C)

Ethanal
(b.p. 20.2 °C)

Solution: Both ethanol and ethanal are polar molecules with simple molecular structures. The lower boiling point of ethanal as compared to ethanol indicates that the pd–pd interactions are weaker than hydrogen bonding. Thus, less energy is required to overcome the attractive forces between ethanal molecules as compared to the hydrogen bonding between ethanol molecules.

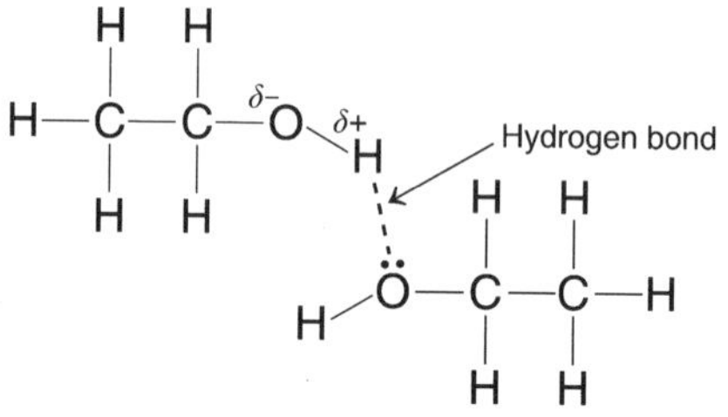

1.4.3.2 *Miscibility of organic compounds*

Organic compounds dissolve more readily in non-polar solvents (e.g. benzene and tetrachloromethane) than in polar solvents such as water.

Q: Why does an organic compound dissolve in one solvent but not in another?

A: For an organic solute to dissolve, energy that is released from the solvation process must be sufficient to offset the energy that is needed to overcome the attractive forces holding the solute molecules together and hence separate them.

Q: The solvation energy is the energy released when the solvent and solute molecules interact. We could call this "future" energy, as the energy can only be released after the solute molecules have separated from each other and at the same time, "spaces" are created between the solvent molecules to accommodate the incoming solute molecules. The separation of the molecules needs energy. So how can "future" energy be used to compensate for the energy that is currently required for the breaking of bonds?

A: One needs to understand how substances actually mix or dissolve. When two substances mix or dissolve, it does not mean that all bonds between the solvent molecules or solute molecules have to be broken first at one go before both the solvent and solute molecules interact. We can imagine the process of an ionic solid dissolving to be like peeling an onion. The positive end of the dipole of a water molecule is attracted to the anions on the surface of the ionic solid, and the negative end of the dipole is attracted towards the cations. The formation of the various ion–dipole interactions releases energy. The energy released is then transferred to the cations and anions, which increases the vibrational energy of these ions. As more ion–dipole interactions occur, releasing more energy, the greater amount of vibrational energy would enable the ions to be freed from the lattice. For the mixing of two liquids, when the boundaries of these two liquids come in contact with each other, the molecules at the two boundaries would interact with each other. If this interaction is favorable, then a sufficient amount of energy would be released, and this energy would be transferred to their neighboring molecules as vibrational energy. With an increase in vibrational energy, the molecules at the boundaries would be able to "break" away from their bulk. This process is similar to the dissolution of ionic compound in water.

To ensure that substantial energy is released from solvation, there must be favorable interactions between the solute and solvent molecules that are stronger than or, at the very least, of similar magnitude to, those between the solute molecules themselves. This is the driving force for the solute molecules to be separated. Likewise, to have solvent molecules attracted to the solute molecules, the interactions between them must be stronger than or similar to those between the solvent molecules. Such explanation is the reasoning behind the saying "like dissolves like." The same reason is used to account for the miscibility, or lack thereof, of two liquids.

The ability to form hydrogen bonds accounts for the high solubility of substances such as ammonia and short-chain alcohols (e.g. methanol, ethanol) in water. Although the id–id interactions among alkane molecules are relatively weak and easily overcome, you do not find alkanes to be miscible with water. This is because it is not "worthwhile" for both the alkane and water molecules to "give up" the "strong" interactions among their own kind in return for less favorable interactions between the two groups of molecules. We thus expect the solubility of alcohol in water to decrease with increasing carbon chain length due to the increasingly hydrophobic nature of the carbon chain. We will touch upon the physical properties of each homologous series in the corresponding chapters.

My Tutorial

1. (a) (i)　The enthalpies of formation of ethane, ethene, ethyne and benzene are -84.7, $+52.3$, $+227$, and $+82.9\,\text{kJ mol}^{-1}$, respectively. With the given bond energies of $H-H$ and $C-H$ as $+436$ and $+412\,\text{kJ mol}^{-1}$, respectively, calculate values for the carbon–carbon bond energies in the four hydrocarbons.

　　(ii)　Account for the differences in carbon-carbon bond length in the compounds ethane, ethene, ethyne and benzene.

　　(iii)　Draw the dot-and-cross diagrams of ethane, ethene and ethyne.

　(b)　Explain why $CH_2=C=CH_2$ is not a flat molecule and the carbon skeleton of the following molecule is not planar.

(c) Explain in molecular terms the differences between the boiling points of:
- (i) methane and *n*-butane
- (ii) *n*-butane and methylpropane
- (iii) *n*-hexane and cyclohexane
- (iv) propene and ethane
- (v) *cis*-but-2-ene and *trans*-but-2-ene
- (vi) *n*-butane and propanone
- (vii) propanone and propanal
- (viii) propanone and propan-1-ol
- (ix) ethylamine and dimethylamine
- (x) ethylamine and ethanol
- (xi) ethanol and ethanoic acid
- (xii) ethanoic acid and amino ethanoic acid

(d) Predict whether each of the compounds below will be miscible or immiscible with water. Give reasons for the predictions.
- (i) *n*-butane
- (ii) chloroethane
- (iii) propanone
- (iv) propan-1-ol
- (v) propylamine

(e) 2,2'-Biquinolyl is an important complexation agent used in the extraction of copper.

2,2'-Biquinolyl

    (i)   Identify the type of hybridization of each of the two N atoms.

    (ii) Circle the atoms that make the compound able to act as a complexation agent.

    (iii) The N atom of 2,2'-biquinolyl is more basic than the N atom of phenylamine but less basic than the N atom of an aliphatic amine. Explain.

    (iv) Explain why the carbon skeleton of 2,2'-biquinolyl is planar.

(f) Use the following bond energies to account for why the following germinal diol undergoes spontaneous intramolecular dehydration to form the aldehyde functional group.

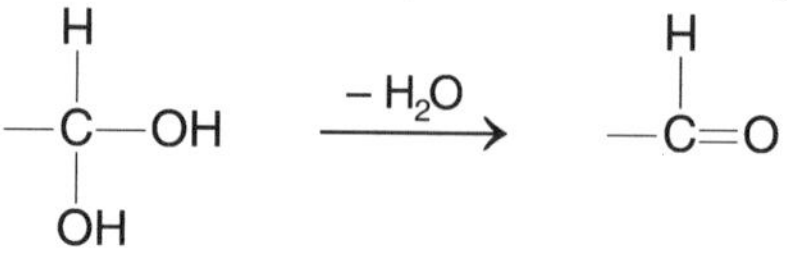

[Bond energies (in kJ mol^{-1}): C$-$O (358); C$-$H (413); C=O (736); H$-$O (464)]

(g) The enthalpy changes of hydrogenation of benzene and cyclohexatriene, both to cyclohexane, are -208 and -360 kJ mol^{-1}, respectively. Discuss the difference between these two values.

2. Morphine is a chemical able to act on the central nervous system to relieve pain, whereas heroin, a synthetic derivative of morphine, is an illegal drug. Both the structures of the compounds are shown below:

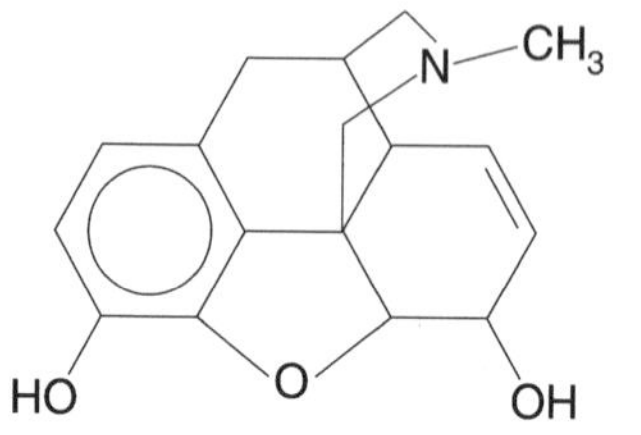

Morphine 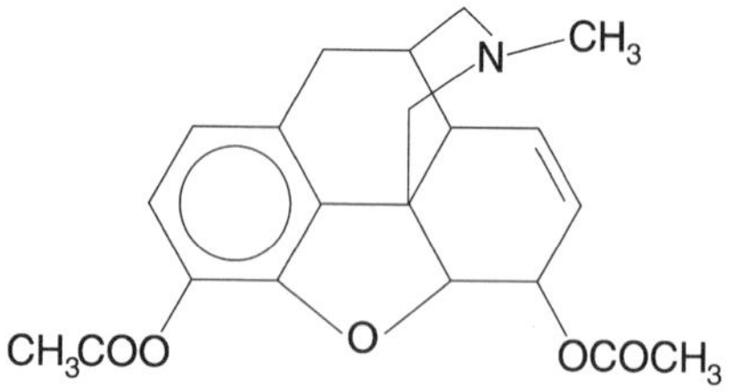   Heroin

(a) Write the condensed structures (i.e. $C_wH_xN_yO_z$) for both morphine and heroin. Thus, find the relative molecular mass of these two compounds.

(b) Identify each type of hybridization used by the carbon atoms in each of the molecule and give the number of carbon atoms exhibiting each type of hybridization.

(c) Describe the bonding represented by the circle in the benzene ring of each of the molecule.

(d) Identify a pair of atoms involved in a pi bond in each of the molecule. Sketch the pi bond and describe briefly how it is formed.

(e) Excluding the benzene ring from consideration, determine the number of pi bonds in each of the molecule.

(f) In each part of the molecule linking the benzene ring to the rest of the molecule,

   (i) name the groups, $C-C(O)-O$ and $C-O-C$

   (ii) predict the value of the $C-C-O$ bond angle and explain your answer.

(g) Identify two different elements in each of the molecules that have lone pairs of electrons and determine the number of such lone pairs in the molecule.

(h) Mark the chiral carbons in each molecule with asterisks and hence predict the number of stereoisomers.

(i) Predict whether morphine or heroin would be more soluble in water. Explain your prediction.

CHAPTER 2

Isomerism in Organic Compounds

2.1 Introduction

If a molecule has the molecular formula C_4H_8, does it imply that all the C_4H_8 molecules are identical? The alkene, but-1-ene, whose molecule is shown below, has the molecular formula C_4H_8.

$$
\begin{array}{ccccccc}
 & \overset{\displaystyle H}{|} & \overset{\displaystyle H}{|} & \overset{\displaystyle H}{|} & \overset{\displaystyle H}{|} & \\
H- & C & - & C & - & C = C & -H \\
 & \underset{\displaystyle H}{|} & & \underset{\displaystyle H}{|} & & &
\end{array}
$$

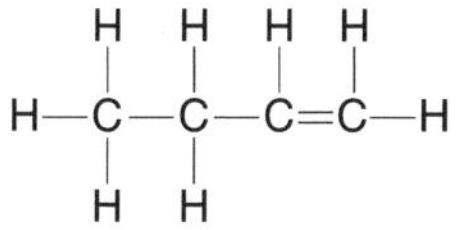

Yet, C_4H_8 also represents the formula for cyclobutane, which belongs to the cycloalkane family:

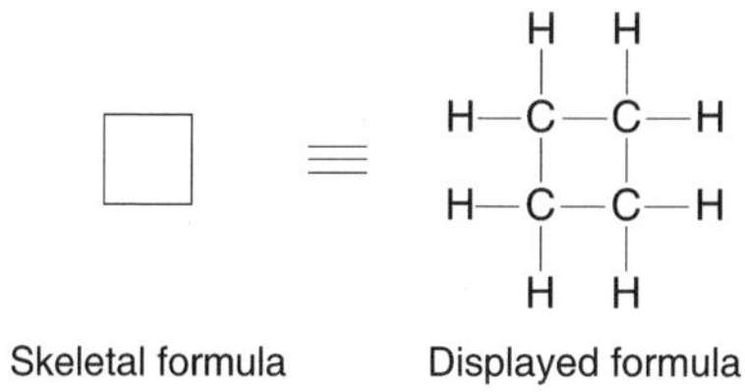

$$
\square \equiv
\begin{array}{ccc}
\overset{\displaystyle H}{|} & & \overset{\displaystyle H}{|} \\
H-C & - & C-H \\
| & & | \\
H-C & - & C-H \\
\underset{\displaystyle H}{|} & & \underset{\displaystyle H}{|}
\end{array}
$$

Skeletal formula   Displayed formula

Compounds that have the same molecular formula but different structures are known as *isomers*. This phenomenon is known as isomerism. The two main types of isomerism are constitutional/ structural isomerism and stereoisomerism. These are further divided into subclasses of which some are discussed in this chapter. Isomers generally have different physical and chemical properties, but they can also have similar chemical properties if they contain the same functional groups. Each specific functional group possesses a characteristic set of chemical reactions.

2.2 Constitutional/Structural Isomerism

Constitutional/structural isomers are compounds with the same molecular formula but different structures or structural formulae. Both but-1-ene and cyclobutane constitute a pair of constitutional/structural isomers. The difference in structures can be attributed to either a difference in the arrangement of atoms or due to the presence of different functional groups.

Based on the above definitions, constitutional/structural isomerism can be classified into three main types:

- chain isomerism;
- positional isomerism; and
- functional group isomerism.

2.2.1 *Chain Isomerism*

Compounds that exhibit chain isomerism with each other have the same functional group but differ in the way the carbon atoms are connected in the mainskeletal carbon chain of their molecules. In other words, these molecules differ in the degree of branching, hence the term *chain isomers.* The length of the main carbon skeletal structures for chain isomers are not the same. An example would be pentane and dimethylpropane.

Pentane   Dimethylpropane

Q: Which isomer, pentane or dimethylpropane, has a higher boiling point?

A: If we are to look at the space-filling model of these molecules, we find that the surface area for contact of the pentane molecule is greater than that of the seemingly spherical dimethylpropane molecule.

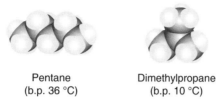

Pentane
(b.p. 36 °C)

Dimethylpropane
(b.p. 10 °C)

The greater surface area for contact translates to more extensive instantaneous dipole–induced dipole (id–id) interactions formed between the pentane molecules than for the dimethylpropane molecules. Since more energy is needed to overcome the more extensive id–id interactions between pentane molecules, pentane has a higher boiling point than dimethylpropane.

Isomers need not necessary be limited to only one pair of compounds. For instance, methylbutane is another chain isomer to both pentane and dimethylpropane.

$$
\begin{array}{c}
\text{H} \\
| \\
\text{H}-\text{C}-\text{H} \\
\end{array}
$$

Methylbutane

Q: Do chain isomers have the same chemical properties?
A: Yes, they do. This is because chain isomers contain the same functional group and hence the same chemical properties.

2.2.2 *Positional Isomerism*

This type of isomerism is used to classify constitutional/structural isomers that have the same functional groups located at different positions along the same carbon chain. That is, the carbon skeletons of these isomers must be of the same length. For instance, both but-1-ene and but-2-ene have a chain length consisting of four carbon atoms. What makes them different is the location of the C=C bond. For but-1-ene, the double bond is between C1 and C2. For but-2-ene, the double bond is between C2 and C3.

But-1-ene   But-2-ene

The name of an alkene informs us of the position of the C=C bond. As such, a "butene" molecule does not exist. The possible non-cyclic molecules, with the molecular formula C_4H_8 and a chain of four carbon atoms, are either but-1-ene or but-2-ene.

Example 2.1: Draw all possible constitutional/structural isomers with the molecular formula C_4H_9Br.

Approach: Start off with structures containing the longest carbon chain, which in this case means a four-carbon chain length.

Identify positional isomers by placing the Br atom on different carbon atoms to obtain unique structures. For instance, placing the Br atom on either C1 or C2 gives structure A and B, respectively. On the other hand, relocating the Br atom on to either C4 or C3 does not produce any new isomers. Instead, by doing so, the two "new" structures are actually identical to structures A and B respectively. Try flipping each of the two "new" structures from left to right and you will be able to see that they are indeed the same as either structure A or B.

Structure A   Structure B
(1-bromobutane)   (2-bromobutane)

After exhausting all possibilities with the longest chain, shorten the chain by one carbon atom. This brings in branching, which means that we are dealing with chain isomers.

Repeat the earlier step of placing the Br atom on different carbon atoms to obtain unique structures, of which there are two, as follows:

Structure C
(1-bromomethylpropane)

Structure D
(2-bromomethylpropane)

Q: Is there any difference between the following two structures?

A: No, they represent the same molecule. The bonding pattern in both structures is exactly the same. In particular, the C atom that the Br atom is connected to is bonded to the same substituents in both structures — a Br atom, two H atoms and a $-CH(CH_3)_2$ group. On the two-dimensional (2-D) paper, the Br atom is oriented differently to the carbon atom, but since this carbon atom is sp^3 hybridized with a tetrahedral geometry and there is free rotation about each of the single bonds, these two structures actually represent the same molecule.

Q: How can there be free rotation about a single bond?

A: When two orbitals come together to form a sigma bond through head-on overlap of orbitals, the electron cloud of the covalent bond is cylindrical in shape. Thus, the two atoms forming the bond can rotate freely at normal room temperature without the sigma bond being broken.

Q: Do we have another unique structure if the Br atom is placed on the top most methyl substituent?

A: If we were to rotate this molecule, we would actually get back structure C. Take note that the 2-D structures do not depict actual bond angles or shape. They just show the connectivity of atoms. One way of checking whether a pair of structures is identical or not is to name them.

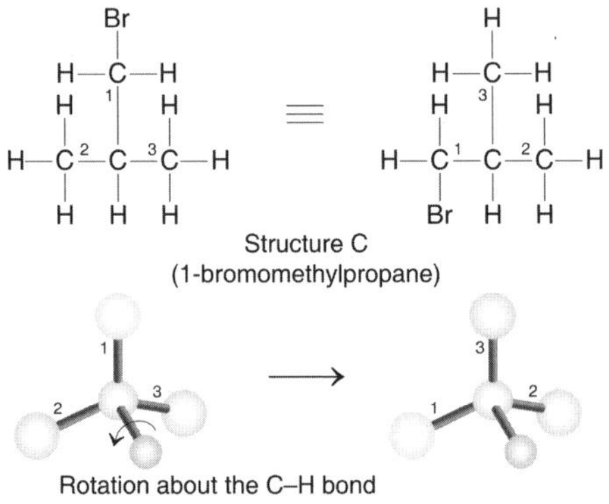

Structure C
(1-bromomethylpropane)

Rotation about the C–H bond

In effect, there are a total of four different possible structural isomers with the same molecular formula C_4H_9Br.

2.2.3 *Functional Group Isomerism*

As the term implies, this type of isomerism is exhibited by compounds that have different functional groups. This means that functional group isomers, unlike the previous two types of constitutional/structural isomers, do not belong to the same homologous series. As such, their chemical and physical properties would certainly be different.

With the same molecular formula, an alcohol and ether constitutes one pair of functional group isomers, e.g. ethanol and dimethyl ether.

Ethanol
(b.p. 79 °C)

Dimethyl ether
(b.p. –23 °C)

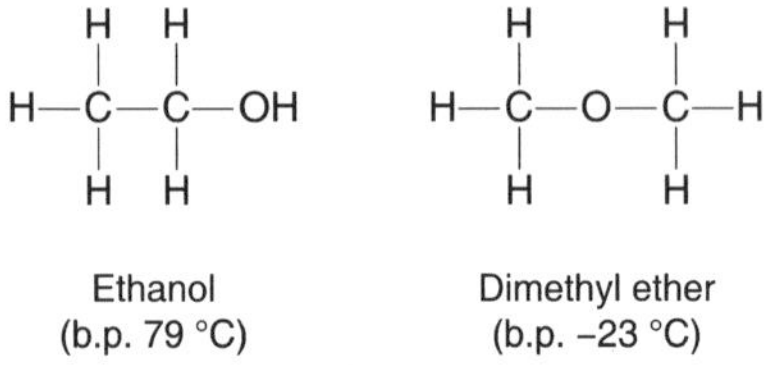

Both ethanol and dimethyl ether are polar molecules. The higher boiling point of ethanol is attributed to the need for a greater amount of energy to overcome the stronger hydrogen-bonding between the ethanol molecules than that needed to overcome the permanent dipole–permanent dipole (pd–pd) interactions that exist between the ether molecules. Other common pairs of compounds that are functional group isomers of each other include:

- alkenes and cycloalkanes, e.g.

But-1-ene   Cyclobutane

- aldehydes and ketones, e.g.

Propanal   Propanone

- carboxylic acids and esters, e.g.

Butanoic acid   Ethyl ethanoate

There are constitutional/structural isomers that can be classified according to one or more of these types of isomerism. Knowing these classifications helps us to figure out the possible types of isomers for a given molecular formula (see Example 2.1).

Exercise: Draw all the possible constitutional/structural isomers that have the molecular formula $C_4H_{10}O$.

Solution: There are seven constitutional/structural isomers, and these are:
(i) $CH_3CH_2CH_2CH_2OH$, (ii) $CH_3CH_2CH(OH)CH_3$, (iii) $(CH_3)_2$ $CHCH_2OH$, (iv) $(CH_3)_3COH$, (v) $CH_3CH_2CH_2OCH_3$, (vi) CH_3CH_2 OCH_2CH_3 and (vii) $CH_3OCH(CH_3)_2$.

2.3 Stereoisomers

Let us startoff by discussing 1,2-dibromoethene. If we were to draw the structure of the molecule, we have this:

$$\underset{H}{\overset{Br}{\diagdown}}C=C\underset{H}{\overset{Br}{\diagup}} \qquad \text{Structure F}$$

The structure of 1,2-dibromoethene can also be drawn as:

$$\underset{H}{\overset{Br}{\diagdown}}C=C\underset{Br}{\overset{H}{\diagup}} \qquad \text{Structure G}$$

If we consider the connectivity of the atoms in the molecules, both structures F and G are the same. But does this mean that they are actually identical and hence represent the same compound?

In fact, structures F and G represent two distinct compounds with different boiling points. The difference in these two structures surfaces when we consider the 3-D orientation of the atoms relative to one another:

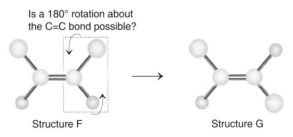

Structure F   Structure G

above the plane of the C=C bond

below the plane of the C=C bond

If we take the doubly bonded carbon atoms to be sitting on a horizontal plane, we have for structure F two of the Br atoms positioned above this plane. In structure G, the two Br atoms are each situated on opposite sides of the plane.

Both structures seem to be interconvertible if we rotate one half-portion of one structure about the C=C bond, i.e.

Is a 180° rotation about
the C=C bond possible?

Structure F   Structure G

If both structures could readily be interconverted, they would represent the same compound. However, the interconversion is not spontaneous at normal room conditions. In fact, the process requires an energy input of about $260\,\mathrm{kJ\,mol^{-1}}$.

Q: What is this energy input required for? Isn't free rotation possible about any bond?

A: In an attempt to convert structure F to structure G, the π bond between the carbon atoms has to be cleaved, and this requires energy. Hence, the process is not spontaneous. The approximated value of $260\,\mathrm{kJ\,mol^{-1}}$ is obtained from calculation using bond energy (BE) values: $\mathrm{BE(C{=}C)-BE(C{-}C)} = 610-350 = 260\,\mathrm{kJ\,mol^{-1}}$. Free rotation about a single bond is spontaneous at normal room conditions because the energy barrier for such rotation is much smaller.

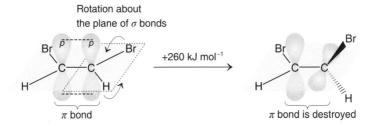

Therefore, structures F and G represent two distinct compounds classified as *cis-trans (geometrical) isomers.*

Cis-trans (geometrical) isomers are a subclass of what is known as *stereoisomers.* These isomers are compounds that have the same molecular and constitutional/structural formulae but different spatial orientation of the atoms within their molecules. Thus, from the above example, "1,2-dibromoethene" does not exist. With the constitutional/structural formula CHBrCHBr, it is either *cis*-1,2-dibromoethene or *trans*-1,2-dibromoethene. We will discuss this in the following section.

2.3.1 *Cis–trans Isomerism (Geometrical Isomerism)*

As shown in the earlier example, *cis-trans* (geometrical) isomerism arises from the restricted rotation about the C=C bond. The *cis-trans* (geometrical) isomers differ in the relative spatial orientation of the substituents with respect to the C=C bond.

Structure F represents the *cis* isomer as the two Br atoms are on the same side of the C=C bond, hence its name *cis*-1,2-dibromoethene. In structure G, the two Br atoms are on opposite sides of the C=C bond, and hence it represents the *trans* isomer, i.e. *trans*-1,2-dibromoethene.

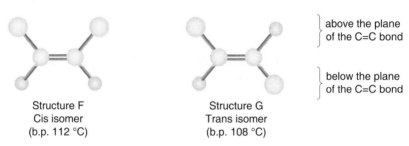

$\left.\rule{0pt}{14pt}\right\}$ above the plane of the C=C bond

$\left.\rule{0pt}{14pt}\right\}$ below the plane of the C=C bond

Structure F
Cis isomer
(b.p. 112 °C)

Structure G
Trans isomer
(b.p. 108 °C)

The effect of the seemingly small difference in orientation is exemplified by the difference in physical properties of these two isomers, such as their boiling points. Chemical reactions and rates may also be affected.

Q: Why does the *cis* isomer have a higher boiling point than the *trans* isomer?

A:

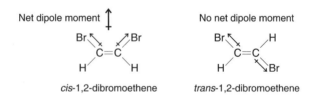

The *cis* isomer is a polar molecule as it has a net dipole moment. The *trans* isomer is non-polar as the dipole moments cancel each other out. More energy is required to overcome the stronger pd–pd interactions between molecules of the *cis* isomer than the id–id interactions between molecules of the *trans* isomer. Hence, the *cis* isomer has the higher boiling point.

Q: But is it always true that the dipole moments of the *trans* isomer will cancel out each other?

A: This is not always true. For instance, if the four groups that are bonded to the C=C double bond are all dissimilar, then there is no way for all the dipole moments to cancel out each other.

Q: Is it right to say that if a compound contains a C=C bond, it would exist as a pair of *cis-trans* (geometrical) isomers?

A: No, this is not always the case. For *cis-trans* (geometrical) isomerism to be possible, each of the doubly bonded carbon atoms must be attached to two different substituents, i.e.

| Cis isomer | Trans isomer | where A, B and D are atoms or groups of atoms |

When one of the doubly bonded carbon atoms is bonded to two identical substituents, *cis-trans* (geometrical) isomerism is not possible. The following structures are not a pair of *cis-trans* (geometrical) isomers:

In fact, they are identical. We can just flip one structure from top to bottom to get to the other structure. It is not necessary for both substituents that are bonded to each of the carbon atoms of the C=C bond to be similar to each other. For example, the following two structures constitute another pair of *cis-trans* (geometrical) isomers:

where A, B, D and E are atoms or groups of atoms

The nomenclature for this pair of isomers does not go by the *cis–trans* designation since there are no two identical substituents in the molecule to bear reference to. Rather, the E, Z notation is used to distinguish between them.

For the E, Z notation, substituents at each of the doubly bonded carbon atom are assigned either a "higher" or "lower" priority. If a structure has both higher priority groups on the same side of the C=C bond, it is labeled the Z isomer (from the German word *zusammen*, which means "together"). When the higher priority groups are on opposite sides of the C=C bond, it is termed the *E* isomer (from the German word *entgegen*, which means "opposite").

Z isomer *E* isomer	where *A* and *B* are higher priority groups; and *D* and *E* are lower priority groups

The assignment of priority is based on the Cahn–Ingold–Prelog rules, which will not be covered in this book. Take note that due to the dissimilarities of the four groups, the net dipole moment of the *trans* isomer may not be zero.

Cis-trans (geometrical) isomerism is not exclusive to the family of alkenes. It can also arise from the restricted rotation about a single bond in a ring structure. Take, for instance, the *cis-trans* (geometrical) isomers *cis*-1,2-dichlorocyclopropane and *trans*-1,2-dichlorocyclopropane. In the *cis* isomer, the two Cl atoms residing on adjacent C atoms are on the same side of the cycloalkane ring, whereas in the *trans* isomer, these two atoms are on the opposite sides of the ring structure.

cis-1,2-dichlorocyclopropane   *trans*-1,2-dichlorocyclopropane

However, if each of the two Cl atoms is bonded to each of the carbon atoms of the C=C bond in a ring structure, we would not get a pair of *cis-trans* (geometrical) isomers. This is because the immediate geometry about each C atom is trigonal planar, and this means that both the Cl atoms lie on the same plane as the doubly bonded C atoms. Any attempt to form the *trans* isomer would cause a puckered (or distorted) ring structure, which is highly unstable due to the presence of ring strain. In addition, the overall molecular geometry of such a puckered molecule would be very much different from the other molecule. Thus, these two molecules would be inherently different due to the overall molecular geometry rather than simply due to a difference in the spatial orientation of two groups of substituents.

Q: Why would ring strain result in non-stability of the molecule?

A: If ring strain is present, this would cause the orbitals that are involved in covalent bond formation NOT to overlap effectively. Such covalent bonds are weaker, hence resulting in lower stability. In addition, there would also be inter-electronic repulsion between groups of atoms, which we will be discussing later.

Restricted rotation about a double bond between nitrogen atoms also gives rise to a pair of *cis-trans* (geometrical) isomers:

Cis isomer   Trans isomer

The number of *cis-trans* (geometrical) isomers that are possible for a given molecular formula depends on the number of bonds wherein restricted rotation arises. In addition, there must be different substituents attached to the bonding atoms at each of these bonds.

Example 2.2: Draw all the possible *cis-trans* (geometrical) isomers of 1-bromopenta-1,3-diene.

Solution: There are a total of four *cis-trans* (geometrical) isomers of 1-bromopenta-1,3-diene.

cis,trans-1-bromopenta-1,3-diene   *trans,trans*-1-bromopenta-1,3-diene

trans,cis-1-bromopenta-1,3-diene   *cis,cis*-1-bromopenta-1,3-diene

Exercise: Identify the non-cyclic *cis-trans* (geometrical) isomers for C_4H_7Cl.

Solution:

One pair of geometric isomers   A second pair of geometric isomers

A third pair of geometric isomers

In summary, *cis-trans* (geometrical) isomerism arises from either a restricted rotation about a double bond or a restricted rotation about a bond in a ring structure. With respect to the bonds concerned, the bonding atoms must be attached to two different substituents.

Q: If there are two pi bonds in a molecule, such as in an alkyne molecule, is there restricted rotation about the $C{\equiv}C$ bond?

A: No. This is because the two pi bonds are formed with two *p* orbitals contributed by each of the carbon atoms. These two *p* orbitals are perpendicular to each other. As a result, the overall pi electron cloud is cylindrically surrounding the carbon–carbon sigma bond. Hence, the pi bonds are not broken even when rotation about the $C{\equiv}C$ bond occurs.

Q: Why is there restricted rotation about single bonds in a ring structure? Aren't they free to rotate?

A: This question brings us to the next theme on ring strain.

The term *ring strain* indicates some form of tension in a ring structure. In certain ring geometries, the atoms are forced to be close together, and this brings about greater repulsion between their electron clouds. Ring strain tends to destabilize the molecule, and indications of such stresses are reflected in the bond angles and the bond energies. Thus, ring strain would be more severe for lower-membered rings such as cyclopropane and cyclobutane.

Take, for instance, cyclopropane. For a carbon atom bonded to four substituents, repulsion between electron clouds is minimal at bond angles of 109.5° (as in a tetrahedral configuration). But when three of such carbon atoms are forced together into a ring, the C–C bond angles are acute and are less than the ideal 109.5° (What is the internal angle of an equilateral triangle?). In order to assume the triangular ring structure, the orbitals of two adjacent carbon atoms would not be able to overlap exactly head-on with each other to form the sigma bond. In chemical bonding terms, we say that the overlap is less effective. Consequently, the bond energy is less endothermic, indicating a weaker bond strength.

Even when we talk about saturated alkanes such as ethane, rotation about a single bond is not smooth sailing.

2.3.2 *Conformational Isomerism*

Conformational isomerism is best understood when one acquires the skill of visualising the spatial orientation of groups of atoms around a central carbon atom using the Newman projection method. In this scheme, the spatial orientation of substituents about a carbon–carbon bond is visualized from the front-to-end view about this bond. Take ethane, for instance. Its Newman projection has C1 depicted as a point of intersection of three C–H bonds, and C2 is represented by the circle at the rear (see Fig. 2.1).

Rotating only one of the C atoms of the carbon–carbon bond will bring the H substituents into different spatial orientation and hence different proximity to one another. On the left side of Fig. 2.1 is the staggered conformation and on the right, the eclipsed conformation. There are other intermediate conformations between these two, but our attention centers on the staggered and eclipsed conformations. These different but specific spatial orientations, obtained through rotation about a single bond, represent what is known as *conformational isomers*. The Newman projection is one approach for illustrating such isomers. Another approach is the Sawhorse representation, as shown in Fig. 2.1.

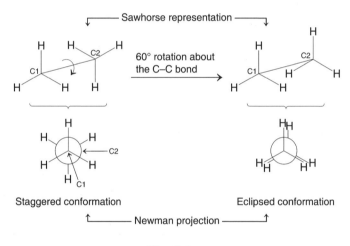

Fig. 2.1.

Notice that the H atoms on both C atoms are much closer together in the eclipsed conformation? The inter-electronic repulsions between these H atoms are the greatest due to their close proximity. This structure is in a "tension" state and has a higher energy state than the staggered conformation. The staggered conformation is thus more stable than the eclipsed conformation, as shown in Fig. 2.2.

Q: Why doesn't the C–C single bond break when the carbon atoms rotate?

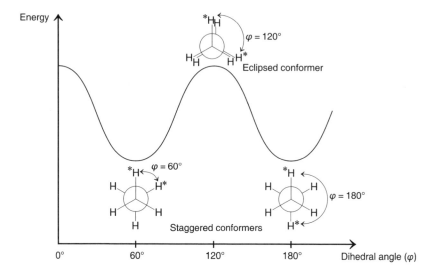

Fig. 2.2. Relative energies of the ethane conformers.

A:

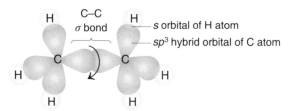

The C–C single bond is a sigma bond formed from the head-on overlap of two sp^3 hybrid orbitals. As a result, the electron density that accumulates within the inter-nuclei region is near to cylindrical in shape. Rotation about the C–C single bond does not change the degree of effective overlap, thus it does not cause the bond to break.

Turning our attention to the middle two carbon atoms in the butane molecule, the Newman projection has C2 depicted as the front carbon and C3, the rear carbon represented by the circle (see Fig. 2.3).

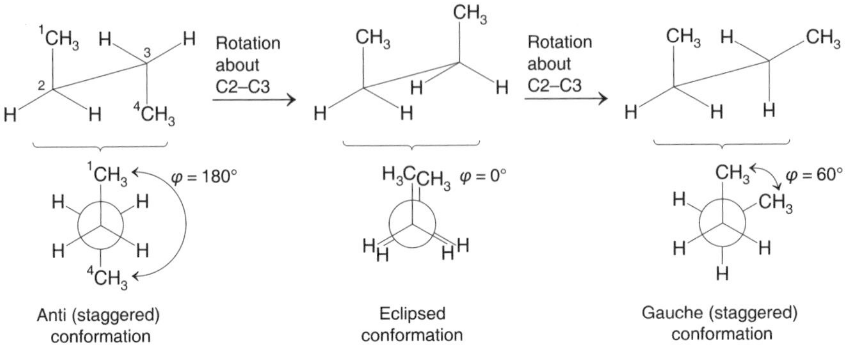

Anti (staggered) conformation   Eclipsed conformation   Gauche (staggered) conformation

Fig. 2.3.

Now imagine rotating the butane molecule about the bond between C2 and C3 clockwise, starting with the leftmost staggered conformation. In no time, we will get to the eclipsed conformation. The two methyl groups are closer together and the greater repulsive forces, especially between these bigger electron clouds, cause the structure to have a higher energy state than the first. If we continue rotating the bond, we will arrive at another staggered conformation, shown on the right. The repulsive forces are now reduced, but this structure still has a higher energy state than the very first structure. This example shows that not all staggered conformations have the same energy, and the same can be said of the eclipsed conformations (see Fig. 2.4). The rotation from one conformation to the next is best described in terms of the dihedral angle, which in this case is the angle between the two methyl groups when viewed in the Newman projection. The anti conformation is the most stable one because the two bulky methyl groups are furthest apart from each other, at a dihedral angle of 180°. In contrast, the eclipsed conformer with the methyl groups aligned together is the least stable of all, as it has the maximum inter-electronic repulsion.

Cis-trans (geometrical) isomers and conformational isomers are actually under the sub-class of stereoisomers known as *diastereomers*. Both diastereomers and *enantiomers* make up the two main types of stereoisomers. Enantiomers consist of molecules which are mirror images of each other, and these two molecules are

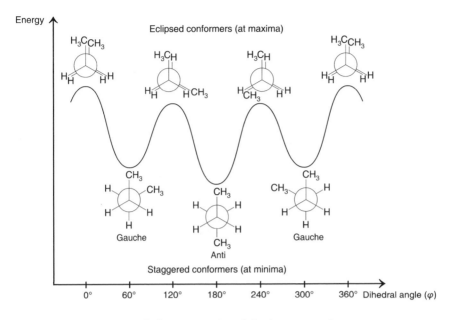

Fig. 2.4. Relative energies of the butane conformers.

non-superimposable, where as diastereomers are not mirror images of each other, and they are non-superimposable.

2.3.3 *Enantiomerism*

Q: What is meant by the term "superimposable mirror images?"

A: If we place a mug in front of a mirror, we get an image that is laterally (side-way) inverted. But when these two mugs are placed together, they are indistinguishable with the handle facing the same direction. We say that both mugs are superimposable mirror images of each other.

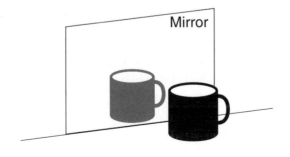

If we place a star on the original mug, and try to rotate its "mirror image" in an attempt to get an exact replica of the first, it cannot be done. The orientation of the star relative to the handle will never be the same for both objects. We term these pair of mugs to be non-superimposable mirror images of each other. Other examples that highlight this notion of non-superimposition include our trusted pair of right and left hands.

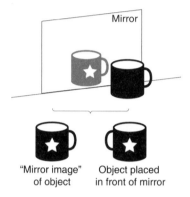

In the molecular world, there are pairs of molecules which are "mirror images" of each other. If the mirror images of this pair of molecules are superimposable on each other, then essentially these two mirror images are identical. But if the mirror images are non-superimposable, then these two molecules are actually two different, unique molecules. We term such a pair of non-superimposable molecules, which are mirror images of each other, enantiomers. But how can we tell whether a pair of molecules is actually a pair of enantiomers?

In Fig. 2.5, we have a molecule of the general formula CX_3Y, where X and Y can be any atoms or groups of atoms attached to the central C atom. Structures A and B are mirror images of each other. To determine if they are enantiomers, they must be non-superimposable mirror images. If we rotate structure A on the vertical axis, structure B is obtained. Hence, they are superimposable images. They are not enantiomers but are actually the same representations of the same compound.

In Fig. 2.6, we have a molecule with a sp^3 hybridized C atom attached to four different atoms or groups of atoms. We term such a carbon atom, with four different groups bonded to it, a *chiral* carbon.

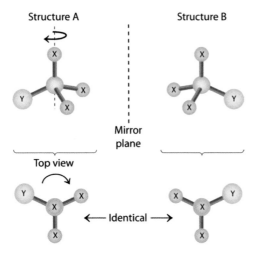

Fig. 2.5.

Regardless of whatever way we rotate structure C, it is not possible to superimpose it onto structure D. For both structures, there are always two substituents that do not match up. Since they are non-superimposable, structures C and D constitute a pair of enantiomers. Thus, due to the difference in spatial orientations, enantiomers are stereoisomers.

The following diagram is one common way of depicting the relationship between a pair of enantiomers:

Mirror
plane   * chiral center

For a molecule with n chiral centers, there can be a maximum of 2^n stereoisomers. Why the term "maximum?" This is due to the presence of a plane of symmetry which may result in superimposable mirror images.

For 2-bromo-3-chlorobutane, the presence of two chiral centers gives rise to a total of four stereoisomers. Structures A and B are a pair of enantiomers while structures C and D represent another pair.

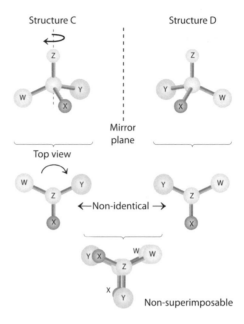

Fig. 2.6.

The two pairs of enantiomers are diastereomers of each other. Thus, we see that diastereomers are actually stereoisomers with the same constitutional/structural formula but different spatial orientation of the various groups of substituents in such a way that the molecules are non-superimposable with each other.

However, for 1,2-dibromoethane-1,2-diol, which has two chiral centers, there are only a total of three stereoisomers — a pair of enantiomers and a diastereomer.

$$HO-\overset{Br}{\underset{H}{C^*}}-\overset{H}{\underset{Br}{C^*}}-OH$$

* chiral center

Structure W   Structure X

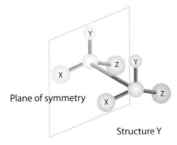

Mirror plane

Front view

←Non-identical→

Enantiomers:
Non-superimposable
mirror images

Structures W and X are a pair of enantiomers. Neither rotation about the C–C bond nor perpendicular to it in structure W can produce an identical structure to that of structure X. They are thus non-superimposable mirror images.

Another possible stereoisomer Y can be obtained by aligning identical substituents on the same planes. Y is said to be a diastereomer of W and X.

Plane of symmetry

Structure Y

Does structure Y have a non-superimposable mirror image which could possibly be the fourth stereoisomer? Notice that structure Y has an internal plane of symmetry.

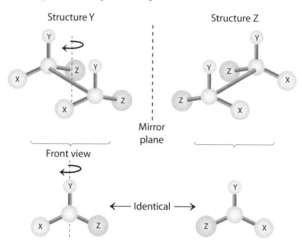

Rotating structure Y about the axis through the C–C bond gives us structure Z. Thus, they are superimposable mirror images and represent the same compound. Such a compound is known as a *meso* compound — an achiral molecule that contains chiral atoms. Meso compounds are optically inactive, which we will discuss later. Looking back at the examples above, you will find that a chiral molecule has no plane of symmetry. On the other hand, an achiral molecule can have an internal plane of symmetry. Of course, a molecule can be achiral if it does not contain any chiral centers.

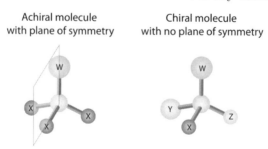

Due to the difference in spatial orientation, when each of the pair of enantiomers interacts with another chiral molecule, the reaction rate is different, and the products formed are obviously dissimilar

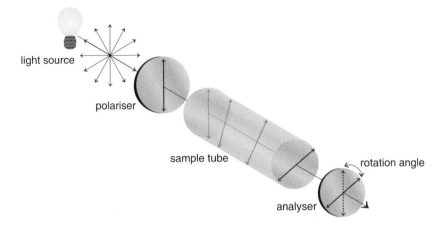

Fig. 2.7. Schematic representation of a polarimeter.

too. But apart from their different behavior towards another chiral molecule, a pair of enantiomers has identical chemical properties towards achiral molecules. They have similar physical properties such as the same densities, melting and boiling points. The only physical property that is dissimilar is the behavior towards plane-polarized light. A pair of enantiomers rotates plane-polarized light in the opposite directions, but to the same extent. The difference in this property helps us to distinguish the enantiomers and hence the use of a polarimeter for this purpose. So, how does a pair of enantiomers response differently to plane-polarized light?

Light is a form of electromagnetic radiation with numerous planes of changing electric field, each plane of electric field with a corresponding plane of changing magnetic field perpendicular to it. An ordinary beam of light is unpolarized because of the presence of different planes of changing electric field perpendicular to the direction of movement. Quite the opposite, plane-polarized light consists of only one plane of changing electric field perpendicular to the direction of movement. Unpolarized light can be made plane-polarized by passing it through a polarising filter, which only allows a particular plane of electric field to pass through and absorbed all the other planes of electric field. As a result, plane-polarized light is "dimmer" than unpolarized light.

The workings of a polarimeter make use of this idea. The schematic set-up of a polarimeter consists of two polarising filters, one is called the polarizer and the other is the analyser (see Fig. 2.7). A sample tube is placed in between them. A beam of light is passed through a polarizer to become plane-polarized. It then passes through the sample tube containing a substance of interest. If this substance is optically active, it rotates the plane-polarized light and this is captured by the analyser. The analyser is rotated until it "sees" the maximum intensity of the rotated polarized light. The amount of this rotation corresponds to the amount of rotation the plane-polarized light has undergone after it emerged from the sample tube. If the initial plane-polarized light is rotated clockwise, the optical rotation is regarded as positive (+) and the substance is said to be *dextrorotatory* (turning to the right). If an anticlockwise rotation (−) is noted, the substance is *levorotatory* (turning to the left). The ability of enantiomers to rotate plane-polarized light means they are optically active and hence are known as optically active compounds. A chiral molecule is optically active whereas an achiral molecule is optically inactive.

Q: But how does a molecule rotate plane-polarized light?

A: As explained before, a plane-polarized light consists of only one plane of changing electric field. When this plane-polarized light encounters a molecule, the electric field of the light interacts with the electron cloud. If the distribution of electron density is homogeneous, which means that there is no potential difference within the homogeneous electron cloud, the plane-polarized light is not rotated. But if the distribution of electron density is non-homogeneous at every point in space, for example in a chiral molecule, then the plane-polarized light has different spatial interaction as it passes through the molecule. This results in a rotation of the plane-polarized light.

Q: Then why does a pair of enantiomers rotate light in opposite directions but to the same degree?

A: A molecule of an enantiomeric pair has a non-homogeneous distribution of electron density at every point in space. The other member of the same enantiomeric pair also has this same non-homogeneous distribution of electron density, but it is actually a mirror image of the former one. Thus, it accounts for the same degree of rotation but in opposite directions.

For one pair of enantiomers, there is one (+)-enantiomer and one (−)-enantiomer. If a sample containing equal proportions of this enantiomeric pair is placed in the polarimeter, no net rotation of the plane-polarized light would be observed. The rotating effect of one enantiomer is exactly cancelled out by the opposite rotation, of the same magnitude, by the other member of the enantiomeric pair. This 50–50 mixture is optically inactive and is known as the *racemic-mixture* or *racemate*.

The configuration of enantiomers at a specific chiral carbon is denoted using the R, S notation. The designations of optical rotation do not correlate with the R, S configurations, i.e. an (R)-enantiomer of a chiral compound may be the (+)-enantiomer, but for another compound, it may be the (−)-enantiomer. The reverse holds true. A chiral carbon is labeled R or S depending on the spatial orientation of the four different groups (see Fig. 2.8). Each group is assigned a *priority*, according to the Cahn–Ingold–Prelog priority rules, which is based on atomic numbers; an atom with higher atomic number would have higher priority. If the chiral carbon is positioned in such a way that the lowest-priority (labeled number 4) of the four groups is pointed away from a viewer, the viewer would see two possibilities: If the priority of the remaining three groups decreases in a clockwise direction $(1 \rightarrow 2 \rightarrow 3)$, it is labeled R (for *Rectus*, Latin for "right"), but if it decreases in a counter clockwise direction $(1 \rightarrow 2 \rightarrow 3)$, it is S (for *Sinister*, Latin for "left").

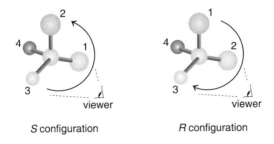

S configuration   R configuration

Fig. 2.8.

The use of molecular models would be useful in visualising a pair of enantiomers. Nonetheless, we can make use of some *general* indicators. The existence of non-superimposable isomers is said to exist:

- if a molecule contains only one chiral center; or
- if there is more than one chiral center and the molecule has no internal plane or center of symmetry.

A chiral center is an atom that is bonded to four different atoms or groups of atoms and such a configuration results in a pair of non-superimposable molecules which are mirror images of each other. And here we have it, chiral compounds exist as a pair of non-superimposable mirror images (enantiomers) which exhibits optical activity (optically active compounds). Let us now look into the imposing criterion of having chiral center(s).

Q: In view of the term "general indicators" used, does it mean that a molecule can still be chiral, i.e. optically active, even though it does not possess any chiral centers?

A: Yes, indeed. The whole idea about a chiral center is that it would certainly give rise to a non-homogeneous distribution of electron density in such a way that there is also a lack of an internal plane of symmetry within the molecule. What do we mean by that? Take CH_2BrCl for instance. It is a polar molecule. This means that there is already a non-homogeneous distribution of electron density in the molecule. But if you look at the molecule closely, there is also a plane of symmetry. This means that the plane of symmetry would cut this non-homogeneous distribution of electron density into two parts which are exactly mirror images of each other. Hence, such a molecule would not be optically active. There are molecules that do not possess any chiral centers at all, and yet there is a non-homogeneous distribution of electron cloud with the absence of a plane of symmetry — such molecules are still optically active. One such molecule is the molecule discussed in Question 3 under the My Tutorial section.

Q: Is chirality and optical activity synonymous?

A: No! Chirality is a molecular property (same for enantiomerism) that is restricted to a particular molecule in consideration whereas optical activity is a BULK property (or macroscopic) of a collection of chiral molecules.

Q: So, we cannot say that an enantiomer is an optical isomer?
A: Yes, it is inappropriate to do that as enantiomerism is the consideration at the molecular level. You cannot measure the optical activity of a molecule alone. So, you cannot say that a molecule is an optical isomer.

Example 2.3: Identify the chiral carbons in the following molecule.

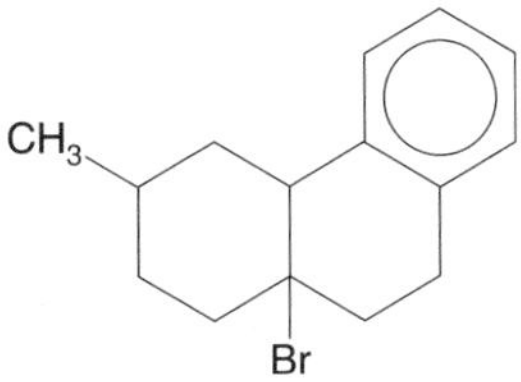

Approach: A chiral carbon is bonded to four different atoms or groups of atoms. Look out in a systematic order for the existence of such carbon atoms. Take note of skeletal structures, such as this one here, where C–H bonds are not explicitly shown and a C–C bond is indicated by a "line." There are a total of three chiral carbons:

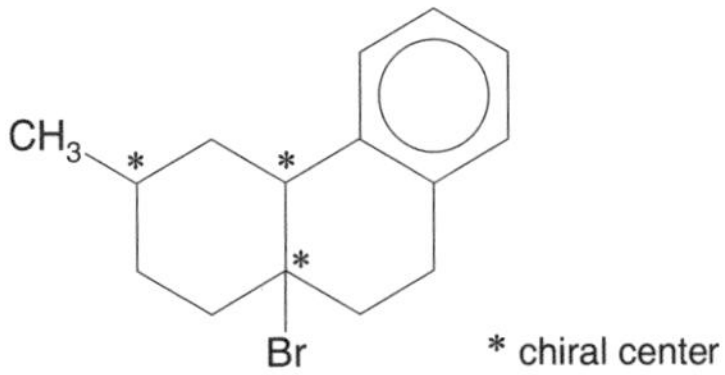

Sometimes, the differences between the substituents may not be obvious. For instance, moving away from C1 are two similar $-CH_2$ groups, but these are further attached to C atoms with different bonding patterns. Thus, these two substituents are not considered identical. The same goes for the chiral carbon bonded to the Br atom — the difference between two of its substituents only surfaces at the third carbon atom away from the chiral center. Note that the circled C atoms are not chiral as they are sp^2 hybridized, i.e. they are bonded to only three other substituents.

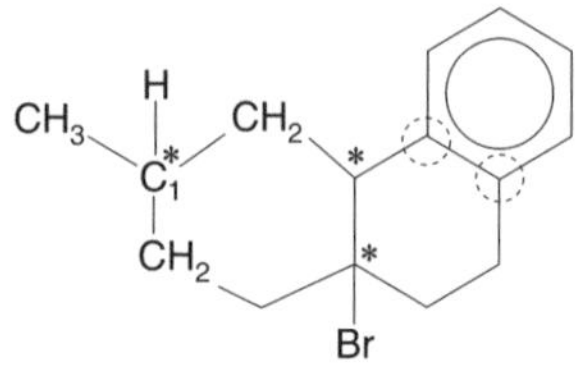

Exercise: Identify the chiral carbons in the following molecules.

(1)   (2)

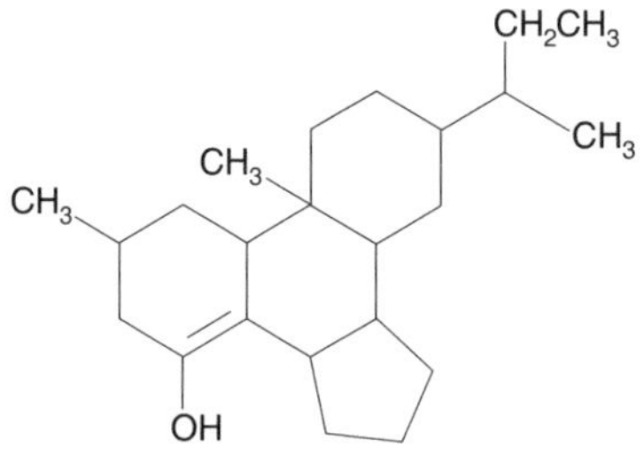

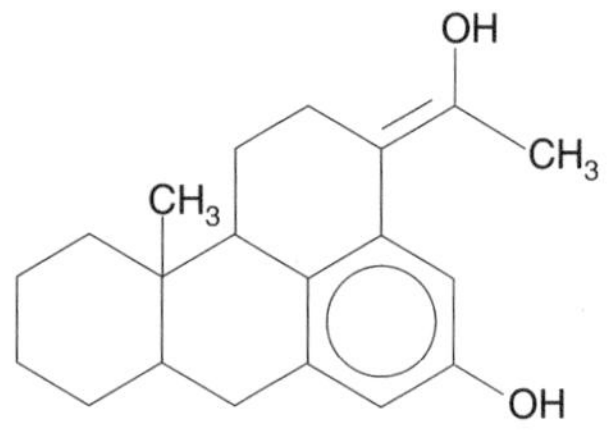

Solution:

(1)   (2)

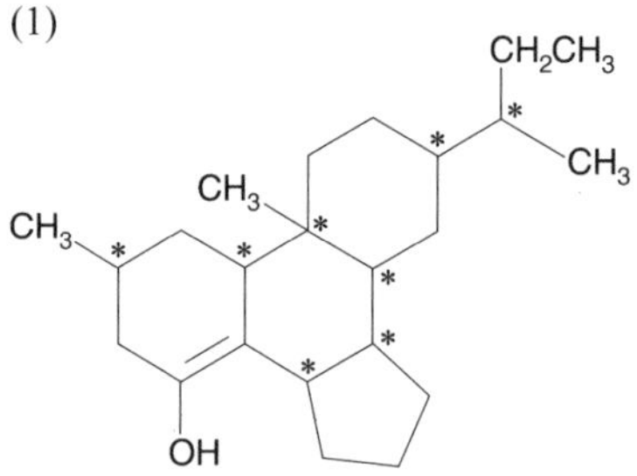

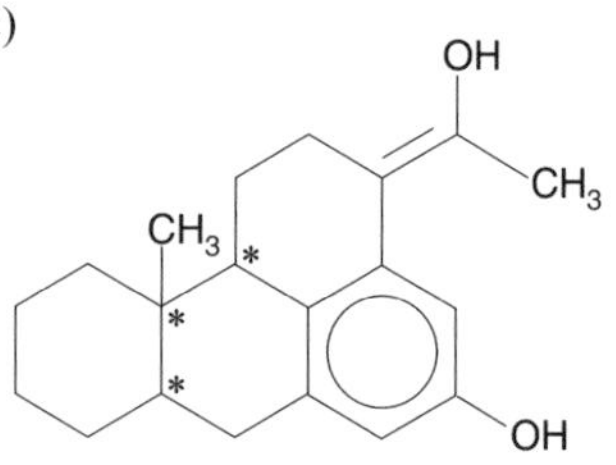

* chiral center

Exercise: How many non-cyclic compounds with the formula $C_5H_{12}O$ exhibit optical activity?

Solution: There are four such compounds, i.e.

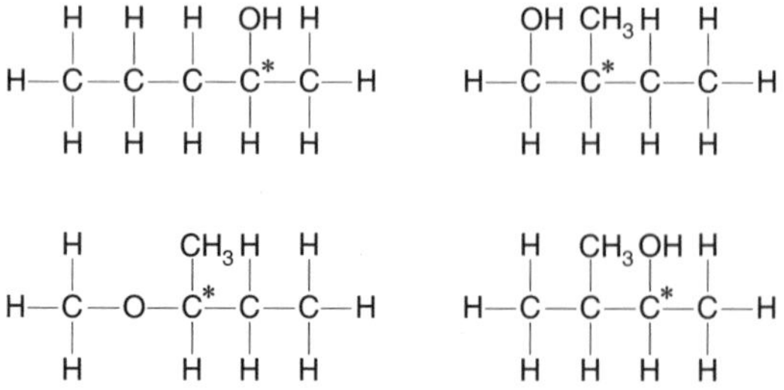

* chiral center

Q: Why is it so important to know about the concept of enantiomers?

A: Just like how a left-handed person cannot play with a golf club meant for right-handed people, requiring instead a custom-made club, there are reactions that utilize one enantiomer and not its mirror image. When in the presence of another enantiomer, a pair of enantiomers may react differently. This is an important point especially when dealing with drugs for consumption. As our biological system involves many molecules that are enantiomeric in nature, the introduction of a pair of enantiomers may have different outcomes — one gives beneficial results while the other may cause adverse effects. For instance, one enantiomer of ethambutol treats tuberculosis, while the other, in stark contrast, causes blindness.

My Tutorial

1. Discuss the various types of isomerism present in the following molecules and draw the constitutional/structural formulae of all possible constitutional/structural isomers and/or stereoisomers:

   (a) C_5H_{12}
   (b) C_2H_6O
   (c) alkenes of formula C_4H_8
   (d) cycloalkane of formula C_4H_8
   (e) 1,2-dibromo-1,2-dichloroethene
   (f) 2-bromobutane
   (g) 1,3-diethylallene

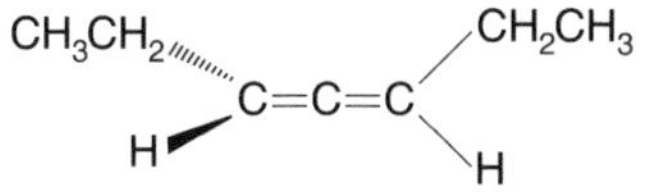

   (h)

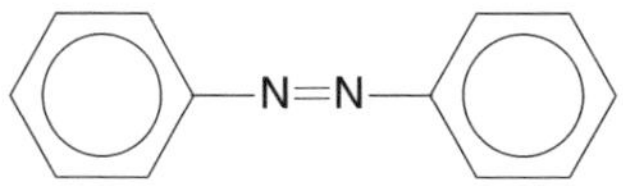

   (i)

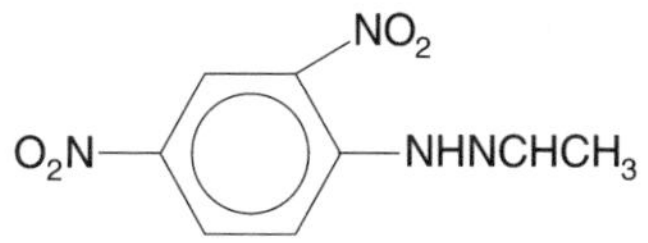

   (j) $HON{=}CHCH(OH)CH_3$

2. Using the isomeric amino acids $C_3H_7NO_2$ as examples, explain the meaning of the terms *constitutional/structural* and *enantiomerism*.

3. Explain how the following structure can occur as enantiomers.

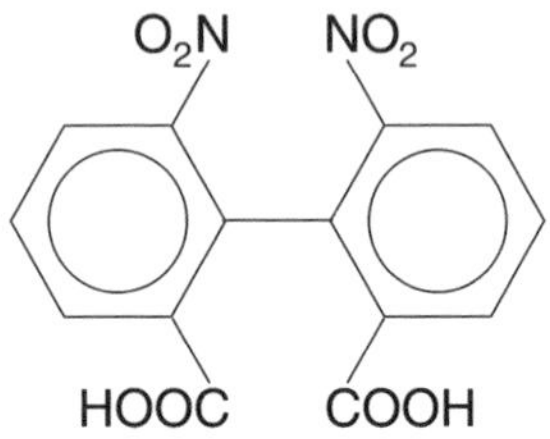

4. The synthesis of ephedrine, a decongestant medicine, is shown below:

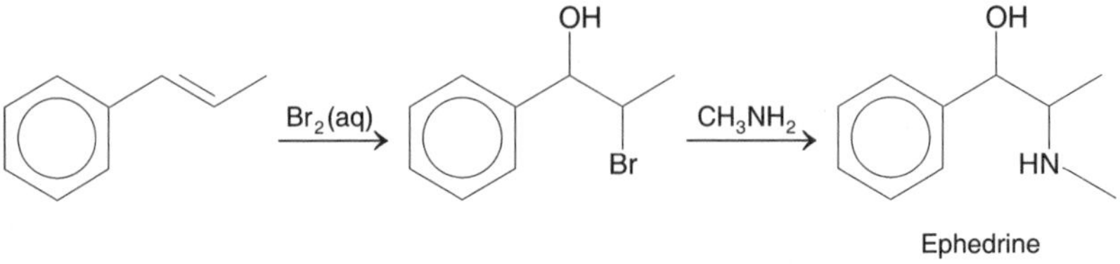

Ephedrine

When the above synthetic route is used, four products with the same constitutional/structural formula are produced. Out of these four products, two have different melting points and solubilities in water.

(a) Explain how four such isomers with the same constitutional/structural formula can arise and draw the three-dimensional structures to show the differences. Name the type of isomerism present.

(b) With the aid of labeled diagrams, explain the reasons for the differences in melting point and solubility of the four isomers.

(c) With reference to the constitutional/structural formula of ephedrine, explain why it can act as a decongestant.

5. Explain clearly each of the following terms:

(a) empirical formula
(b) molecular formula
(c) constitutional/structural isomerism
(d) chain isomerism

(e) positional isomerism
(f) functional group isomerism
(g) stereoisomerism
(h) *cis-trans*/geometrical isomerism
(i) enantiomerism

CHAPTER 3

Organic Reactions and Mechanisms

3.1 Introduction

What is a chemical reaction? You would say that it is a process whereby substances interact and transform into new ones, usually with different properties. How does the transformation occur? Do the reactant molecules simply collide with each other and therefore combine to form new compounds? If this is so, why do certain reactions occur only when appropriate conditions such as heating or catalyst are imposed?

A classical view of chemical reactions involves the rearrangement of particles (atoms, ions or molecules), and for this to occur, "old" bonds are broken before "new" bonds are formed. But these rearrangements do not happen by simply bumping molecules around, even though collision is a precursor for a fruitful reaction. When two reacting particles approach each other, there is repulsion between their negatively charged electron clouds. This repulsive force decreases the speed of the approaching particles while they are colliding with each other. If they do not have sufficient kinetic energy to overcome the repulsive force that tries to keep them apart, by the time they "touch" each other, they would simply be "pushed" apart. In such an instance, bonds are neither broken within a particle nor formed between them.

Let us try a different scenario. This time round, when the approaching particles have sufficient kinetic energy, effectively known as the **activation energy** (E_a), they can overcome the repulsive force, which then leads to the inter-penetration of their electron clouds. At this juncture, there can be rearrangement of valence electrons, with the partial breaking of old bonds and the partial formation of new ones — this fuzzy picture is aptly described as the formation of an **activated complex** in the transition state. In the energy profile diagram (see Fig. 3.1), the transition state is located at the point of maximum potential energy. This is because part of the kinetic energy that the reacting particles have initially has been converted to electrostatic potential energy and stored within the bonds of the activated complex.

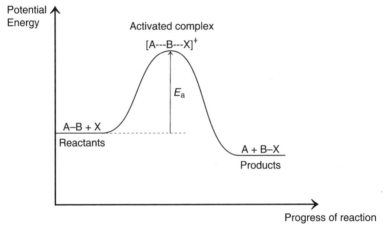

Fig. 3.1.   Energy profile diagram for a single-step reaction.

The preceding description is the essence of what is known as the transition state theory, formulated to account for what specifically has happened to the initial kinetic energy of reactant particles as the reaction progresses, using the law of the conservation of energy. It focuses on the activated complex that is formed at the point where the reactants are just about to change into products. Once formed, the activated complex can transform into either products or "fall back" to regenerate the reactants. This explains the reversible nature of reactions and why certain reaction conditions have to be utilized to counteract this "issue" so as to increase the desired yield of the product.

More often than not, reactions occur in more than one step. Take, for instance, Reaction 1, represented by Eq. (3.1):

Reaction 1: $H_2O_2(aq) + 2H^+(aq) + 2I^-(aq)$

$$\longrightarrow 2H_2O(l) + I_2(aq). \tag{3.1}$$

There are five reactant particles. What is the probability of having these collide instantaneously with one another at one go? It is as improbable as trying to roll the same number of billiard balls across the table in the hope that they would be striking altogether at the same time. It is at this point that we would like to introduce the concept of the reaction mechanism — to get an insight on how reactants combine to form products.

A reaction can be thought to proceed in a series of elementary steps. An elementary step is a fundamental step that cannot be reduced further into other sub-steps. Each of these elementary steps produces new substances which, in order to differentiate them from the final product, are called **intermediates**. There is a transition state pertaining to each elementary step with an associated activated complex for that step. But unlike an activated complex, an intermediate can be isolated as it is much more stable than the activated complex. As its name implies, an intermediate once formed subsequently reacts in a later step. A series of intermediate reactions will then lead to the final step that produces the desired product.

The set of elementary steps is collectively known as the reaction mechanism. Referring to Eq. (3.1), the reaction mechanism proposed for Reaction 1 consists of the following elementary steps:

$$\text{Step 1: } H_2O_2 + I^- \longrightarrow H_2O + OI^- \quad \text{(slow)}$$
$$\text{Step 2: } H^+ + OI^- \longrightarrow HOI \quad \text{(fast)}$$
$$\text{Step 3: } HOI + H^+ + I^- \longrightarrow H_2O + I_2 \quad \text{(fast)}.$$

OI^- and HOI are the intermediates or reaction intermediates. The **molecularity** of an elementary step is the number of *reacting species* that react in that particular step. For instance, both Steps 1 and 2 are bimolecular in nature. Step 3 is termolecular, involving three reacting species. A unimolecular step consists of only one reacting species.

Q: Why is the term "reacting species" being used and not "reactants?"

A: The term "reactant" refers to the species that we put into the reaction vessel in the beginning. The reacting species that appear in the intermediate steps may not be a reactant but an intermediate instead. Thus, to prevent the confusion, we have chosen to use the term "reacting species" to describe the particles involved in the intermediate reactions.

When all the elementary steps of a proposed reaction mechanism are summed up, you get the overall chemical equation.

Step 1: $H_2O_2 + I^- \longrightarrow H_2O + OI^-$ (slow r.d.s.)

Step 2: $H^+ + OI^- \longrightarrow HOI$ (fast)

Step 3: $HOI + H^+ + I^- \longrightarrow H_2O + I_2$ (fast)

Overall equation: $H_2O_2 + 2H^+ + 2I^- \longrightarrow 2H_2O + I_2$.

In any particular reaction mechanism, there is bound to be one particular step that is the slowest step. This slowest step is known as the **rate-determining step** (r.d.s.). It is usually indicated by the word "slow" in the elementary step. The overall rate of reaction is equal to (or controlled by) the rate of the rate-determining step, i.e. the rate of formation of the final products, H_2O and I_2, depends on how fast Step 1 proceeds and not how quickly Step 2 or 3 proceed. Take the analogy of moving out of a crowded place like a cinema or a concert venue. Even if you are the fastest walker on the planet, the rate at which you can exit depends not on you but on the persons in front of you, and how fast they exit in turn depends on those further up-front closer to the doorways.

A rate equation is usually obtained first before formulating the mechanism for a reaction. It informs us of the quantified relationship between the rate of reaction and the concentration of the reactants. It is expressed as:

$$\text{Rate} = k[\text{A}]^x[\text{B}]^y,$$

where:

- rate has the units of mol dm^{-3} time^{-1},
- x and y are known as the order of reaction, and
- k is the rate constant for a given reaction. It is temperature-dependent and its units depend on the overall order of reaction.

The order of reaction with respect to a given reactant is the power to which the reactant's concentration is raised in the rate equation. The sum of the orders of the reaction (i.e. $x + y$) gives the overall order of reaction. These are experimentally determined quantities; they may not be equivalent to the stoichiometric coefficients of the reactants in the balanced chemical equation.

But how do we obtain the rate equation? We first have to determine the order of reaction, i.e. we experimentally measure the change in rate when the concentration of a reactant changes. These concentration changes are indirectly monitored through a related physical property whose changes can be measured during the progress of reaction. Examples include visible changes such as formation of product precipitate, color changes, volume of gas evolved, pressure of gas and electrical conductivity. For Reaction 1, we can monitor the intensity of the brown color of iodine, since this is the only colored species present in the system.

Based on experimental kinetic studies, the rate equation for Reaction 1 is:

$$\text{Rate} = k[H_2O_2][I^-].$$

It tells us that the reaction rate is directly proportional to $[H_2O_2]$, i.e. rate $\propto [H_2O_2]$. When $[H_2O_2]$ is doubled and all else kept constant, the reaction rate is twice as fast. Reaction rate is also directly proportional to $[I^-]$, i.e. rate $\propto [I^-]$.

We have to take note that when we try to determine the rate equation, the only variables that we can alter are the concentrations of the respective reactants. Hence, the rate equation must always be expressed in terms of the concentration of reactants and not the intermediates of the elementary step. The reactants that appear in the rate equation are the ones that appear in the rate-determining step. But what happens when an intermediate appears in the rate-determining step? Then all we need to do is to find a relationship between the concentration of this intermediate and the concentrations of the reactants that formed this particular intermediate in the preceding step, using the steady-state approximation method. For the case of Reaction 1, the stoichiometric ratio in the rate-determining step is denoted by the power of the concentration terms in the rate equation. In this case, H_2O_2 and I^- react in the mole ratio of 1:1 in the rate-determining step.

The reaction mechanism is depicted pictorially by the energy profile diagram shown in Fig. 3.2. We have only shown the reacting species for each of the elementary steps and not the products of the elementary steps in the energy profile.

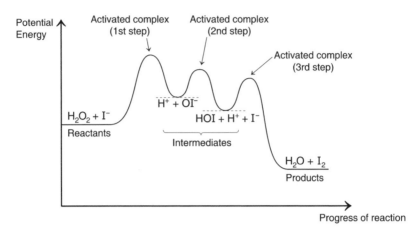

Fig. 3.2.   Energy profile diagram for a three-step reaction (not drawn to scale).

Q: Can there be more than one proposed mechanism for a given reaction?

A: Yes, it is possible as long as the following two criteria are met: (i) the reacting stoichiometry of the reactants in the rate-determining step and the other preceding steps leading to it coincide with the experimentally determined order of reactions in the rate equation and (ii) the summation of the elementary steps should give the overall reaction chemical equation.

Q: What is steady-state approximation?

A: Since the rate-determining step controls the overall rate of the reaction, we can assume that each of the preceding steps will reach a state of dynamic equilibrium or steady-state "while waiting in the queue to clear through the jam." Thus, we can use the equilibrium expression to find the concentration of any intermediate with respect to the concentrations of the reactants that we put into the system. The following example could help to clarify how we use steady-state approximation. Consider the overall reaction:

$$2A + B \longrightarrow C + D.$$

Its mechanism is:

Step 1: $A + B \rightarrow \beta$ (fast)

Step 2: $\beta + A \rightarrow C + D$ (slow).

For Step 1, the rate equation is $\text{rate}_1 = k_1[A][B]$.

For Step 2, the rate equation is $\text{rate}_2 = k_2[\beta][A]$.

(Note: We can only write the rate equation straight away if it is an elementary step.)

Assuming steady-state for Step 1, then forward rate equals to backward rate, i.e.

$\text{rate}_1 = k_1[\text{A}][\text{B}] = \text{rate}_b = k_b[\beta]$

$\Rightarrow k_1[\text{A}][\text{B}] = k_b[\beta]$.

Rearranging, we get $[\beta] = k_f[\text{A}][\text{B}]/k_b$.

Substituting $[\beta] = k_f[\text{A}][\text{B}]/k_b$ into the rate equation for Step 2 gives

$$\begin{aligned}
\text{rate}_2 &= k_2[\text{A}] \times k_f[\text{A}][\text{B}]/k_b \\
&= (k_2 \times k_f/k_b)[\text{A}]^2[\text{B}] \\
&= k[\text{A}]^2[\text{B}]
\end{aligned}$$

where

$$k = k_2 \times k_f/k_b.$$

So experimentally, we would find that the above reaction is second-order with respect to A and first-order with respect to B.

3.2 Types of Bond Cleavage

One of many reasons that cause an elementary step to be the rate-determining step is the need to break chemical bonds within the reacting species. And the stronger the bond to be broken, the slower would be this particular elementary step. The breaking of chemical bonds is a typical phenomenon that happens in a chemical reaction before new bonds can be formed to generate the products. There are two ways that bond cleavage or fission can occur:

- Homolytic bond cleavage
- Heterolytic bond cleavage

The type of bond cleavage is affected by whether the bond is polar or not, which in turn is affected by the difference in electronegativity of the two bonding atoms.

Q: What is electronegativity?

A: It denotes the ability of an atom to polarize (distort) the electron cloud that is shared between the two atoms. The higher the effective nuclear charge, the higher the electronegativity value. Thus, we expect electronegativity to increase across a period and decrease down a group in the periodic table.

When the bonding atoms have the same electronegativity, the bonding electrons are shared equally between the two nuclei, forming a pure covalent bond.

A difference in electronegativity creates a permanent separation of charges (known as dipole) in the bond. The more electronegative atom has a stronger hold on the bonding electrons. It is considered "electron-rich" and gains a partial negative charge ($\delta-$). The other bonding atom, which is less electronegative, partially loses its hold on the electrons and acquires a partial positive charge ($\delta+$) — an indication of it being "electron-deficient." Such covalent bonds with unequal sharing of electrons are termed polar bonds.

3.2.1 *Homolytic Bond Cleavage*

If the electronegativity of the bonding atoms, X and Y, are similar, each of them has comparable attractive forces for the bonding electrons. When the bond between them is cleaved, the shared pair of electrons is distributed equally among the two atoms. As X and Y have one unpaired electron each, they are known as **free radicals** — which refer to species that have one or more unpaired electrons.

$$X—Y \longrightarrow X\cdot + Y\cdot$$

Take, for instance, the dot-and-cross diagram that illustrates the distribution of electrons in a Br_2 molecule before and after homolytic bond cleavage:

$$:\overset{\cdot\cdot}{Br}\cdot_\times\overset{\times\times}{\underset{\times\times}{Br}}{}_\times^\times \xrightarrow[\text{cleavage}]{\text{bond}} :\overset{\cdot\cdot}{Br}\cdot + {}_\times\overset{\times\times}{\underset{\times\times}{Br}}{}_\times^\times$$

Bonding electrons   2 Br• free radicals

To differentiate between the types of bond cleavages, we make use of curved arrows. The "half-headed" arrow indicates the transfer of a single electron, from the tail of the arrow to the head of the arrow that points at the recipient.

$$Br—Br \longrightarrow Br\cdot + Br\cdot$$

Q: If a radical has two unpaired electrons, why wouldn't the two unpaired electrons simply pair up? Then it would no longer be a radical and would hence more stable, wouldn't it?

A: It is incorrect to think that pairing electrons up always leads to stability. Pairing electrons up would only lead to stability if the pairing is necessary compared to other alternatives. Take, for instance, the oxygen atom, which has eight electrons. There are four electrons to be distributed into the $2p$ subshell. It is necessary to pair up the electrons in one of the $2p$ orbitals (although there is inter-electronic repulsion by doing this), as the pairing up is more viable than putting one electron in the next $3s$ orbital, which has a higher energy. Now for the two remaining electrons in the $2p$ subshell, it is not viable to pair them up as compared to distributing them equally into two unused $2p$ orbitals which are degenerate in energy. The distribution into the degenerate orbitals would have less inter-electronic repulsion than pairing them up into an orbital. Thus, the oxygen atom is a radical with two unpaired electrons.

3.2.2 *Heterolytic Bond Cleavage*

If Y is more electronegative than X, it has a stronger hold on the bonding electrons than X. When the bond between them is cleaved, Y acquires both the shared pair of electrons. Overall, Y gains an extra electron and acquires a negative charge, while X lose an electron and becomes positively charged. Hence, the products of a heterolytic bond cleavage are charged species — a cation and an anion.

$$^{\delta+}X—Y^{\delta-} \longrightarrow X^+ + Y^-$$

Take, for instance, the dot-and-cross diagram that illustrates the bonding in a HBr molecule before and after heterolytic bond cleavage:

$$H \cdot {\times} \overset{\times\times}{\underset{\times\times}{Br}} {\times} \xrightarrow[\text{cleavage}]{\text{bond}} H^+ + \left[\overset{\times\times}{\underset{\times\times}{\times Br \times}} \right]^-$$

Bonding electrons   Cation   Anion

When using curved arrow notation, a "full-headed" arrow is used to show the movement of two electrons or a pair of electrons, originating from the tail-end of the arrow to its head.

$$^{\delta+}H—\overset{\frown}{Br}^{\delta-} \longrightarrow H^+ + Br^-$$

The various products of bond cleavage, i.e. the free radical, the cation and the anion, are aptly known as reaction intermediates.

Q: The Br–Br bond is non-polar. Does this mean that it cannot undergo heterolytic cleavage?

A: No. The Br–Br bond can undergo heterolytic cleavage to give Br^+ and Br^- too. It all depends on the amount of energy that is present at the time of cleavage. If the right amount of energy is present for heterolytic cleavage to occur, then heterolytic cleavage will occur and not homolytic cleavage. Similarly, the polar H–Br bond can also undergo homolytic cleavage with the right amount of energy present.

3.2.3 *Type of Reaction Intermediates*

Intermediates common to organic reactions include the following three:

(1) A free radical is a highly reactive species that has at least one unpaired electron. An example is the methyl radical ($CH_3^{\bullet}$), which contains an electron-deficient carbon atom. This carbon atom is sp^2 hybridized with a trigonal planar geometry; it lies on the same plane as the three hydrogen atoms bonded to it with bond angles of $120°$. The carbon atom has an unhybridized p orbital, containing a single electron, perpendicular to the three sp^2 hybrid orbitals.

Q: If we look at the number of electron densities surrounding the carbon atom of the methyl radical, there are four regions. Based on VSEPR Theory, shouldn't the electron pair geometry be tetrahedral?

A: Based on VSEPR Theory, you would postulate that the electron pair geometry to be tetrahedral in shape. Unfortunately, this postulate cannot be validated by experimental result, which shows that it is trigonal planar. It is easy to understand why the preferred geometry should be trigonal planar instead of tetrahedral. For the trigonal planar geometry, the lone electron is perpendicular to three bond pairs of electrons while these three bond pairs of electrons are 120° apart. In the tetrahedral geometry, no doubt the lone electron is further apart from the bond pairs of electrons (~109.5°) as compared to the trigonal planar geometry, hence smaller inter-electronic repulsion between the lone electron and the three bond pairs of electrons in the tetrahedral geometry. But now the three bond pairs of electrons are also much closer than they would be when compared to the trigonal planar geometry, meaning greater inter-electronic repulsion between the three bond pairs of electrons in the tetrahedral geometry. The fact that experimentally the electron pair geometry is trigonal planar instead, is a clear indication that the overall inter-electronic repulsion within the trigonal planar geometry must be smaller than that in the tetrahedral geometry, right? Later on, you will see that the geometry for a carbanion is different from that of a methyl radical.

(2) A carbocation contains a positively charged carbon atom that is electron-deficient. It too is sp^2 hybridized, but with an empty p orbital perpendicular to the plane of sp^2 hybrid orbitals. The geometry around the sp^2 hybridized carbon atom is trigonal planar. An example is the methyl carbocation (CH_3^+):

(3) A carbanion contains a negatively charged carbon atom that is electron-rich. It is sp^3 hybridized with three of the hybrid orbitals used for covalent bond formation while the other hybrid orbital contains a non-bonding pair of electrons. An example is the methyl carbanion (CH_3^-):

Q: Why can't the carbanion be trigonal planar in shape just like the free radical?

A: If the carbanion were trigonal planar in shape, then the inter-electronic repulsion would be too great due to the presence of three 90° lone pair–bond pair repulsion. But for the methyl free radical, if its lone electron "sits" in an unhybridized p orbital, the trigonal planar geometry would still be able to "withstand" the three 90° lone electron–bond pair repulsion. This is because the bond pair–bond pair repulsion in the methyl radical would be much lesser if the geometry is trigonal planar in shape. Thus, Nature seems to know how to strike a good balance!

Q: Why are these intermediate species reactive?

A: Both the free radical and the carbocation are electron-deficient. They seek out species to minimize their electron-deficiency, consequently forming a covalent bond and achieving octet configuration. Thus, these electron-deficient species look out for electron-rich species that are willing to share electrons with them. On the other hand, a carbanion is "too" electron-rich, and one way to minimize the inter-electronic repulsion between the electron pairs surrounding the carbon atom is to use the lone pair of electrons for bond formation, whereby the lone pair of electrons get to spread out over a bigger space.

Q: Are you saying that as long as it is a charged particle, it will be reactive?

A: No! This is not the implication. Take, for instance, the Na^+ and Cl^- ions; they are charged particles. Do they continue to be reactive after the formation? No, they don't. So, the important thing to take note of is that the reactivity of a charged species after its formation is relative to the stability before its formation. If the energy state before the formation of the charged particle is a more stable state than the energy state of the charged particle, then we would expect the charged particle to be reactive, and vice versa.

This brings us next to the classification of reactants especially useful in discussing organic reactions. A reactants can be labeled as:

• an electrophile; or
• a nucleophile.

An electrophile is "electron-loving" and tends to seek out electron-rich sites in other molecules. These electrophilic species include electron-deficient free radicals and cations (carbocations included). It also includes molecules that contain an atom that can act as a Lewis acid (e.g. $AlCl_3$), accepting electron-pairs from donor species.

In contrast, a nucleophile is "nucleus-loving," and it is drawn towards electron-deficient sites in other molecules. Nucleophilic species are electron-rich, like anions (e.g. OH^-, CN^-, and Br^-) and neutral molecules (e.g. H_2O and NH_3) that have at least a lone pair of electrons that can be used for dative covalent bond formation.

In a typical organic reaction, it is basically the reaction between an electrophile and nucleophile in which electron transfer occurs from the nucleophile to the electrophile.

Q: Can we use the terms "electrophile" and "free radical" interchangeably?

A: No. Although both free radicals and electrophiles seek out electrons, the driving force for this is different. An electrophile is a much more electron-deficient species than a free radical, and it usually possess a charge or partial charge, whereas a radical is an electrostatically neutral species.

Q: Since a nucleophile is electron-rich, does it mean that it is also a base?

A: Yes, you are right. The Lewis theory of acids and bases defines a base to be an electron-pair donor. Hence, a nucleophile is indeed a base too. In essence, a typical organic reaction involves the movement of electron density from an electron-rich species (nucleophile) to an electron-deficient species (electrophile). From the Lewis acid–base perspective, we can perceive the nucleophile as a Lewis base, while the electrophile as a Lewis acid. Thus, a typical organic reaction is a Lewis acid–base reaction!

3.3 Types of Organic Reactions

There are countless organic reactions, given the myriad of organic compounds in existence. But if we take a closer look at these, we can find similar characteristics among them and, hence, group the reactions into a few general types. Some of these reactions are:

- rearrangement
- redox (oxidation–reduction)
- addition
- elimination
- substitution

3.3.1 *Rearrangement Reactions*

A rearrangement reaction, as its name implies, consists of the shifting of atoms and bonds in a molecule to form a structural isomer of the original molecule.

Notice that in the above rearrangement reaction, the number of double bonds and single bonds is still the same; i.e. the degree of unsaturation or saturation has not changed. The main difference lies in the length of the carbon skeletal structure.

3.3.2 *Redox Reactions*

A redox reaction consists of the simultaneous oxidation of one reactant and the reduction of another reactant. For organic redox reactions, since the focus is on the "transformation" of the organic compound, formulae of oxidising agents and reducing agents are usually left out in the chemical equations. But in order to balance such chemical equations, the symbols [O] and [H] are used respectively. For instance, when writing the chemical equation for the oxidation of ethanal using hot acidified $KMnO_4$, the latter is not depicted. Instead, the symbol [O] is used to highlight the fact that ethanal undergoes oxidation.

Notice that the oxidation state of the carbon atom of the aldehydic functional group ($-CHO$) has increased from -1 to $+3$ in the carboxylic acid ($-COOH$) functional group.

3.3.3 *Addition Reactions*

An addition reaction involves two reactants coming together to form a single product, i.e., "$1 + 1 = 1$".

$$A + B \longrightarrow C$$

This type of reaction is generally observed for unsaturated compounds such as alkenes and carbonyls. An example is the addition reaction between ethene and Br_2:

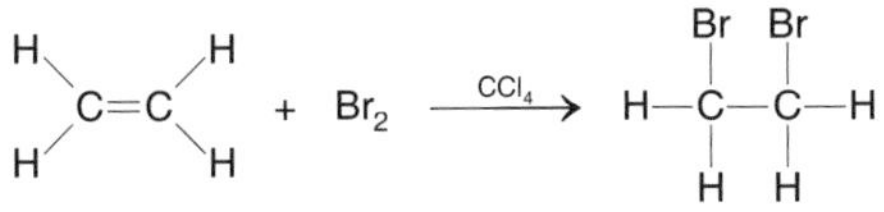

$$H_2C=CH_2 + Br_2 \xrightarrow{CCl_4} H_2BrC-CBrH_2$$

Q: Can we term the above reaction as redox?

A: Although it is indeed a redox reaction in nature, you cannot name it as redox because it would lose the specificity of the reaction.

Another example is the addition reaction between ethanal and HCN:

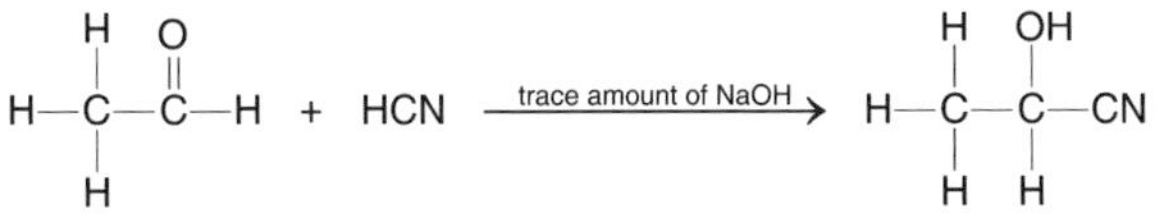

$$CH_3CHO + HCN \xrightarrow{\text{trace amount of NaOH}} CH_3CH(OH)CN$$

Notice that what characterizes an addition reaction is the decrease in the degree of unsaturation as reactants form products, or an increase in the degree of saturation. Put simply, the number of double or triple bonds in the product is lower than that in the reactants.

3.3.4 *Elimination Reactions*

On the other hand, an elimination reaction involves the removal of an atom or group of atoms from a single reactant molecule, i.e., "1 = 1 + 1". This type of reaction generally results in the formation of an unsaturated compound.

$$C \longrightarrow A + B$$

An example is the elimination reaction between bromoethane and a base to form ethene:

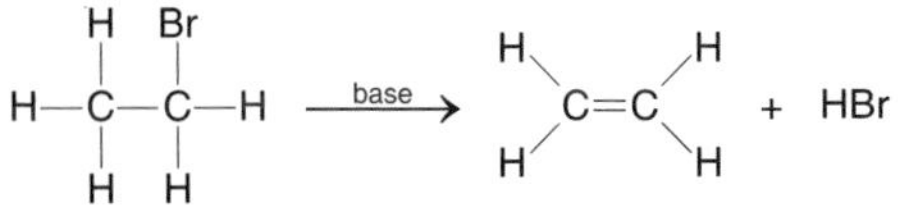

$$CH_3CH_2Br \xrightarrow{\text{base}} H_2C=CH_2 + HBr$$

Notice that what characterizes an elimination reaction is the increase in the degree of unsaturation as reactants form products, or a decrease in the degree of saturation. Put simply, the number of double or triple bonds in the product is greater than that in the reactants.

3.3.5 *Substitution Reactions*

A substitution reaction involves the exchange of an atom or group of atoms between two reactants, i.e., "$1 + 1 = 1 + 1$".

$$A\text{—}B + X\text{—}Y \longrightarrow A\text{—}X + B\text{—}Y$$

An example is the substitution reaction between ethane and Br_2:

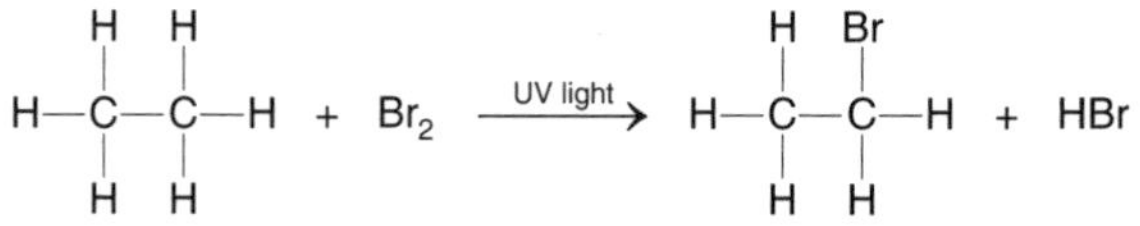

Another example is the substitution reaction between benzene and Br_2:

A third example is the substitution reaction between bromoethane and NaCN:

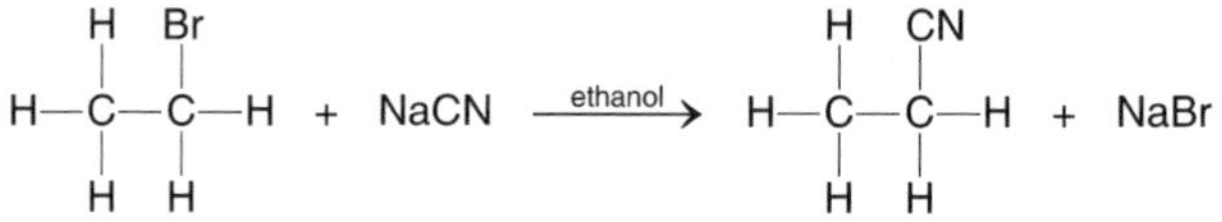

Notice that what characterizes a substitution reaction is that, there is no change in the degree of unsaturation as reactants form products, or there is no change in the degree of saturation.

In subsequent chapters, we will revisit these in detail, along with other types of organic reactions pertaining to each homologous series.

3.4 Types of Reaction Mechanisms

If we analyze the addition reactions of ethene and ethanal in Section 3.3.3, questions that come to mind include the following:

- How do these reactions proceed? What bonds are broken and what bonds are formed?
- Do they proceed via the same mechanism?
- If so, what common feature between them leads to the same reactivity?
- If not, what are the distinguishing characteristics that set them apart?

In order to answer these questions, we have to compare the structures of ethene and ethanal:

$$\underset{\text{Electron-rich site}}{\overset{\displaystyle \underset{H}{\overset{H}{>}}C=C\underset{H}{\overset{H}{<}}}{}} \qquad \underset{\text{Electron-deficient site}}{\overset{\displaystyle H-\underset{H}{\overset{H}{C}}-\overset{\delta^+}{C}\overset{\overset{\delta^-}{O}}{\|}-H}{}}$$

In ethene, the non-polar C=C double bond is an electron-rich site (i.e. nucleophilic in nature) and electrophiles are therefore attracted to it. As the pi bond is much weaker than the sigma bond, it cleaves much more easily, allowing the π electrons to form covalent bonds with the attacking electrophiles. In contrast, for ethanal, although there is also a weak pi bond, but the carbonyl (C=O) bond is polar with the electron-deficient carbon atom bearing the partial positive charge. Thus, nucleophiles would be more likely to be attracted to this electron-deficient carbon atom.

Q: Since the C=C double bond of ethene is non-polar, can it attract an attack from an even more electron-rich species?

A: Yes, certainly. In the presence of an even more electron-rich species, the C=C double bond becomes electrophilic in nature. Hence, take note that both nucleophilicity and electrophilicity are relative to each other. For a chemical reaction to proceed, as long as there are attractive forces between the reacting species and the conditions are right, it will proceed.

And here we have the answer. Based on electronic factors, reactions of alkenes and carbonyl compounds proceed differently. We shall now look into the mechanism for the addition reaction of alkenes, exemplified by ethene.

3.4.1 *Electrophilic Addition Mechanism*

Q: The electron-rich C=C double bond attracts electrophiles. Why is the non-polar Br_2 being considered an electrophile? Is there an electron-deficiency site on the molecule?

A: A Br_2 molecule by itself is a non-polar molecule. But in close proximity to the ethene molecule, the π electron cloud interacts with the electron cloud of the Br_2 molecule, causing the Br–Br bond to be polarized, i.e. a permanent dipole is created as long as the Br_2 molecule is near to the C=C double bond. The Br atom that is nearer to the C=C double bond acquires a partial positive charge ($\delta+$) while the other acquires a partial negative charge ($\delta-$). It is the $\delta+$ end of the molecule ($^{\delta+}$Br–Br$^{\delta-}$) that acts as an electrophile.

Q: Why can't the Br$_2$ polarize the electron cloud of the C=C double bond instead?

A: The lone pair of electrons on the Br$_2$ molecule is comparatively easier to polarize as it is only being attracted to a nucleus of an atom. Whereas the pi electrons of the C=C double bond are being attracted by two nuclei, hence less polarizable.

The mechanism for the reaction between ethene and bromine proceeds as follows:

- In close proximity to the ethene molecule, the Br$_2$ molecule is polarized with the electrophilic $^{\delta+}$Br atom drawn towards the electron-rich C=C bond.
- In a *concerted* move, the following can be thought to occur:
  - The Br–Br bond undergoes heterolytic cleavage forming a Br$^-$ ion and a Br$^+$ ion.
  - The π bond cleaves with the two π electrons used to form a bond between one of the C atoms in the alkene double bond and the Br$^+$ ion (a Lewis acid). Accordingly, the other C atom loses an electron and is thus positively charged.

  The concerted step is depicted using the curved arrow notation that shows the transfer of two electrons from the pi bond to the $^{\delta+}$Br atom and the two bonding electrons in the Br–Br bond are acquired by the $^{\delta-}$Br atom (see Step 1). The positively charged intermediate is known as a carbocation and it is formed together with a Br$^-$ ion.
- The reactive carbocation (a Lewis acid) then readily combines with the bromide ion (a Lewis base) to form the final addition product 1,2-dibromoethane.

In this electrophilic addition reaction, there is a change in the hybridization of the doubly bonded carbon atoms from sp^2 to sp^3 in the halogenoalkane product.

Step 1: Electrophile attacks electron-rich C=C bond to form a carbocation intermediate

carbocation

Step 2: Carbocation is readily attacked by Br⁻ to form the final product

halogenoalkane

Q: Why is Step 1 indicated as the slow step and Step 2 the fast step?

A: Energy is needed to cleave the alkene pi bond and the Br–Br sigma bond. Although at the same time a C–Br sigma bond is formed, the net energy change is still endothermic in nature, resulting in this reaction step being the rate-determining step. In Step 2, there is natural attraction of oppositely charged particles that leads to bond formation (there is no bond cleavage involved at all), which is exothermic in nature.

Q: What is the "incentive" involved in cleaving the pi bond to form the addition product?

A: The driving force for a reaction is the formation of a more stable compound, and such reactions are energetically favorable. The greater stability of the product as compared to the reactants comes from the formation of two C–Br sigma bonds in exchange for the breaking of one C–C pi bond and one Br–Br sigma bond. The net energy change is exothermic in nature. You can do a bond energy calculation to verify it!

Mathematically, using bond energy values to calculate the enthalpy change of reaction, we have:

$$\Delta H_{rxn} = \Sigma \text{ (energy } absorbed \text{ in bond breaking)}$$
$$- \Sigma \text{ (energy } released \text{ in bond formation)}$$

$$= [BE(C{=}C) - BE(C{-}C) + BE(Br{-}Br)]$$
$$\quad - 2 \times BE(C{-}Br)$$
$$= [610 - 350 + 193] - 2(280)$$
$$= -107\,\text{kJ}\,\text{mol}^{-1}.$$

The electrophilic addition reaction is an exothermic one with an overall energy release of $107\,\text{kJ}\,\text{mol}^{-1}$. The following energy profile diagram depicts the electrophilic addition mechanism:

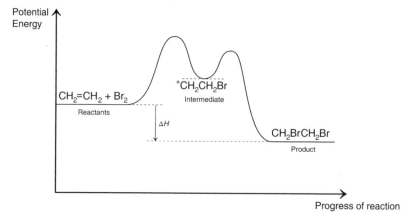

Q: Since Step 2 involves the attraction of two *oppositely* charged particles, why is there still an activation energy barrier present in Step 2?

A: It is a common misconception to think that a cation "carries" a "+" sign when it moves around whereas an anion "carries" a "−" sign. The net electrostatic attractive force between a cation and an anion is the sum of the attractive forces each nucleus has for the electron cloud of the other ion (i.e. the nucleus of the cation attracting the electron cloud of the anion and vice versa) minus the inter-electronic repulsion between the two electron clouds and the inter-nuclei repulsion between the two nuclei. When both oppositely charged ions attract each other, their electron clouds usually do not interpenetrate each other. Although Step 2 involves the attraction of two oppositely charged particles, in order for the product to be formed, the electron clouds of the two oppositely charged particles still have to interpenetrate each other so that covalent bonds can be formed. This close contact of the two oppositely charged particles results in strong inter-electronic repulsion between the electron clouds, which needs energy to be overcome. This accounts for the observed activation barrier.

Q: Since the Cl–Cl bond is stronger than the Br–Br bond, when the Cl_2 acts as an electrophile, shouldn't the reaction be slower as compared to when Br_2 acts as an electrophile?

A: Yes, cleaving the Cl–Cl bond is more difficult than cleaving the Br–Br bond. But do not forget that the C–Cl bond formed is stronger than the C–Br bond. You can do the bond energies sum to verify whether when Cl_2 acts as an electrophile, the reaction would be slower.

You may have noticed that the carbocation intermediate has a trigonal planar geometry around the sp^2 hybridized carbon (the C atom bearing the positive charge). This means that the site of attack is a flat plane at which the nucleophile (Br^-) can approach from either side of the plane. If we start out with an appropriate alkene, there is the possibility of getting a pair of enantiomers. For example, in the reaction between $CH_3CH=CH_2$ and Br_2, a pair of enantiomers is obtained in the second step of the electrophilic addition mechanism:

* chiral center

3.4.2 *Nucleophilic Addition Mechanism*

Nucleophilic addition reaction is a common reaction for carbonyl compounds, such as aldehyde (–CHO) and ketone (C–CO–C) functional groups. A typical nucleophilic addition reaction is the addition of a HCN molecule to the carbonyl functional group.

Electron-deficient site

Take, for instance, ethanal; the C=O bond is polar with the C atom bearing the partial positive charge. Hence, nucleophiles are attracted to this electron-deficient site.

Q: How is H–C≡N considered a nucleophile?

A: The nucleophile is actually the ⁻C≡N ion and not HCN. So although we normally represent the cyanide ion as CN⁻, it does not mean that the negative charge (which represents an extra electron in the species) resides on the nitrogen atom. Thus, take note that the attacker is actually the more electron-rich carbon atom (as it is negatively charged) and not the nitrogen atom. In addition, although the nitrogen atom also possesses a lone pair of electrons (which means that it is also nucleophilic in nature), since the C atom is less electronegative than N, it is still more willing to donate its lone pair of electrons. In essence, HCN is just a source of the nucleophile. However, as HCN is a weak acid that partially dissociates in water, the [CN⁻] is considerably low:

$$\text{HCN} \rightleftharpoons \text{H}^+ + \text{CN}^- \; K_\text{a} = 5 \times 10^{-10}\,\text{mol}\,\text{dm}^{-3}.$$

Q: Why is HCN a weak acid? Isn't the electronegativity difference between H and C small?

A: In HCN, the most electronegative N withdraws electron density away from the C atom, which in turn withdraws from the H atom. This makes the H atom relatively electron-deficient and hence susceptible to extraction by a base.

Q: But shouldn't the CN⁻ formed be highly unstable as the negative charge resides on the C atom which has a lower electronegativity value?

A: Yes, you are right that the electronegativity of C is not "good enough" to hold the negative charge. But do not forget that there is a highly electronegative N atom bonded to the C atom. Through inductive effect, the N atom would withdraw electron density away from the C atom and help to stabilize the CN⁻. In addition, the orbital that holds the extra electron on the C atom is a *sp* hybridized one, containing 50% *s*-character. Since an *s*-orbital is relatively closer to the nucleus than a *p*-orbital, an extra electron in a *sp* hybridized orbital is more strongly attracted by the nucleus than in a *sp²* one, which in turn is more strongly attracted than in a *sp³* hybridized one. This accounts for the stability of the CN⁻ ion.

From kinetics study of reaction rate, the derived rate equation is: rate = k[ethanal][CN⁻].

Since reaction rate is dependent on [CN⁻], a low [CN⁻] means a slow reaction.

To increase the reaction rate, [CN⁻] has to be increased without totally removing all the HCN in the system. There are two ways to do this:

(1) Add a small amount of strong base such as NaOH(aq) and KOH(aq).

The added OH⁻ ions react with H⁺ ions and cause [H⁺] to decrease. According to Le Chatelier's principle, the system will attempt to increase the [H⁺] by favoring the dissociation reaction. The equilibrium position will shift to the right and, as a result, [CN⁻] is increased.

(2) Add a strong electrolyte containing CN⁻ such as NaCN(aq) and KCN(aq).

The complete ionization of the electrolyte provides sufficient [CN⁻] to start-off the reaction.

$$NaCN(s) + aq \longrightarrow Na^+(aq) + CN^-(aq)$$

Q: Why can't we just simply allow the ethanal to react with NaCN(aq) or KCN(aq)? What is the purpose of ensuring that there is still the undissociated HCN present in the system?

A: HCN serves two roles in this reaction. First, it provides the nucleophile CN⁻, albeit in low concentrations. Second, it acts as a proton donor in the second step of the mechanism to generate the final product (see Step 2 of the mechanism).

The mechanism consists of two steps as follows:

- CN⁻, from the dissociation of HCN, acts as the nucleophile (or a Lewis base) which is attracted to the electron-deficient carbonyl carbon atom (a Lewis acid). The attack of the CN⁻ on the $^{\delta+}C$ constitutes the slow rate-determining step.
- In a *concerted* move, the following can be thought to occur:
  - The lone pair of electrons on the carbon atom of the CN⁻ ion (a Lewis base) is donated to form a bond with the carbonyl carbon atom (a Lewis acid).

○ The pi bond in the C=O double bond cleaves, with the two π electrons being acquired by the carbonyl O atom which becomes negatively charged.

The concerted step is depicted using the curved arrow notation as shown in Step 1.

• The anionic intermediate (a Lewis base) then extracts a proton from either the undissociated HCN molecule (a Lewis acid) or a solvent molecule to form the final addition product, a cyanohydrin. As seen in Step 2, CN^- and OH^- can be regenerated. This means that only small amounts of the strong electrolyte (KCN) or base (NaOH) are required since they behave as catalysts.

Step 1: Nucleophile attacks electron-deficient carbonyl carbon to form an anionic intermediate.

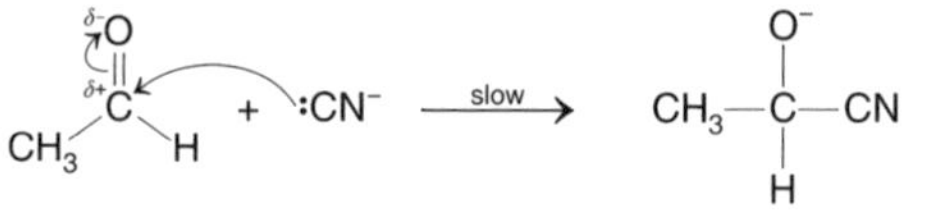

anionic intermediate

Step 2: The anionic intermediate is protonated to form the final cyanohydrin product.

A proton can be extracted from the undissociated HCN molecule,

cyanohydrin

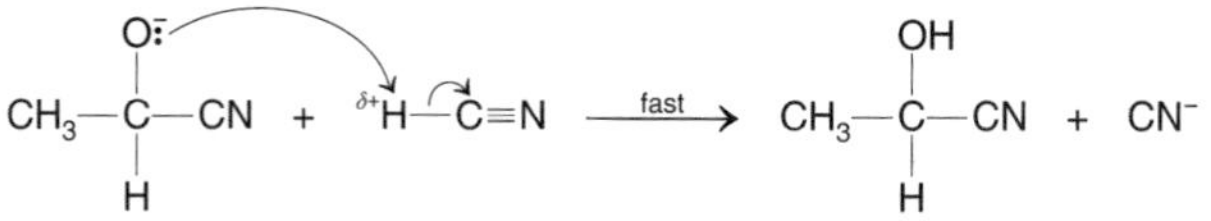

or it can be extracted from the H_2O solvent molecule,

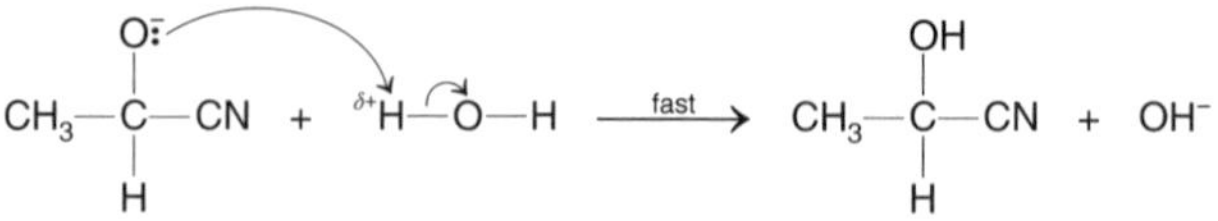

cyanohydrin

Q: If the solvent can act as a proton donor and NaCN as a source of CN^-, can't we forego the use of HCN?

A: Now, the extraction of a H^+ ion from a water molecule is much more difficult than from a H–CN molecule. This is because the H–O (BE $= +460\,\text{kJ mol}^{-1}$) bond is much stronger than the H–C bond (BE $= +410\,\text{kJ mol}^{-1}$). Such an explanation also provides us with the understanding of why HCN, though it is a weak acid, is still a stronger acid as compared to H_2O. So, in reality, extracting a H^+ from a water molecule to form the cyanohydrins is a minor reaction as compared to the extraction from a HCN molecule.

Q: Since the nitrogen atom of CN^- also possesses a lone pair of electrons, can it make the nucleophilic attack instead of the carbon atom?

A: Although both the carbon and nitrogen atoms of CN^- possess a lone pair of electrons each, the carbon atom is more electron rich than the nitrogen atom because it is negatively charged. This coupled with the fact that the carbon atom is less electronegative than the nitrogen atom, made the lone pair of electrons on the carbon atom more likely to be donated, that is, acting as a Lewis base or a nucleophile.

Q: So, although we commonly write the cyanide ion as CN^-, it does not mean that the negative charge is residing on the nitrogen atom?

A: Indeed, the negative charge is "sitting" on the carbon atom. If you know that CN^- originates from HCN after a H^+ ion has dissociated from the H-CN, then it is logical to remember that the negative charge must reside on the carbon atom. Of course, it should be more appropriate to write the cyanide ion as ^-CN or NC^- instead.

The following energy profile diagram depicts the nucleophilic addition mechanism:

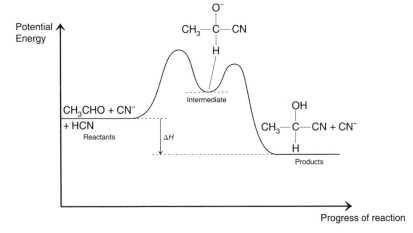

Ethanal is an example of a prochiral molecule — an achiral molecule that can be transformed into a chiral molecule in one step. Just like a carbocation, the geometry around the sp^2 hybridized carbonyl carbon atom is trigonal planar. The nucleophile can attack the carbonyl carbon from either side of the plane to form a racemic mixture (or racemate) containing equal proportions of the two optically active enantiomers.

* chiral center

If we look back at the two mechanisms that we have discussed, both types of addition reactions effectively consist of the cleavage of a pi bond within the unsaturated reactant molecule. At the end, two sigma bonds are formed with the other reactant in obtaining the final saturated product.

So, we see that addition reaction can involve the attack by different species that maybe either electron-rich (nucleophilic addition) or electron-deficient (electrophilic addition). Similarly, in an organic substitution reaction, the attacking species may also differ. As a result, what appears on the surface to be just a simple substitution reaction can actually proceed via different mechanisms. Of course, the nature of the mechanism would depend on the type of reactants involved.

Take, for instance, both ethane and benzene undergoing the substitution reaction with Br_2. The reaction conditions are totally different. This difference can be explained by looking into the mechanisms entailed.

3.4.3 *Free Radical Substitution Mechanism*

Alkanes, although unreactive, with seemingly few reactions, do undergo substitution, but only in the presence of stringent conditions.

The mechanism involved is known as the free radical substitution reaction, which is a chain reaction comprising of initiation, propagation and termination steps:

- To initiate the chain reaction, free radicals are generated from the homolytic cleavage of the Br–Br bond in the bromine molecule. The energy needed for this process comes from either ultraviolet (UV) light or the heat supplied.
- The highly reactive Br$^\bullet$ radical then reacts with the ethane molecule, seeking out electrons through bond formation. For this to occur, a C–H bond in the ethane molecule needs to be cleaved homolytically so as to furnish the electron that is needed for covalent bond formation with the Br$^\bullet$ radical.
- In a *concerted* move, the following can be thought to occur:
  - Homolytic cleavage of the C–H bond occurs with the help of Br$^\bullet$ to produce the ethyl $CH_3CH_2^\bullet$ and H$^\bullet$ radicals.
  - The Br$^\bullet$ and H$^\bullet$ radicals each donate an electron to form the H–Br bond.
  - The equally reactive ethyl radical then reacts with a Br_2 molecule (if it encounters it). Homolytic cleavage of the Br–Br bond occurs, with one Br$^\bullet$ combining with $CH_3CH_2^\bullet$.
  - The other Br$^\bullet$ subsequently reacts with another ethane molecule (if it encounters it) and thus the entire cycle repeats itself with the generation of free radicals, bond breaking and bond formation. And there you have it, a chain reaction.

Q: Why can't an alkane molecule react with an electrophile or nucleophile?

A: Although a carbon atom is slightly more electronegative than a hydrogen atom, there is a lack of prominent sites on an alkane molecule that are especially electron-rich or electron-deficient enough to "invite" an attack from an electrophile or nucleophile.

Step 1: Chain initiation with the generation of free radicals

$$\text{Br—Br} \xrightarrow{\text{UV light}} 2\ \text{Br·}$$

Step 2: Chain propagation

(a) $\text{Br}^{\bullet} + \text{CH}_3\text{CH}_3 \longrightarrow \text{CH}_3\text{CH}_2^{\bullet} + \text{HBr}$

(b) $\text{CH}_3\text{CH}_2^{\bullet} + \text{Br}_2 \longrightarrow \text{CH}_3\text{CH}_2\text{Br} + \text{Br}^{\bullet}$

and the cycle repeats itself.

Take note that the repeated cycle of getting the free radicals and their "blind" attack on any of the molecules that "cross" their path can lead to the substitution of all the H atoms in the alkane molecule by the Br atoms, i.e.

$$\text{H—C(H)(H)—C(H)(H)—H} \longrightarrow \text{Br—C(Br)(Br)—C(Br)(Br)—Br}$$

In fact, various degrees of substituted alkanes or even the formation of polymeric alkane molecules are obtained from this chain reaction that lacks controllability. The chain reaction is terminated when two free radicals collide and react to form a stable compound.

Step 3: Chain termination

$\text{Br}^{\bullet} + \text{CH}_3\text{CH}_2^{\bullet} \longrightarrow \text{CH}_3\text{CH}_2\text{Br}$

$\text{Br}^{\bullet} + \text{Br}^{\bullet} \longrightarrow \text{Br}_2$

$\text{CH}_3\text{CH}_2^{\bullet} + \text{CH}_3\text{CH}_2^{\bullet} \longrightarrow \text{CH}_3\text{CH}_2\text{CH}_2\text{CH}_3,$

etc.

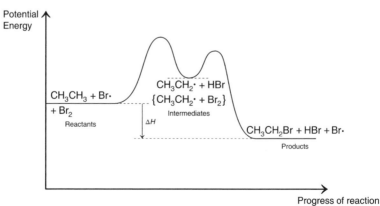

Q: If a free radical reacts so indiscriminately, does it mean that even an alkene can also undergo free radical reaction?

A: Yes, indeed. There is free radical addition reaction for alkene or even for benzene molecules. So ultimately, the reaction mechanism for a reaction really depends very much on the nature of the reactants and the reaction conditions involved.

3.4.4 *Electrophilic Substitution Mechanism*

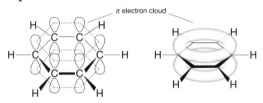

Benzene, unlike alkene, does not undergo an addition reaction, although both of them are unsaturated hydrocarbons with electron-rich sites. They behave very differently towards electrophiles.

Benzene has an unexpectedly stable structure that is attributed to the extensive network of delocalized π electrons. We say that the benzene ring is resonance stabilized (we will discuss this in greater detail in Chapter 6). When a compound is said to be stable, it means that it can exist as it is without being attacked by other substances under certain conditions. But if the so-called stable molecule "faces" a highly reactive species, it can still undergo a reaction. So, take note that stability is rather relative!

Since both benzene and ethene are electron-rich species, they are susceptible to attack by electrophiles, with ethene being more so. The electron density is high at the ethene C=C bond, which is the target of electrophiles. As for benzene, the electron density is spread out uniformly over six carbon atoms — there is a greater degree of charge dispersal brought about by the ability of the π electrons to delocalize over the entire network of carbon atoms. As a result, if one compares the carbon–carbon bond of an alkene to that of a benzene molecule, one would discover that the carbon–carbon bond of a benzene molecule is intermediate between a single bond and a double bond. The average is about one and a half bond. Hence, in reality, the carbon–carbon bond of a benzene molecule is actually less electron-rich than the C=C double bond of an alkene! Therefore,

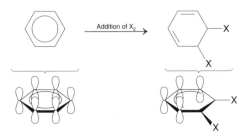

Fig. 3.3. (Left) delocalization of electrons over the entire ring and after addition reaction, (right) there is lesser extent of delocalization, involving only four carbon atoms.

the benzene molecule would need a much stronger electrophile to attack the benzene ring in order to destroy the resonance stability. In addition, if benzene were to undergo an addition reaction rather than substitution, then the resonance stability would be destroyed once the product is formed.

Benzene undergoes mostly substitution reactions with the aromatic character still retained. As π electrons are available, benzene is an attractive site for strong electrophiles to attack.

Q: So, does that mean that benzene cannot undergo addition reactions?

A: As a matter of fact, it can; if the reaction condition is harsh enough, addition reaction would take place. One typical reaction would be a free-radical addition reaction.

We have come across Br_2 as a poor electrophile that needs to be polarized by an alkene to serve as an effective electrophile. In the case of benzene, Br_2 is made a better electrophile by subjecting it to reaction with a Lewis acid such as anhydrous $FeBr_3$ or $AlBr_3$. These Lewis acids polarize the Br_2 molecule to such a great extent that the electrophile, Br^+, is generated through heterolytic cleavage of the Br–Br bond forming a $[FeBr_4]^-$ ion and a Br^+ ion. In effect, the Lewis acid also serves the role of a catalyst for the reaction as it is being regenerated at the end of the reaction.

$$Br{-}Br + FeBr_3 \;\rightleftharpoons\; \overset{\delta^+}{Br}{-}\overset{\delta^-}{Br}\cdots FeBr_3 \;\rightleftharpoons\; Br^+ + [FeBr_4]^-$$
polarised

Q: What characteristic features do these reagents have that enable them to function as Lewis acids?

A: A Lewis acid is an electron-pair acceptor, and in this way, it reduces the electron density on the donor atom, i.e. a Br atom in the Br_2 molecule. As a result, the Br–Br bond is weakened to an extent that it results in heterolytic cleavage. Both $FeBr_3$ and $AlBr_3$ are able to function as Lewis acids due to the presence of a low-lying vacant orbital which is able to accept a lone pair of electrons from the donor atom. In addition, both the metal centers have high charge density and thus are highly electron deficient.

Q: Why must the reagents be anhydrous?

A: Due to the high charge density of the metal ion, both the $FeCl_3$ and $AlCl_3$ will undergo hydrolysis (reaction with water), producing ions that would decrease the electron deficiency of the metal center, rendering it incapable of functioning as a Lewis acid.

$$AlCl_3(s) + 6H_2O(l) \longrightarrow [Al(H_2O)_6]^{3+}(aq) + 3Cl^-(aq)$$

$$FeCl_3(s) + 6H_2O(l) \longrightarrow [Fe(H_2O)_6]^{3+}(aq) + 3Cl^-(aq)$$

The two-step electrophilic substitution mechanism is as follows:

- In a *concerted* move, the following can be thought to occur:
  - A pi bond cleaves with the two π electrons being donated to form a bond between one of these "doubly bonded" C atoms and the Br^+ ion (a Lewis acid). Accordingly, although the other C atom lost an electron, the positive charge is actually dispersed over the remaining five sp^2 hybridized C atoms in the carbocation intermediate (also known as benzonium ion).

- In the subsequent fast step, the carbocation (a Lewis acid) loses a proton which is taken up by the Lewis base $[FeBr_4]^-$. Bromobenzene is obtained along with the acid HBr and the catalyst $FeBr_3$ being regenerated.

Step 1: Electrophile attacks electron-rich benzene to form a carbocation intermediate.

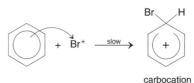

carbocation

Step 2: The carbocation is deprotonated to regain the resonance stabilized benzene ring.

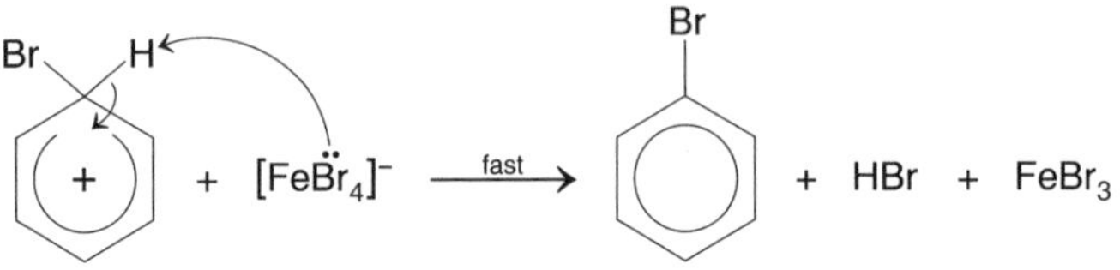

In this electrophilic substitution reaction, there is a change in the hybridization of one of the carbon atoms as shown below:

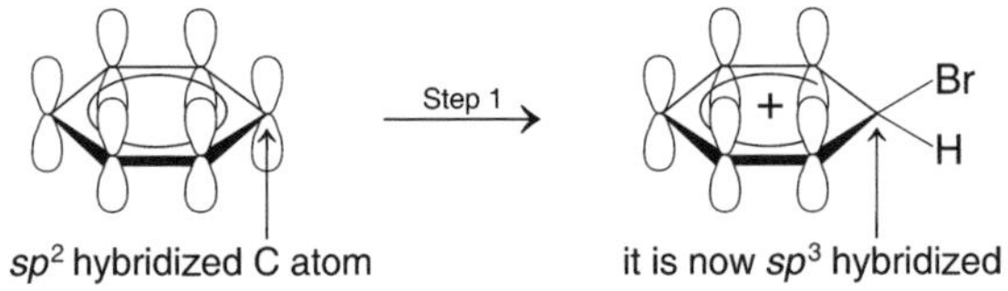

sp^2 hybridized C atom   it is now sp^3 hybridized

The resonance stability of the ring is destroyed in Step 1 because the sp^3 hybridized carbon atom disrupts the network of delocalized electrons. As a result, Step 1 is the rate-determining step. However, the five remaining sp^2 hybridized carbon atoms can still form a π network consisting of four delocalized electrons, hence dispersing the positive charge on the carbocation. Resonance stability is regained in the product when a proton is removed from the sp^3 hybridized carbon in the second step. This results in Step 2 being a fast step as the driving force is the regeneration of the resonance stabilized benzene ring.

sp^3 hybridized C atom   it is sp^2 hybridized again

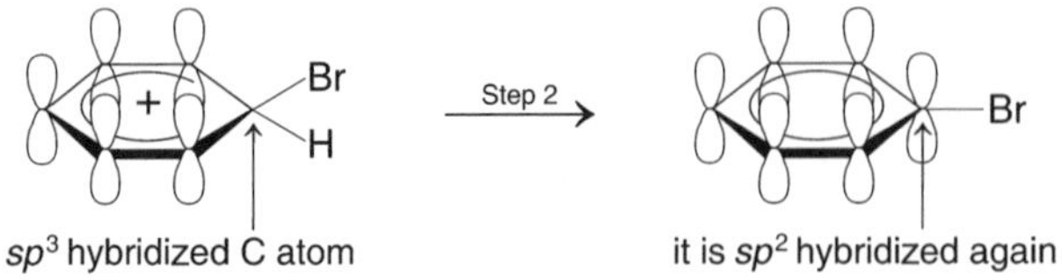

The following energy profile diagram depicts the electrophilic substitution mechanism:

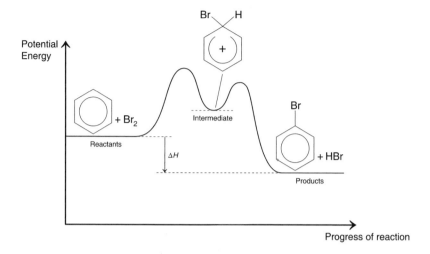

Q: Can we simply say that the reason for the Step 1 being the rate-determining step is because of the need to break bonds rather than to "destroy" the resonance stability?

A: No, it is not enough. The fact that a stronger electrophile is needed would mean that the high activation energy is not just to break bonds alone as compared with the electrophilic addition of an alkene. The benzene molecule is resonance stabilized as compared to an alkene.

3.4.5 *Nucleophilic Substitution Mechanism*

$$
\begin{array}{ccc}
\text{H} & \text{Br} & \\
| & | & \\
\text{H}-\text{C}-\text{C}-\text{H} & + \text{ NaCN} & \xrightarrow{\text{ethanol}} \\
| & | & \\
\text{H} & \text{H} &
\end{array}
\quad
\begin{array}{c}
\text{H} \;\; \text{CN} \\
| \;\;\;\; | \\
\text{H}-\text{C}-\text{C}-\text{H} \;\; + \;\; \text{NaBr} \\
| \;\;\;\; | \\
\text{H} \;\; \text{H}
\end{array}
$$

As you would have predicted, the substitution reaction between bromoethane and NaCN occurs via yet another mechanism, which will be discussed next.

$$
\begin{array}{c}
\text{H} \;\; {}^{\delta-}\text{Br} \\
| \;\;\;\; | \\
\text{H}-\text{C}-\!\!\overset{\delta+}{\text{C}}\!-\text{H} \\
| \;\;\;\; | \\
\text{H} \;\; \text{H}
\end{array}
$$

In bromoethane, the C–Br bond is polar with C bearing the partial positive charge. Hence, nucleophiles are attracted to this electron-deficient site. Halogenoalkanes are saturated molecules, unlike their carbonyl counterparts. This rules out the possibility of an addition reaction. What we have instead is nucleophilic substitution, whereby a nucleophile attacks the $^{\delta+}$C atom and displaces another atom.

Q: Why can't there be an addition reaction?

A: A nucleophile brings with it a lone pair of electrons for bond formation with the $^{\delta+}$C atom. As this C can only accommodate a maximum of eight electrons in its valence shell, the C–Br must undergo heterolytic cleavage with both bonding electrons acquired by the Br atom that subsequently parted as the Br$^-$ ion.

Q: Can the H atom that is bonded to the $^{\delta+}$C atom be as easily displaced as the Br atom?

A: The C–H bond is much stronger than the C–Br bond and hence requires a greater amount of energy to cleave it. This stronger bond arises from more effective orbital overlap between that of C and H atoms since the $1s$ orbital which the H atom used in bond formation is smaller and less diffuse than the valence orbital of the Br atom. This brings to mind again the reason behind the lack of reactivity of alkanes.

There are two different mechanisms for nucleophilic substitution as observed from the kinetics studies on such reactions — it may proceed via either a one-step mechanism (S_N2) or a two-step mechanism (S_N1).

3.4.5.1 *Nucleophilic Substitution (S_N2 Mechanism)*

pentavalent
activated complex

The one-step electrophilic substitution mechanism is as follows:

- The nucleophile CN^- attacks the $^{\delta+}C$ from the side opposite to that of the C–Br bond in what is known as the "backside" attack.
- In a concerted move, the following can be thought to occur:

  ○ C–Br bond undergoes heterolytic cleavage and a Br^- ion is formed, which is known as the leaving group;
  ○ C–CN bond is formed.

The simultaneous bond breaking and bond formation is depicted in the pentavalent activated complex, which is trigonal bipyramidal in shape. As the nucleophile is a negatively charged species, the pentavalent activated complex would also possess a net negative charge, as depicted by the two $\delta-$. But what happens if the attacking nucleophile is a neutral species, such as a NH_3 molecule? Then take note that the pentavalent activated complex would also be electrostatically neutral, as shown below.

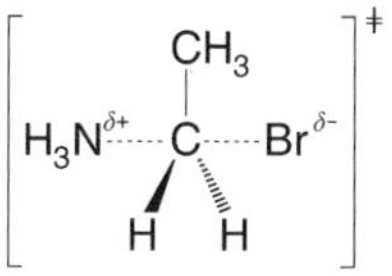

$$\left[\begin{array}{c} CH_3 \\ | \\ H_3N^{\delta+}\text{----}C\text{----}Br^{\delta-} \\ \diagup \quad \backslash\!\!\!\backslash \\ H \quad H \end{array} \right]^{\ddagger}$$

If the halogenoalkane molecule has a chiral $^{\delta+}C$ atom — due to the direction of "backside" attack — there will be an inversion of stereochemistry known as the Walden inversion.

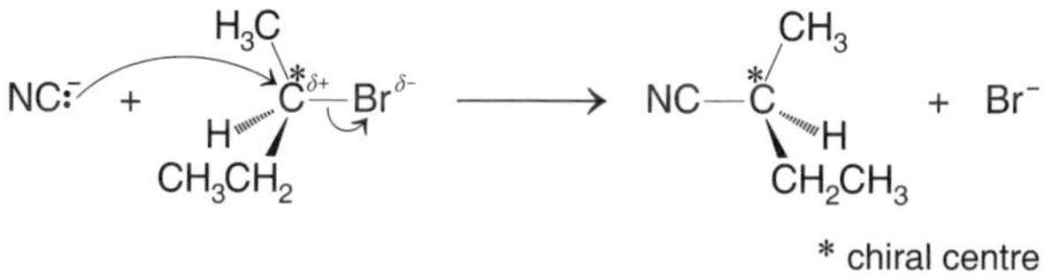

$$NC\!:^{\curvearrowright} + \overset{H_3C}{\underset{CH_3CH_2}{\overset{|}{\underset{H}{\diagup}}}}\!\!\overset{*}{C}^{\delta+}\!\!-\!\!Br^{\delta-} \longrightarrow NC\!-\!\overset{CH_3}{\underset{CH_2CH_3}{\overset{|}{\underset{H}{\diagup}}}}\!\!\overset{*}{C} + Br^-$$

* chiral centre

Q: Why does the nucleophile approach from the "backside" route?
A: Due to the large size of the Br atom, the "backside" approach poses less steric hindrance (which is inter-electronic repulsion in nature) and hence greater ease of accessibility to the $^{\delta+}C$ atom by the nucleophile. Furthermore, the repulsion between the electron clouds of the negatively charged CN^- and $^{\delta-}Br$ can be minimized. In addition, in order for the nucleophile to form a bond with the electron-deficient carbon atom, the donating pair of electrons has to "enter" the small lobe of the sp^3

hybrid orbital from the side that is opposite to that of the Br group.

From kinetics study of reaction rate, the derived rate equation is: rate = k[RX][nucleophile]. Rate is dependent on both the concentrations of the halogenoalkane (RX) and the nucleophile (Nu). The rate equation indicates a bimolecular rate-determining step consisting of two species, RX and Nu, reacting with each other. Hence, the acronym S_N2 that stands for nucleophilic substitution mechanism with a bimolecular rate-determining step.

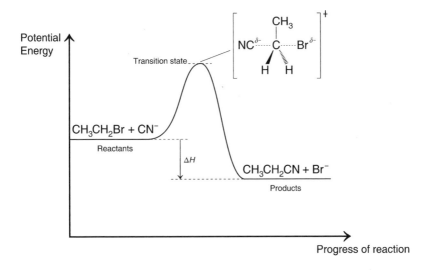

Considering the same nucleophile reacting with different halogenoalkanes, the factors that influence the reaction rate center upon:

(i) The ease of cleavage of the C–X bond (X stands for halogen);
(ii) The ease of forming the C–Nu bond.

(i) The ease of cleavage of the C–X bond

The ease of cleavage of the C–X bond depends of course on the strength of the bond. The stronger the bond, the harder it is to cleave, and this slows down the reaction.

Bond	Bond Energy/kJ mol^{-1}
C–F	485
C–Cl	340
C–Br	280
C–I	240

The C–X bond strength decreases in the order: C–F > C–Cl > C–Br > C–I.

As the size of the halogens gets bigger from F to I, the valence orbital used for bonding is larger and more diffuse. As a result, the overlap of the orbital with that of the carbon atom becomes less effective, and this accounts for the weaker bond strength, which is reflected in the bond energies.

So we would expect iodoethane to react the fastest followed by bromoethane, chloroethane and fluoroethane, with the same nucleophile. In fact, due to the very strong C–F bond, fluoroalkanes are generally unreactive towards nucleophilic substitution or any reactions that involve cleavage of the C–F bond.

Indeed, the low reactivity of chlorofluorocarbons (CFCs) contributes to their environmentally destructive nature. Since they can't be broken down easily, they have enough time to get to the stratosphere, where the UV radiation is strong enough to convert these to free radicals that attack the ozone layer.

Q: Since electronegativity decreases down a group, the electron deficiency of the carbon atom is greater in a C–F bond than a C–Cl bond, and so on and so forth. Then shouldn't the more electron deficient carbon atom in the C–F bond be more susceptible to nucleophilic attack, thus making it more reactive?

A: You are right. Indeed the carbon atom of C–F is the most electron-deficient of all. And yes, it is most "attractive" to a nucleophile. But unfortunately, this does not make it the most reactive. This is because kinetics studies have shown that fluoroalkane is the least reactive of all, thus we cannot make use of this factor to account for our observation. The only feasible explanation would be to use the C–X bond strength to account for it.

(ii) The ease of forming the C–Nu bond.

The ease of forming the C–Nu bond depends in part on the accessibility of approach by the nucleophile to the electron-deficient site ($^{\delta+}$C atom).

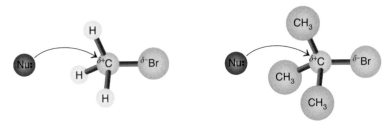

As seen from the diagram above, the larger methyl substituents cause steric hindrance (which is inter-electronic repulsion in nature) to the "backside" approach of the nucleophile towards the $^{\delta+}$C atom in $(CH_3)_3CBr$ as compared to the case in CH_3Br. The less the steric hindrance present, the faster the reaction rate. The greater the number of alkyl substituents, the more problematic is the steric hindrance.

Apart from steric hindrance, the alkyl substituents also affect the electrophilic nature of the electron-deficient $^{\delta+}$C atom. Alkyl groups are considered to be electron-releasing groups. As such, the greater the number of alkyl groups being attached to the $^{\delta+}$C atom, the lower the electron-deficiency on the C atom. Thus, the carbon atom would be less attracted by the nucleophiles.

In consideration of the points mentioned, we will expect the reaction rate to increase in the order for:

$$CH_3X \quad > \quad RCH_2X \quad > \quad R_2CHX \quad > \quad R_3CX$$

1° halogenoalkane   2° halogenoalkane   3° halogenoalkane

where R stands for alkyl groups.

It may seem that tertiary (3°) halogenoalkanes tend to be less susceptible towards nucleophilic attack, but this is not true. They do undergo nucleophilic substitutions as readily, but they do so via a different way — the S_N1 mechanism.

3.4.5.2 *Nucleophilic Substitution ($S_N 1$ Mechanism)*

In the rate-determining step, we have the heterolytic cleavage of the C–Br bond in the halogenoalkane that produces a carbocation intermediate. The carbocation (a Lewis acid) is then attacked by a nucleophile (a Lewis base) attracted to the electron-deficient carbon atom.

Step 1: Formation of carbocation.

$$CH_3\text{---}\overset{\overset{\displaystyle CH_3}{|}}{\underset{\underset{\displaystyle CH_3}{|}}{C}}{}^{\delta+}\text{---}Br^{\delta-} \xrightarrow{\text{slow}} \overset{\overset{\displaystyle CH_3}{\diagdown}}{\underset{\underset{\displaystyle CH_3}{\diagup}}{C}}{}^{+}\text{---}CH_3 \ + \ Br^-$$

Step 2: Nucleophile attacks the carbocation to form final product.

$$\overset{\overset{\displaystyle CH_3}{\diagdown}}{\underset{\underset{\displaystyle CH_3}{\diagup}}{C}}{}^{+}\text{---}CH_3 \ + \ :CN^- \xrightarrow{\text{fast}} CH_3\text{---}\overset{\overset{\displaystyle CH_3}{|}}{\underset{\underset{\displaystyle CH_3}{|}}{C}}\text{---}CN$$

Q: Step 1 of the S_N1 mechanism gives me the impression that the C–Br bond of the halogenoalkane simply cleaves on its own. Am I right about this?

A: No! The C–Br bond does not cleave "automatically." It is actually mediated by the approach of a nucleophile, quite similarly to an S_N2. But the proximity of the nucleophile in an S_N1 reaction is further than in an S_N2. As a result, the effect of the nucleophile is not great enough to make its "impact" in the rate equation. In addition, you could also imagine that as the nucleophile approaches the halogenoalkane from the "backside" route, it would repel against the electron clouds of the bulky alkyl groups. This inter-electronic repulsion would "push" the electron clouds of the bulky alkyl groups towards the Br atom, thus "helping" the C–Br bond to cleave "on its own."

As we have seen in earlier mechanisms, this carbocation intermediate has a trigonal planar geometry around the sp^2 hybridized carbon that allows the nucleophile to approach from two directions. If we start off with a halogenoalkane that has a chiral C atom attached to the halogen atom as shown, we will get a racemic mixture containing equal proportions of the two optically active isomers. For example,

$CH_3CH_2CH(CH_3)Br$

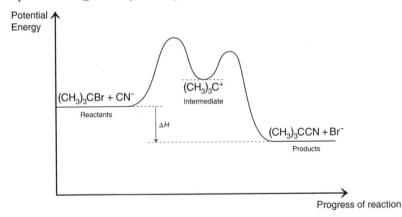

* chiral center

Q: Wouldn't the Br^- ion re-attack the carbocation since it is a nucleophile too?

A: Yes, it can. But you would just get back the original reactant.

From kinetics study of reaction rate, the derived rate equation is: rate = k[RX].

Rate is dependent only on the concentration of the 3° halogenoalkane (RX). The rate equation indicates a unimolecular rate-determining step consisting of only one species, RX. Hence, the acronym S_N1.

Taking into account what happens in the rate-determining step, the factors that influence the reaction rate center upon the ease of cleavage of the C–X bond:

(i) the C–X bond strength

As we have discussed earlier, a slower rate is attributed to a stronger bond that is harder to be cleaved. With the C–X bond strength decreasing in the order:

$$C{-}F > C{-}Cl > C{-}Br > C{-}I.$$

Considering halogenoalkanes that only differ in the type of halogen substituent, the reaction rate is the fastest for compounds that contain the C–I bond, followed by C–Br, C–Cl and lastly, C–F, which would be the slowest.

There must be a good reason for the C–X bond to cleave, and that is the formation of a stable species — the carbocation intermediate. Hence, we will expect that a more stable carbocation will be formed faster since its formation is favored over a less stable carbocation.

With this said, the stability of a carbocation increases in the order:

$$\underset{\substack{\text{1° carbocation} \\ \text{(primary)}}}{CH_3^+} \quad < \quad \underset{\substack{\text{1° carbocation} \\ \text{(primary)}}}{RCH_2^+} \quad < \quad \underset{\substack{\text{2° carbocation} \\ \text{(secondary)}}}{R_2CH^+} \quad < \quad \underset{\substack{\text{3° carbocation} \\ \text{(tertiary)}}}{R_3C^+}$$

and the reaction rate increases in the order of:

$$\underset{\text{1° halogenoalkane}}{CH_3X} \quad < \quad \underset{\text{2° halogenoalkane}}{RCH_2X} \quad < \quad \underset{\text{2° halogenoalkane}}{R_2CHX} \quad < \quad \underset{\text{3° halogenoalkane}}{R_3CX}$$

Q: What determines the relative stability of the carbocations?

A: The stability of charge particles depends on the extent of charge dispersal. For a carbocation, if the positive charge can be made less concentrated on the carbon bearing it, there will be a lower tendency for it to be attacked by nucleophiles, rendering it more stable. The way to disperse the positive charge is to neutralize it, and the presence of electron-releasing groups, such as alkyl groups, will do the trick. Further details are given in Chapter 4 on Alkanes.

Q: The stability of a species is a thermodynamic property — why should a more stable carbocation be formed faster? How can we link up both thermodynamic stability and kinetic feasibility?

A: The reason why a 3° carbocation is more stable than a 2° carbocation is due to a greater amount of charge dispersal on the electron-deficient carbon atom brought about by a greater number of alkyl groups present. The activated complex of a reaction is a species that is intermediary between the reactant and the product or the intermediate if the reaction has multiple elementary steps. For example, this would mean that the activated complex for the formation of a 3° carbocation would also possess an electron-deficient carbon center. Now, factors that stabilize the intermediate would also be the factors that stabilize the activated complex. As a result, we would expect the activated complex for a 3° carbocation to be more stable than that for the 2° carbocation due to the greater amount of charge dispersal. This would translate into a smaller activation energy barrier for the formation of the 3° carbocation. Thus, when the reaction temperatures for both the 3° and 2° halogenoalkanes are the same, this would mean that the average kinetic energies of both the 3° and 2° halogenoalkanes are also going to be the same (similar Maxwell–Boltzmann distribution of kinetic energy). With the same average kinetic energy, the 3° halogenolkane would thus have a higher rate of reaction.

Q: If nucleophilic reactions of 1° halogenoalkanes proceed via the S_N2 mechanism, and for 3° halogenoalkanes, the S_N1 mechanism, what about 2° halogenoalkanes?

A: The type of nucleophilic substitution mechanism for a 2° halogenoalkanes depends on the type of the C–X bond to be broken and also on the bulkiness of the alkyl groups. Thus, it is quite difficult to pin-point whether it would be predominantly S_N1 or S_N2. In reality, both mechanisms are likely for a particular type of 2° halogenoalkane.

In summary, factors that affect nucleophilic substitution reaction are:

- *Electronic factor*
  - Polarity of the C–X bond would make C–F more susceptible to nucleophilic attack, especially for S_N2.
  - Strength of the C–X bond to be broken would make C–I more susceptible to cleavage for both S_N1 and S_N2.
  - Strength of the C–Nu bond to be formed would make C–X bond more susceptible to cleavage for S_N2.
  - Inter-electronic repulsion between the bulky alkyl groups and the halogen atom would favor the cleavage of the C–X bond for both S_N1 and S_N2.
  - Relative stability of the carbocation formed would favor S_N1.

o Polarity of the solvent used for reaction would favor S_N1, as it would help to stabilize the carbocation if the solvent used is a polar solvent.

- *Steric factor*

  o Increased steric hindrance for the approaching nucleophile would favor S_N1.

3.4.5.3 *Nucleophilic Acyl Substitution (Condensation) of Carboxylic Acid and its Derivatives*

In addition to halogenoalkanes, carboxylic acid and its derivatives also undergo nucleophilic acyl substitution reaction with an energy profile diagram (see below) similar to that for S_N1 (refer to Section 3.4.5.2). But take note that the intermediate is tetrahedral in shape for carboxylic acid and its derivatives, rather than a trigonal planar intermediate.

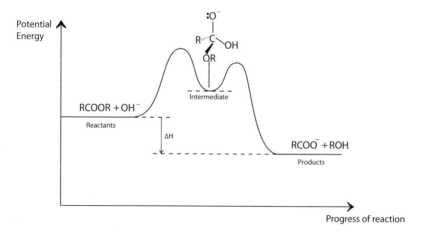

The following shows the different functional groups:

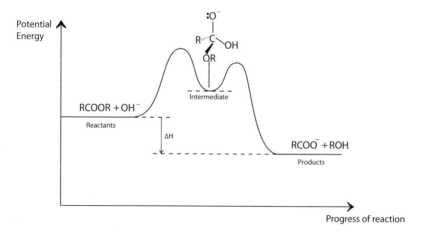

A typical nucleophilic acyl substitution reaction of an ester by a base is shown below:

intermediate

If you compare the nucleophilic acyl substitution of carboxylic acid and its derivatives with that of halogenoalkane, one would find that nucleophilic acyl substitution of carboxylic acid and its derivatives occurs much more readily. This is because the electron-deficient carbon atom for carboxylic acid and its derivatives is less sterically hindered. This is because the carbon center is sp^2 hybridized and is trigonal planar in shape. In addition, the highly electron deficient carbon atom of the carbonyl group (C=O) attracts the nucleophile very readily. The additional electron deficiency is brought about by the presence of two highly electronegativity atoms, such as O, Cl or N.

The tetrahedral intermediate that is formed in the course of the reaction is highly unstable due to the presence of three highly electronegative atoms. This instability results in a driving force to regenerate the trigonal planar geometry, which is resonance stabilized. In this process, a leaving group is "expelled."

Carboxylic acid and its derivatives undergo nucleophilic acyl substitution rather than nucleophilic addition, which is typical for carbonyl compounds. This is due to the ease of leaving groups, such as –OH, –Cl, –OR and –NH$_2$, to be replaced by a nucleophile. Nucleophilic acyl substitution of an aldehyde or ketone is difficult as the alkyl group or hydrogen atom is a bad leaving group. In addition, the carbon-leaving group bond in carboxylic acid and its derivatives is further weakened due the presence of highly electronegative atom such as O, Cl or N (refer to Chapter 10 for more details).

My Tutorial

1. Discuss the various types of reaction mechanism between the following pairs of molecules:

   (a) butane and bromine in the presence of UV light
   (b) ethylbenzene and bromine in the presence of UV light
   (c) propene and bromine
   (d) propene and aqueous bromine
   (e) propene and aqueous bromine in the presence of sodium cyanide (NaCN)
   (f) benzene and bromine in the presence of $AlBr_3$ or $FeBr_3$
   (g) chlorobenzene with concentrated nitric and sulfuric acids
   (h) methylbenzene and chlorine in the presence of $AlCl_3$ or $FeCl_3$
   (i) 1-bromobutane and aqueous barium hydroxide with heating
   (j) 2-bromo-2-methylpropane and aqueous barium hydroxide with heating
   (k) 1-bromobutane and alcoholic potassium hydroxide with heating
   (l) propanone and HCN with sodium hydroxide
   (m) ethanal and KCN with sulfuric acid
   (n) butanone and HCN with NaCN

2. (a) Describe the meaning of the term *homolytic fission,* using an organic reaction as an example.
   (b) Describe the meaning of the term *heterolytic fission,* using an organic reaction as an example.
   (c) Why do halogenoalkanes react with nucleophiles at the carbon center but with electrophiles at the halogen atom?
   (d) What do you understand by the *polarity* of a bond and what determines whether a bond is polar or not?

CHAPTER 4

Alkanes

4.1 Introduction

Hydrocarbons are a class of organic compounds containing only two types of elements — carbon and hydrogen (see Fig. 4.1). These include members of the homologous series of alkanes, alkenes, alkynes and arenes. The difference between these groups of compounds lies in the structure of the carbon skeleton.

Alkanes are saturated hydrocarbons, which can be either straight-chain or branch-chain molecules. Each of the carbon atoms in the molecule is singly-bonded to four atoms — carbon or hydrogen. All these molecules share the same general molecular formula C_nH_{2n+2}, where n stands for the number of carbon atoms. For instance, butane, with $n = 4$, has the formula C_4H_{10}. If we join the two carbon atoms at each end of a butane molecule, and in so doing remove one hydrogen atom from each of these terminal carbon atoms, we will end up with a ring structure known as cyclobutane.

Cycloalkanes are also saturated hydrocarbons since no further hydrogen atoms can be added to the ring structure; they cannot undergo hydrogenation, which is the addition of hydrogen. Cycloalkanes have the general formula C_nH_{2n}. Sharing this same general formula are the alkenes, which are constitutional/structural isomers of the corresponding cycloalkanes.

Alkenes are unsaturated hydrocarbons as they contain a C=C double bond. Alkynes, of the general formula C_nH_{2n-2}, are also unsaturated hydrocarbons identified by the C≡C triple bond. Both alkenes and alkynes can be hydrogenated to form saturated hydrocarbons. With all this discussion thus far, it should not be difficult to see that two hydrogen atoms is equivalent to the presence of a double bond or a ring.

Q: So, are alkane and cycloalkane considered isomer?

A: No! Isomers have the same molecular formula. But alkane and cycloalkane do not share the same molecular formula.

Another group of hydrocarbons are the aromatic hydrocarbons, also known as arenes, which contain the benzene ring. The term "aromatic" arises from the sweet-smelling aroma of benzene, which has the formula C_6H_6.

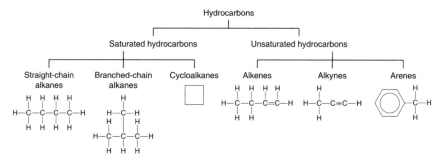

Fig. 4.1.

4.2 Nomenclature

Unlike in earlier times, when the names of compounds were given by their discoverers, the myriad of compounds that were discovered soon called for a systematic naming procedure — this comes in the form of a set of rules convened by the International Union of Pure and Applied Chemistry (IUPAC) in the 1950s. Based on the IUPAC nomenclature, a chemical name comprises three components, namely the root, suffix and prefix(es):

Prefix(es) – Root – Suffix

The root indicates the number of carbon atoms in the main carbon skeleton (see Table 4.1). The suffix indicates the principal functional group or homologous series. The prefix indicates the substituent or functional group attached to the main carbon skeleton. As such,

Table 4.1

Number of C atoms	1	2	3	4	5	6	7	8	9	10
Root	meth-	eth-	prop-	but-	pent-	hex-	hept-	oct-	non-	dec-

there can be more than one prefix according to the number of substituents present.

For saturated alkanes, their chemical names end with the suffix — *ane*. Table 4.2 lists the names of the first four members of the alkane family.

Table 4.2

Alkane	Methane	Ethane	Propane	Butane
Structural formula	H—C—H with H above and below	H—C—C—H with H H above and H H below	H—C—C—C—H with H H H above and H H H below	H—C—C—C—C—H with H H H H above and H H H H below

If you notice, each successive member of the alkane family differs by a methylene ($-CH_2$) group. This is in fact true when we descend through each homologous series.

To differentiate them from straight-chain alkanes, ring structures are named with the word *cyclo* attached to the front of the root. The first member of the cycloalkane family is cyclopropane, since a minimum of three carbon atoms are needed to form a closed loop.

For alkanes with four carbon atoms onwards, there exists different ways in which the carbon atoms can be connected to each other, hence giving rise to the notion of constitutional/structural isomerism (see Chapter 2). Take, for instance, an alkane with the formula C_6H_{14}. Hexane, a straight-chain alkane, is but one possible structure that we can draw:

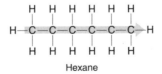

Hexane

The other possible structures arise when we consider branching along the main carbon skeleton — when the sequence of carbon atoms splits in two or more directions. With branching, we get the following branched-chain alkanes, which are isomeric to hexane:

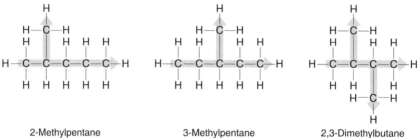

2-Methylpentane   3-Methylpentane   2,3-Dimethylbutane

A chemical name is unique, and it can only represent one structure. For branched-chain alkanes, the root indicates the maximum number of carbon atoms in the main carbon skeleton, while the branches are indicated as alkyl prefixes. Using 2-methylpentane as an example, the following are the steps used to name it:

Step 1: Identify the longest carbon chain. This is regarded as the main chain.

o Count the number of carbon atoms in the main chain to get the corresponding root word of its name, i.e. since there are five carbon atoms, its name ends with — **pentane**.

Step 2: Identify the branches, which in this case, are alkyl substituents. There is a — **CH₃** branch which is called the methyl substituent.

o The methyl substituent is named as a prefix stated in front of the root word, i.e. **methyl** pentane.

Q: What are alkyl groups?

A: Alkyl groups can be thought of as alkane derivatives with one less hydrogen atom. For instance, removing a hydrogen atom from methane, CH_4, leaves us with a methyl, CH_3, group. Alkyl groups have their names end in –*yl*. They have the general formula C_nH_{2n+1}.

Step 3: The locations of substituents on the main carbon chain are denoted by **positional numbers**. Each carbon atom on the main chain is numbered, starting from one end of the chain, i.e.

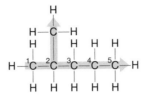

o The positional number is stated in front of the prefix for that particular substituent, i.e. **2**-methylpentane. Note that "hyphens" are used to separate numbers from words.

Q: Which end of the carbon chain do we start counting from?

A: We can start from any end of the chain. The main objective is to indicate the location of the substituents using the **lowest** number possible. As shown here, it is preferred to identify the methyl substituent to be residing on C2 rather than C4.

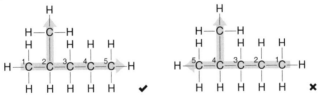

Q: Can the following highlighted sequence of carbon atoms be regarded as the longest chain?

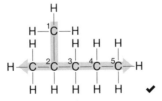

A: Yes, a sequence consists of a continuous series of carbon atoms. It does not consider changes in angles. As long as you can trace a series of carbon atoms without lifting your pencil from the paper, it is considered a sequence. If we apply the rules in naming the structure above, its name is concluded to be 2-methylpentane. This reinforces the fact that a chemical name constitutes one unique structure, and that the reverse holds true.

In another example using 2,3-dimethylbutane, the following are the steps used to name it:

Step 1: Identify the longest carbon chain.

    ○ Number of carbon atoms in main chain: 4 ⇒ root −suffix: **−butane.**

Step 2: Identify the substituents.

    ○ Substituents: methyl groups ⇒ prefix: **methyl−.**

Step 3: Assign the lowest possible positional numbers to each substituent by numbering the carbon atoms on the main chain, i.e.

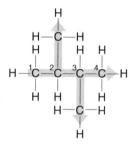

    ○ Position of substituents: at C2 and C3.

Step 4: Cite the name and positional number of each substituent as prefixes before the root, bearing in mind the following rules:

    ○ prefixes are cited in alphabetical order.

    ○ if there are identical substituents, the numeric prefixes *di-*, *tri-*, *tetra-* are attached to the prefix. These numeric prefixes have no bearing on the alphabetical order of prefixes.

    ○ the number of positional numbers must be equal to the number of substituents.

    ○ a "hyphen" is used to separate numbers from words.

    ○ a "comma" is used to separate numbers.

Examples of numeric prefixes:

- "di" means "2"
- "tri" means "3"
- "tetra" means "4"
- "penta" means "5"
- "hexa" means "6"
- "hepta" means "7"

After going through the steps, we get the name **2,3-dimethylbutane**.

Exercise: Name the following alkanes.

(i) CH_2CH_3 (ii) CH_2CH_3

      $CH_3CH_2CH_2\overset{|}{C}HCH_3$ $CH_2\overset{|}{C}HCH_3$

                                    $CH_3CH_2CH_2(CH_3)\overset{|}{C}HCH_2CH_3$

Solution: (i) 3-methylhexane; (ii) 4-ethyl-3,6-dimethyloctane

4.3 Physical Properties

4.3.1 *Melting and Boiling Points*

Alkanes are non-polar molecules, and only weak instantaneous dipole–induced dipole (id–id) interactions exist between their molecules. The melting and boiling points of the alkanes only depend on the strength of the id–id interaction, which is affected by:

the number of electrons in the molecule: the greater the number of electrons (i.e. the bigger the electron cloud), the more polarizable is the electron cloud, the stronger are the id–id interactions;

surface area for contact of molecule: the greater the surface area of contact possible between the molecules, the greater the extent of id–id interactions.

Table 4.3

Alkane	Methane	Ethane	Propane	Butane	Pentane	Hexane	Heptane
	CH_4	C_2H_6	C_3H_8	C_4H_{10}	C_5H_{12}	C_6H_{14}	C_7H_{16}
b.p. /°C	−162	−89	−42	−0.5	36	69	98

As shown in Table 4.3 above, the boiling points of the alkanes increase as the carbon chain gets longer. This is due to the greater number of electrons that contribute to a more polarizable electron cloud leading to stronger id–id interactions.

In comparison to the straight-chain alkane of the same molecular formula, the branched-chain alkane has a lower boiling point. Recall from Chapter 2 the pair of chain isomers pentane and dimethyl propane:

Pentane
(b.p. 36 °C)

Dimethylpropane
(b.p. 10 °C)

They have the same molecular formula, C_5H_{12}, which means they have the same number of electrons. The difference in boiling points is explained by the relative size of the surface areas of the molecules. There are more extensive id–id interactions between pentane molecules due to their greater surface area of contact, and this accounts for the higher boiling point of pentane.

Q: Why isn't dimethylpropane named "2,2-dimethylpropane?"

A: The numbers are being dropped as, for a parent carbon chain of three carbon atoms, there is only one possibility for the methyl groups to be bonded to without changing the length of the carbon skeleton, i.e. on the second carbon. Placing the methyl groups onto the terminal carbon atoms would result in a parent chain consisting of at least four carbon atoms in length. Thus, both names are acceptable here because if we translate each of the names into a molecular structure, we get the same molecule. In this case, there is no ambiguity.

4.3.2 *Solubility*

Alkanes, being non-polar compounds, are soluble in non-polar organic solvents such as carbon tetrachloride (CCl_4) but insoluble in polar solvents such as water. This is due to the similar intermolecular attractive forces, i.e. van der Waals forces, that exist between the non-polar molecules that lead to favorable mixing. In fact, hydrocarbons in the liquid state such as hexane and benzene are commonly used as non-polar solvents.

Q: What is the type of interaction between a non-polar alkane molecule and a polar molecule?

A: As a polar molecule already has a permanent dipole, when this polar molecule is near a non-polar alkane molecule, it would cause an induced dipole on the non-polar alkane molecule. Then the interaction between them would be called permanent dipole-induced dipole interaction. But there is also a possibility that the alkane molecule already "comes" with an induced dipole when being approached by the polar molecule. Thus, from this perspective, it seems like a kind of permanent dipole-permanent dipole interaction. So, to make things simple, we sometimes just consider the interaction between a non-polar and a polar molecule as instantaneous dipole-induced dipole interaction. Meaning? We look at the polar molecule as simply "non-polar" when interacting with another non-polar molecule. Such approach is to simplify the type of interaction between polar-polar (pd-pd), non-polar-non-polar (id-id) and polar-non-polar (id-id or pd-id or id-pd) molecules.

4.4 Preparation Methods

4.4.1 *Catalytic Hydrogenation of Alkenes-Reduction*

Reagents:	$H_2(g)$ with $Ni(s)$ catalyst
Conditions:	Heat
Options:	$Pt(s)$ or $Pd(s)$ as catalyst

Catalytic hydrogenation can be done on an alkene to obtain the alkane with the same number of carbon atoms. An example is the conversion of ethene to ethane. In this reaction, the pi bond is broken to form two $C-H$ sigma bonds. Although it appears to be an addition reaction, it does not proceed via the electrophilic addition mechanism. Rather, it is a heterogeneous catalytic reaction (refer to Chapter 5 for further details on this reaction).

4.5 Chemical Properties

In Chapter 3, we have seen organic compounds being attacked by either nucleophiles (because the organic compounds are electron-deficient) or by electrophiles (because the organic compounds are electron-rich).

Electron-rich pi system   Electron-deficient carbon atom

Both ethene and benzene are sources of pi electrons, making them electron-rich sites that attract electrophiles.

Ethanal and bromoethane have polar C=O and C–Br bonds respectively. With oxygen and bromine being more electronegative than carbon, these molecules contain an electron-deficient carbon atom that attract nucleophiles.

If we look back at the structure of any alkane, there are no electrostatically attractive features that draw in a potential attacking species. C–H bonds are essentially non-polar as both the carbon and hydrogen atoms have a rather small electronegativity difference. There are also no pi electrons as these are saturated molecules, with each carbon atom using all its valence electrons to form sigma bonds with four others. Subtly, it is this unattractive feature of the C–C and C–H bonds that forms the backbone of all organic compounds and is responsible for their existence. With their chemical inertness, reactions that do occur involve some form of energy input.

4.5.1 *Combustion–Oxidation*

Reagents:	$O_2(g)$
Conditions:	Heat at high temperature
Options:	—

All hydrocarbons produce the same products, CO_2 and H_2O, when they undergo complete combustion. As the reaction is very exothermic, alkanes serve as an important industrial fuel.

The general equation depicting the combustion reaction is:

$$C_xH_y + \left(x + \frac{y}{4}\right)O_2 \longrightarrow xCO_2 + \frac{y}{2}H_2O$$

Q: It is said that combustion of an alkane is actually a redox process. How can we prove it?

A: If you count the oxidation number of the C atom in ethane, each has an oxidation number of -3. This oxidation number is derived if we consider that a C atom is slightly more electronegative than an H atom; as a result, with three C–H bonds, the C atom would have a -3 oxidation state. Next, the oxidation number of the C atom in CO_2 is $+4$. Therefore, combustion of alkane to CO_2 is a redox reaction.

Q: Why is combustion of an alkane exothermic? And why is it that the more carbon atoms present in alkane molecule, the more exothermic the combustion?

A: You can use bond energies data to prove this. Essentially, the energy that is released during the formation of the strong bonds in CO_2 and H_2O is much more than the energy that is required to break the C–H, C–C and O=O bonds. Hence, you have an exothermic reaction.

However, if the supply of oxygen is limited, incomplete combustion occurs and products such as non-combusted carbon (in the form of soot) and CO are formed apart from CO_2 and H_2O, i.e.

Complete combustion: $CH_4(g) + 2O_2(g) \longrightarrow CO_2(g) + 2H_2O(l)$

Incomplete combustion: $CH_4(g) + \frac{3}{2}O_2(g) \longrightarrow CO(g) + 2H_2O(l)$

$$CH_4(g) + O_2(g) \longrightarrow C(s) + 2H_2O(l)$$

Q: What is the reaction mechanism like during the combustion process?

A: The combustion process is of the free radical reaction type. In order for free radicals to be generated, first the molecules need to be vaporized. Then homolytic cleavage occurs, generating free radicals, which then react with the oxygen present.

4.5.2 Halogenation — Free Radical Substitution

Reagents:	Halogens such as $F_2(g)$, $Cl_2(g)$ or $Br_2(l)$
Conditions:	UV light or heat
Options:	—

As mentioned in Chapter 3, various types of substituted halogenated products can be obtained. Although the lack of control on the reaction is inevitable, the yield of desired products can still be managed to a certain extent by controlling the relative amount of reactants used. If an excess of ethane is used over bromine, the main product will be monobrominated ethane. When an excess of the halogen is used, further substitutions are more likely to occur with polybrominated ethane as the major product.

Coupled with the fact that the C–H bonds are of about the same energy, there is a high possibility of having a different hydrogen atom substituted such that isomeric products are obtained. For instance, when excess pentane reacts with bromine, three types of monosubstituted products can be formed:

1-Bromopentane   2-Bromopentane   3-Bromopentane

Q: Assuming that only three of the monosubstituted products are formed, what is the ratio in which they might be formed?

A: Considering that there are a total of 12 C–H bonds in a pentane molecule, all with the same bond energy, cleaving any six of the C–H_a bonds produces 1-bromopentane. This gives a 50% probability that a monosubstituted product is likely to be 1-bromopentane. Cleaving any of the four C–H_b bonds gives 2-bromopentane and cleaving the two C–H_c bonds produces 3-bromopentane.

Based on statistical probability, the ratio in which 1-bromopentane, 2-bromopentane and 3-bromopentane are likely to be formed is thus 6 : 4 : 2 which simplifies to 3 : 2 : 1.

Q: So based on what was said, does that mean that when propane is brominated, the ratio of 1-bromopropane to 2-bromopropane would be 3:1?

A: Based on the statistical factor, the ratio would indeed be 3:1. But in addition to the statistical factor, there is also the electronic factor that affects the composition of the products. To form 1-bromopropane and 2-bromopropane, we need the following radicals to be generated, respectively:

$$\dot{C}H_2CH_2CH_3 \quad CH_3\dot{C}HCH_3$$
$$\text{Radical 1} \qquad \text{Radical 2}$$

The stability of the two radicals differs. Radical 2 is more stable than Radical 1 because there are two alkyl groups bonded to the carbon atom with the lone electron whereas Radical 1 has only one alkyl group. An alkyl group is known to be electron-donating (via inductive effect), and thus the electron deficiency on the carbon atom possessing the lone electron would be diminished by a greater extent for Radical 2 than Radical 1. Thus, take note that factors such as the reactants ratio, the statistical factor and electronic factor would all affect the final composition of the products.

Alkyl radicals can be classified as one of four types — methyl radical, primary alkyl radical, secondary alkyl radical or tertiary alkyl radical. These differ in the number of alkyl (R) groups attached to the carbon atom bearing the lone electron. The greater the number of electron-donating alkyl groups, the more stable is the alkyl radical. Thus, in order of increasing stability, we have:

$$
\begin{array}{cccc}
\text{H} & \text{H} & \text{R} & \text{R} \\
| & | & | & | \\
\text{H—C}\cdot & < \quad \text{R—C}\cdot & < \quad \text{R—C}\cdot & < \quad \text{R—C}\cdot \\
| & | & | & | \\
\text{H} & \text{H} & \text{H} & \text{R} \\
\text{Methyl} & \text{Primary (1°)} & \text{Secondary (2°)} & \text{Tertiary (3°)} \\
\text{(least stable)} & & &
\end{array}
$$

Q: Why is an alkyl group able to donate electron density via inductive effect? What is an inductive effect?

A: Inductive effect refers to "transmission" through the sigma bond. Hence, electron-donating via inductive effect refers to the movement of electron density via the sigma bonds. An alkyl group is able to donate electron density via inductive effect only if there is a demand. What do I mean by "if there is a demand?" We cannot say that the central carbon atom of a propane molecule is more electron-rich than the other two carbon atoms because there are two methyl groups bonded to it, which are electron donating via inductive effect. There are actually no great differences between these three carbon atoms. But, if the central carbon atom has a lone electron, like Radical 2 above, then this carbon atom is less electron-rich (or more electron deficient) than the other two carbon atoms. This electron deficiency would then cause the "outflow" of electron density from the more electron-rich methyl groups through the sigma bonds. The fundamental principle is for electron density to be as spread out as possible so as to minimize inter-electronic repulsion.

Q: How do we depict the free radical substitution (FRS) mechanism showing the formation of 1-bromopropane, 2-bromopropane, 1,1-dibromopropane, and hexane from the same starting reagent, propane?

A: The first step of the FRS mechanism to generate each of these alkanes is the same, i.e.

Step 1: Chain initiation with the generation of free radicals

$$Br\text{---}Br \xrightarrow{\text{UV light}} 2\ Br\cdot$$

The differences occur in the subsequent propagation and termination steps.

- For the formation of 1-brompropane:

  Step 2: Chain propagation

  (a) $Br^\bullet + CH_3CH_2CH_3 \longrightarrow CH_3CH_2CH_2^\bullet + HBr$

  (b) $CH_3CH_2CH_2^\bullet + Br_2 \longrightarrow CH_3CH_2CH_2Br + Br^\bullet$
  ... etc.

  Step 3: Chain termination

  $$Br^\bullet + CH_3CH_2CH_2^\bullet \longrightarrow CH_3CH_2CH_2Br$$

○ For the formation of 2-brompropane:

Step 2: Chain propagation

(a) $Br^\bullet + CH_3CH_2CH_3 \longrightarrow (CH_3)_2CH^\bullet + HBr$

(b) $(CH_3)_2CH^\bullet + Br_2 \longrightarrow CH_3CHBrCH_3 + Br^\bullet$

... etc.

Step 3: Chain termination

$$Br^\bullet + (CH_3)_2CH^\bullet \longrightarrow CH_3CHBrCH_3$$

○ For the formation of 1,1-dibrompropane:

Step 2: Chain propagation

(a) $Br^\bullet + CH_3CH_2CH_3 \longrightarrow CH_3CH_2CH_2^\bullet + HBr$

(b) $CH_3CH_2CH_2^\bullet + Br_2 \longrightarrow CH_3CH_2CH_2Br + Br^\bullet$

(c) $Br^\bullet + CH_3CH_2CH_2Br \longrightarrow CH_3CH_2CHBr^\bullet + HBr$

(d) $CH_3CH_2CHBr^\bullet + Br_2 \longrightarrow CH_3CH_2CHBr_2 + Br^\bullet$

... etc.

Step 3: Chain termination

$$Br^\bullet + CH_3CH_2CHBr^\bullet \longrightarrow CH_3CH_2CHBr_2$$

○ For the formation of hexane:

Step 2: Chain propagation

(a) $Br^\bullet + CH_3CH_2CH_3 \longrightarrow CH_3CH_2CH_2^\bullet + HBr$

(b) $CH_3CH_2CH_2^\bullet + Br_2 \longrightarrow CH_3CH_2CH_2Br + Br^\bullet$

... etc.

Step 3: Chain termination

$$2CH_3CH_2CH_2^\bullet \longrightarrow CH_3CH_2CH_2CH_2CH_2CH_3$$

Q: So, generally, what do you need to do if you are asked to show the FRS mechanism for the formation of a particular halogenolkane?

A: (i) The chain propagation steps would only focus on the sequential steps that would lead to the formation of the desired product, for instance, 1,1-dibromoporpane. Basically, we start off with a halogen radical and an alkane in the first step, we end with the desired product and a "different" halogen radical in the last step.

(ii) The termination step only shows the reaction of two radicals that give us our desired product, which, in this case here, involves bromine and 1-bromopropyl radicals.

(iii) To depict the formation of side-products, such as hexane, the chain propagation steps would demonstrate the formation of the alkyl radical which when reacts in the termination steps, would give us our desired side-product.

4.5.3 *Thermal and Catalytic Cracking*

Long-chain alkane $\xrightarrow{\text{strong heating}}$ Smaller-chain alkanes + alkenes

For Thermal Cracking,		For Catalytic Cracking,	
Reagents:	—	Reagents:	silica-alumina catalyst
Conditions:	800–900°C	Conditions:	450–550°C
Options:	—	Options:	—

Thermal cracking involves the breaking of strong C–C bonds, which requires strong heating. Homolytic cleavage of these C–C bonds leads to the formation of free radicals. And just like the free radical substitution reaction, the cracking process is not controllable, and a variety of hydrocarbon products is formed depending on the reaction conditions.

For instance, the cracking of $C_{14}H_{30}$ can give the following products:

$$C_{14}H_{30} \longrightarrow C_8H_{18} + C_6H_{12},$$

or it can break down into more fragments as follows:

$$C_{14}H_{30} \longrightarrow C_6H_{14} + C_5H_{10} + C_3H_6.$$

Nonetheless, the process is useful in converting long-chain alkanes into shorter-chain alkanes which have more uses, as in the cracking of petroleum oil to give simpler hydrocarbons that can be used as lubricants, fuels and so forth. Extremely long-chain alkanes are not very volatile, hence it is not easy to make them react so as to get more useful products. In the worst scenario where these long-chain alkanes cannot be further broken down, they are used as fuel to heat up boilers in power plants.

Catalytic cracking is similar to thermal cracking except that lower heating temperatures are needed and a catalyst is used. The catalyst used is none other than silica-alumina catalyst (SiO_2–Al_2O_3).

My Tutorial

1. (a) When ethane reacts with bromine in light to produce the desired product, bromoethane, small traces of butane and other multiple brominated products are found after the reaction has completed.

   (i) Describe the mechanism for the formation of bromoethane.

   (ii) Describe the mechanism for the formation of 1,1-dibromoethane.

   (iii) Describe the mechanism for the formation of 1,2-dibromoethane.

   (iv) Describe the mechanism for the formation of 1,1,2-tribromoethane.

   (v) Describe the mechanism for the formation of butane.

   (vi) How can you maximize the formation of bromoethane?

   (vii) How can you maximize the formation of hexabromoethane?

   (viii) Suggest a simple way to separate the desired product from the mixture of products.

   (b) When $50\,cm^3$ of gaseous hydrocarbon, **P**, was completely combusted with $400\,cm^3$ oxygen gas, the residual gases occupied $300\,cm^3$ at r.t.p. After passing through barium hydroxide solution, the final volume was $100\,cm^3$.

   (i) Why was there a decrease in volume when residual gases were shaken with aqueous barium hydroxide? Give a balanced equation for the reaction.

   (ii) Determine the molecular formula of **P**.

   (iii) Give the constitutional/structural formulae of six possible compounds with this molecular formula.

   (iv) Which of the six constitutional/structural formulae show compounds which are

   (A) constitutional/structural isomers;

(B) stereoisomers;

(C) enantiomers?

(c) Methane and chlorine react only very slowly in the dark, but in bright sunlight, the reaction takes place readily to give a mixture of organic products. Explain.

(d) When pentane is treated with a solution of bromine in CCl_4 in the presence of sunlight, the reddish-brown coloration disappears and steamy white fumes evolve. Identify the white fumes.

CHAPTER 5

Alkenes

5.1 Introduction

Alkenes are unsaturated hydrocarbons containing the C=C double bond. They have the general formula C_nH_{2n}. Sharing this same general formula are the cycloalkanes, which are constitutional/structural isomers to the corresponding alkenes. Apart from functional group isomerism, constitutional/structural isomerism in alkenes can also arise due to the different degree of branching in the main carbon chain of their molecules (chain isomerism) and the location of the C=C bond (positional isomerism).

Alkenes also exhibit *cis-trans* isomerism, otherwise known as geometrical isomerism (see Chapter 2). This form of stereoisomerism is attributed to the restricted rotation about the C=C bond.

As the doubly bonded carbon atoms are sp^2 hybridized, the geometry about each of them is trigonal planar. To show the trigonal planar geometry, the structure of an alkene is normally drawn in such a way as to depict an angle of 120° about the double bond, as follows:

$$\begin{array}{ccc} H & & H \\ & C{=}C & 120° \\ H & & H \end{array}$$

This form of drawing is especially important in illustrating both the *cis* and *trans* isomers.

5.2 Nomenclature

For alkenes, their chemical names end with the suffix –*ene*. Table 5.1 lists the names of the first few members of the alkene family. For an alkene with four carbon atoms onwards, there exists different ways in which the carbon atoms can be connected to each other, hence giving rise to the notion of constitutional/structural isomerism (see Chapter 2). For instance, both but-1-ene and but-2-ene have a chain of four carbon atoms. The difference between them is the location of the C=C bond.

Table 5.1

Alkene	Ethene	Propene	"Butene"	"Pentene"
Constitutional /Structural formula			**But-1-ene** **"But-2-ene"**	**Pent-1-ene** **"Pent-2-ene"**

The word "butene" does not just wrongfully account for the two constitutional/structural isomers mentioned above, ambiguity is also found in the word "but-2-ene," which actually constitutes a pair of *cis-trans* (geometrical) isomers:

cis-but-2-ene   trans-but-2-ene

Hence, "butene" actually stands for a total of three distinct compounds – but-1-ene, *cis*-but-2-ene and *trans*-but-2-ene. Likewise, the word "pentene" can mean any one of the three different compounds when we include the stereoisomers:

cis-pent-2-ene   trans-pent-2-ene

Apart from these straight-chain alkenes with five carbon atoms, we also have the isomeric branched-chain alkenes:

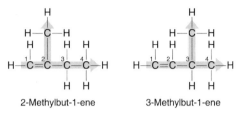

2-Methylbut-1-ene   3-Methylbut-1-ene

Similar to the naming of alkanes, the root indicates the number of carbon atoms in the main carbon skeleton. Prefixes, accompanied by positional numbers, indicate the alkyl branches or other functional groups that are attached to the main carbon skeleton. Positional numbers are also used to indicate the location of the C=C bond, which allows the differentiation of positional isomers from each other. But since there are two carbon atoms in the C=C bond, which means that there is a pair of consecutive positional numbers that identify the position of the C=C bond, only the lower of these consecutive numbers appears in the chemical name, and it is stated in front of the suffix *–ene*. The numbering of the carbon atoms on the main chain is done in such a way as to give the principal functional group, which in this case is the alkene functional group, the lowest possible positional number.

Using 2-methylbut-1-ene as an example, the following are the steps used to name it:

Step 1: Identify the longest carbon chain and the principal functional group.

○ Number of carbon atoms in main chain: 4 ⇒ root: **but–**
○ Principal functional group: alkene ⇒ suffix: **–ene**.

Step 2: Identify the location of the C=C bond by assigning it the lowest possible positional number when numbering the carbon atoms on the main chain.

○ Position of C=C bond: between C1 and C2 ⇒ root–suffix: **–but-1-ene**.

Step 3: Identify the substituents and their locations.

○ Substituent: methyl group
○ Position of substituent: at C2 ⇒ prefix: **2–methyl–**.

Note the following rules when stringing together the various parts of a chemical name:

- prefixes are cited in alphabetical order and *not* in numerical order.
- if there are identical substituents, the numeric prefixes *di-*, *tri-*, *tetra-* are attached to the prefix. These numeric prefixes have no bearing on the alphabetical order of prefixes.
- the number of positional numbers must be equal to the number of substituents.
- a "hyphen" is used to separate numbers from words.
- a "comma" is used to separate numbers.
- if there is more than one C=C bond, the corresponding positional number for each C=C bond and the numeric prefixare stated in front of the suffix *–ene*. In addition, the letter "a" is added to the root. An example is buta-1,3-diene:

Buta-1,3-diene

In certain situations, positional numbers can be omitted only if this does not present any ambiguity. Examples include propene and methylpropene:

Propene   Methylpropene

There is no ambiguity when the positional number for the location of the double bond is omitted from their names. There is also no confusion as to where the methyl group is located in methylpropene. Based on its name, the methyl group must be attached to the middle carbon atom. If it is attached to one of the two terminal carbon atoms, we end up with a "butene" molecule.

5.3 Physical Properties

5.3.1 *Melting and Boiling Points*

As with the alkanes, melting and boiling points increase with increasing carbon chain length and decrease with increasing degree of branching for the alkenes.

But unlike the alkanes, which are non-polar, alkenes can be either non-polar or polar molecules. For an alkene to be polar, it must have a net dipole moment. Since C−H bonds are essentially non-polar, we will focus on the type of substituents attached to the C=C bond and determine if there is a net dipole moment at this site.

Take, for instance, the *cis-trans* (geometrical) isomers of but-2-ene. The methyl substituents are electron-releasing groups that contribute towards an increase in electron density at the C=C bond. Coupled with the fact that the geometry about each of the doubly bonded carbon atoms is trigonal planar, we can conclude there is no cancellation of individual dipoles for the *cis* isomer and hence, it is polar. The *trans* isomer is non-polar since the dipole moments cancel each other out.

net dipole moment

no net dipole moment

cis-but-2-ene
(b.p. 4 °C)

trans-but-2-ene
(b.p. 1 °C)

Q: It was mentioned in the previous chapter that the electron-releasing effect of the alkyl groups only works if there is a "demand." For the case of the above *cis* and *trans* alkenes, the alkyl groups are bonded to carbon atoms that are not electron-deficient, so where is the driving force for the electron-releasing effect?

A: The carbon atoms of the C=C double bond are both sp^2 hybridized, whereas those of the alkyl groups are usually sp^3 hybridized. An s orbital is closer to the nucleus than a p orbital; we term this phenomenon "penetrating power." Therefore, an s electron is more strongly attracted than a p electron by the nucleus. As a result, due to the greater percentage of s character in a sp^2 hybridized ($\sim$33%) orbital than in a sp^3 hybridized

($\sim$25%) one, we would expect a sp^2 hybridized carbon atom to be more electron withdrawing. This accounts for the dipole moment that is created in a sp^2–sp^3 overlapped carbon–carbon bond.

Henceforth, with the presence of permanent dipole–permanent dipole (pd–pd) interactions in addition to the instantaneous dipole–induced dipole (id–id) interactions, the polar *cis* isomer has a higher boiling point than the non-polar *trans* isomer.

Q: But *trans*-but-2-ene (m.p. $-106°$C) has a higher melting point than *cis*-but-2-ene (m.p. $-139°$C). How can we use the relative strength of the intermolecular attractive forces to rationalize this fact?

A: A greater amount of energy is required to overcome the sum of the pd–pd and id–id attractive forces for *cis*-but-2-ene than to overcome the only id–id attractive forces in *trans*-but-2-ene. This accounts for the higher *boiling* point of *cis*-but-2-ene. In the process of *melting*, the intermolecular forces for *cis*-but-2-ene are in fact still stronger than that of the *trans*-but-2-ene. However, there is another more significant factor that we need to consider here — it concerns the lattice structure of these covalent molecules. Due to the more symmetrical shape of the *trans* isomer molecules, these molecules tend to *pack better* together than those of the *cis* isomer. This basically means that for a unit volume of space, there are more *trans* isomer molecules packed closer to one another than there are for the *cis* isomer. With more molecules per unit volume, the extensiveness of the van der Waals forces of attraction is greater, and this contributes to the higher melting point of *trans*-but-2-ene. Henceforth, take note that if you encounter a *trans* isomer having a higher melting point than the *cis* isomer, likely explanatory factor to use would be the packing effect. But if the *trans* isomer has a lower melting point than the *cis* isomer, the factor to attribute would be that the *cis* isomer is polar whereas the *trans* isomer is not. Later on, we will learn that we simply need to know the various factors that affect the physical property and use the factors appropriately based on the observed physical data.

Q: Does the packing factor account for the fact that trans fat tends to clot up the artery than cis fat?

A: Yes, indeed. Due to the better packing of the *trans* isomer, the trans fat that is accumulated in our body at the physiological temperature becomes more difficult to break up, hence less fluid.

Q: Why doesn't the packing factor cause the *trans* isomer to have a higher boiling point than the *cis* isomer?

A: In the liquid state, the molecular particles are not rigidly held in fixed positions as in the solid state. Hence, the packing factor is not influential to the boiling point.

For any given pair of *cis-trans* (geometrical) isomers, the *cis* isomer is polar and the *trans* isomer is non-polar. However, we cannot assume that in all cases, the *cis* isomer will certainly have a higher boiling point than the *trans* isomer. Take, for instance, the *cis-trans* (geometrical) isomers of butenedioic acid:

net dipole moment

cis-butenedioic acid
(b.p. 160 °C)

no net dipole moment

trans-butenedioic acid
(b.p. 290 °C)

Although the *cis* isomer is polar, it has a lower boiling point than the *trans* isomer. The dominant factor that accounts for the difference in boiling points observed is the extensiveness of intermolecular hydrogen bonding. In *cis*-butenedioic acid, the two carboxylic acid functional groups are in close proximity, so they are able to form **intra** molecular hydrogen bond within themselves. Such intramolecular hydrogen bonding is absent in the *trans* isomer. The presence of intramolecular hydrogen bonding within the *cis* isomer molecule limits the number of sites available for **inter** molecular hydrogen bonding.

Less extensive intermolecular hydrogen bonding in *cis*-butenedioic acid

With less extensive intermolecular hydrogen bonding between molecules of *cis*-butenedioic acid, the attractive forces holding them together are weaker than those among the same number of *trans* isomer molecules. Hence, *cis*-butenedioic acid has a lower boiling point. The same factor accounts for the lower solubility of *cis*-butenedioic acid in water since its molecules form less extensive hydrogen bonds with water molecules.

More extensive intermolecular hydrogen bonding in *trans*-butenedioic acid

5.3.2 *Solubility*

Due to their relatively non-polar nature, alkenes are soluble in non-polar organic solvents but insoluble in polar organic solvents.

Q: If the *cis* isomer of an alkene is polar, wouldn't it dissolve in polar solvents?

A: The *cis* isomer is relatively more polar than the *trans* isomer, but it is still not as polar as, let's say, an alcohol. Thus, relative to a polar organic solvent, alkene is still hydrophobic in nature. The interaction that an alkene molecule would have with a polar organic molecule such as ethanol is simply the id–id type of interaction.

5.4 Preparation Methods

5.4.1 *Dehydration/Elimination of Alcohols*

Reagents:	Excess concentrated H_2SO_4
Conditions:	$170°C$
Options:	Al_2O_3, $400°C$ or H_3PO_4, 200–$250°C$

Q: Can we use reflux when using concentrated H_2SO_4 to dehydrate alcohol?

A: No! Reflux refers to prolong heating with condensation. At 170°C, the concentrated H_2SO_4 can decompose to SO_3 gas. Thus, reflux could only cause the H_2SO_4 molecule to "disappear".

This is an example of an elimination reaction that involves the removal of a group of atoms from a single reactant molecule. In this case, a hydrogen atom and an OH–group (or hydroxyl group) on adjacent carbon atoms are removed from the propan-2-ol molecule to form the unsaturated alkene compound (see Chapter 8 for further details on this reaction).

When an unsymmetrical alcohol undergoes dehydration, the products obtained are a mixture of alkenes in unequal proportions as shown in the case for butan-2-ol:

but-1-ene (least yield) + cis-but-2-ene + trans-but-2-ene (most yield)

Q: Why are the products not produced in equal proportions?

A: A reaction proceeds favorably to form stable compounds. There are two factors to consider in the formation of an alkene from an alcohol:

- the greater stability of a more substituted alkene than a less substituted one; and
- the greater stability of the *trans* isomer than the *cis* isomer.

A more stable alkene is formed in larger proportions, and greater stability is found in alkenes with more alkyl (R) groups attached to the doubly bonded carbons. In order of increasing stability, we have:

Monosubstituted alkene (least stable) < Disubstituted alkene < Trisubstituted alkene < Tetrasubstituted alkene

This is known as Saytzeff's rule. When butan-2-ol is dehydrated, $CH_3CH=CHCH_3$ is the major product and $CH_3CH_2CH=CH_2$ the minor product.

Q: Why is a more substituted alkene more stable than a less substituted one?

A: Since alkyl groups are electron-releasing via inductive effect, they contribute electron-density to the pi bond and thus strengthen it. As to the driving force for the movement of electron density, it has been explained in Section 5.3.1.

Q: Why is the *trans* isomer more stable than the *cis* isomer even when they have the same numbers and types of substituents?

A: Here, we need to consider steric factor — alkyl groups are relatively bulky and during the formation process; they are best placed opposite to each other diagonally across the C=C bond, so as to minimize inter-electronic repulsion. Thus, with the *trans* isomer being more stable, it is formed in greater quantity than the *cis* isomer.

5.4.2 *Dehydrohalogenation/Elimination of Halogenoalkanes*

Reagents:	KOH, ethanol
Conditions:	Heat
Options:	NaOH as a base

This is another example of an elimination reaction. A hydrogen atom and a halogen atom on adjacent carbon atoms are removed from the halogenoalkane molecule to form the alkene (see Chapter 7 for further details). Notice that the elimination reactions of both propan-2-ol and 2-chloropropane give the same product, propene. Although both dehydration and dehydrohalogenation are elimination reactions, they do not proceed via the same mechanism. Nonetheless, Saytzeff's rule still applies when an unsymmetrical halogenoalkane undergoes elimination. Can you figure out the major product formed when 2-chlorobutane undergoes elimination?

Q: How are the mechanisms for the dehydration of alcohol and dehydrohalogenation of halogenolkane like?

A: In dehydration, concentrated sulfuric acid plays the catalytic role of protonating the OH–group. As a result, this causes the elimination of a H_2O molecule, which means that the organic intermediate would have the carbocation characteristics. Whereas in dehydrohalogenation, the KOH acts as a base, extracting a H^+ ion from the β-carbon (α-carbon atom is the one that a functional group is bonded to, in this case, the halogen atom. Subsequent, carbon atoms are labeled as β, γ, δ, etc.). The extraction is feasible because the electron-withdrawing effect of the halogen group makes the neighboring hydrogen atoms acidic, hence "attracting" attack from a base. This would also mean that the hydrogen atom becomes less acidic if it is further away from the halogen atom. This accounts for why the hydrogen and halogen atoms come preferably from adjacent carbon atoms. So, one can see that the organic intermediate of the dehydrohalogenation process would have the carbanion characteristics.

Q: How does Al_2O_3 cause dehydration of alcohol to form an alkene?

A: Al_2O_3 is an amphoteric compound. The Al^{3+} is highly electron-deficient, hence very acidic and thus acting as a Lewis acid. It can accept the lone pair of electrons from the OH–group, thus causing the OH–group to "leave" the alcohol. In addition, the O^{2-} ion is highly basic, which can extract the acidic hydrogen atom from the β-carbon atom. Therefore, Al_2O_3 is a good dehydrating agent.

5.5 Chemical Properties

An alkene, being an electron-rich species due to the presence of the $C=C$ double bond, undergoes attack by electron-deficient species, namely electrophiles (a Lewis acid). Due to the relatively strong $C-H$ and $C-C$ bonds present in the molecule, the reactions that an alkene undergoes is mainly addition in nature, which involves the breaking of the weaker pi bond. But still, some alkenes, such as chloroethene, can undergo elimination to give alkyne (e.g. ethyne). In addition, take note that an alkene can also be attacked by a free radical or even a nucleophile, as long as the reaction condition is appropriate.

5.5.1 *Catalytic Hydrogenation — Reduction*

$$CH_3\text{—C}=\text{C—H} \ (\text{+ H}_2) \xrightarrow[\text{heat}]{\text{Ni catalyst}} \ \text{H—C—C—C—H}$$

Reagents:	$H_2(g)$ with $Ni(s)$ catalyst
Conditions:	Heat
Options:	$Pt(s)$ or $Pd(s)$ as catalyst

Catalytic hydrogenation is an example of what is called heterogeneous catalysis, whereby the catalyst and the reactants are in different phases.

Commercial applications include the use of powdered nickel in the hydrogenation of unsaturated fats to the saturated form in the manufacture of margarine:

$$R\text{–CH} = \text{CH–R}'(l) + H_2(g) \longrightarrow R\text{–CH}_2\text{–CH}_2\text{–R}'(l).$$

Q: What is the purpose of hydrogenating unsaturated fats to saturated fats?

A: The presence of C=C bonds make the unsaturated fat more prone to free radical attack at the electron-rich sites, which makes the fat turn rancid easily, meaning shorter shelf-life in the market. But note that saturated fat has a higher melting point than unsaturated fat, which is not very good for our bodies.

Q: Why is the melting point of an unsaturated fat lower than saturated fat?

A: Due to the presence of multiple C=C bonds along the carbon skeleton of an unsaturated fat molecule, the three-dimensional structure of the molecule is relatively more spherical in shape. In contrast, saturated fat is more linear. As a result, the surface area of contact for a saturated fat molecule is greater than for an unsaturated molecule. Hence, more extensive intermolecular forces of attraction account for the higher melting point of saturated fat.

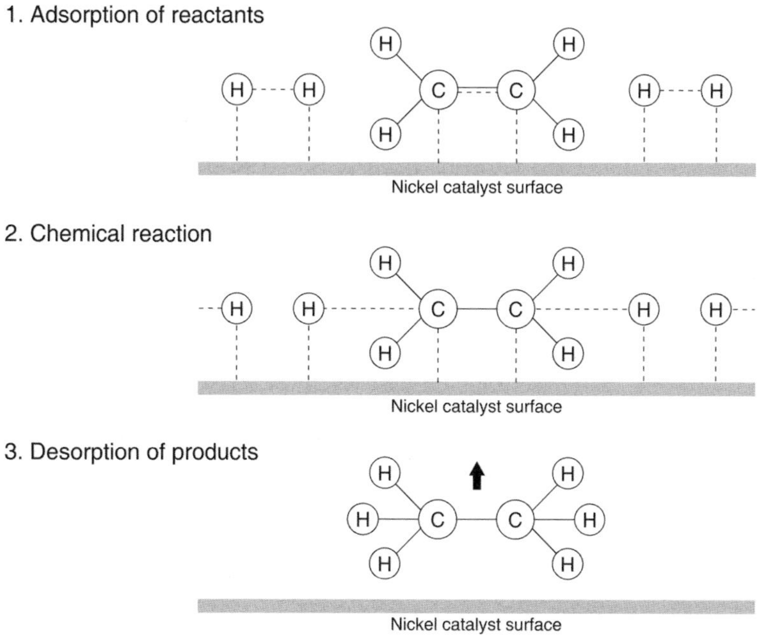

1. Adsorption of reactants

Nickel catalyst surface

2. Chemical reaction

Nickel catalyst surface

3. Desorption of products

Nickel catalyst surface

Fig. 5.1.

This would mean that at the physiological temperature, saturated fat molecules are more likely to solidify than unsaturated fat molecules, which would be more likely to cause arterial blockage.

Three likely steps can be put forward for any heterogeneous catalysis (see Fig. 5.1):

Step 1: Adsorption of the reactant particles onto the active sites on the surface of the catalyst. This is facilitated by the formation of weak bonds between the reactant particles and the catalyst. The adsorption process increases the surface concentration of the reactant.

Step 2: Reaction at the surface.

- The activation energy of the catalyzed process is lowered because the intramolecular bonds within the reactant particles are weakened by the adsorption effects, thereby reducing the energy required to disrupt them.

- The reactant particles are brought into close contact and are properly oriented for reaction.

Step 3: Desorption of the reactants or products from the surface.

5.5.2 *Formation of Halogenoalkane — Electrophilic Addition*

major product

Reagents:	Dry HBr(g)
Conditions:	Room temperature
Options:	Dry HCl(g) or HI(g)

The mechanism involved is that of electrophilic addition. HBr is a polar molecule with the $^{\delta+}$H acting as the electrophile (a Lewis acid) that attacks the alkene. The reaction of ethene with HBr gives us a single product — bromoethane. But for an unsymmetrical alkene such as propene, there are two possible products that are formed. However, note that these two products are not formed in equal proportions:

2-bromopropane
(major product)

1-bromopropane
(minor product)

If we are to look back at the mechanism for the reaction of propene and HBr, either C1 or C2 can form a bond with the $^{\delta+}$H atom in the HBr molecule. This means that it is possible to obtain two different carbocations in Step 1:

2° carbocation   1° carbocation

The reason as to why 2-bromopropane is the major product lies in the characteristics of these two carbocations. Carbocations can be classified as primary (1°), secondary (2°) or tertiary (3°) carbocations based on the number of alkyl (R) groups attached to the carbon atom bearing the positive charge.

The driving force of a reaction is the formation of a stable product. We expect a more stable carbocation to be *formed faster* since its formation is favored over a less stable carbocation.

An alkyl group is known to be electron-releasing (via inductive effect), and thus the electron deficiency on the carbon atom bearing the positive charge would be diminished by a greater extent when it is attached to more of these alkyl groups. Hence, the stability of a carbocation, and its rate of formation, increases in the following order:

$$
\begin{array}{cccc}
\underset{\substack{| \\ H}}{\overset{\substack{H \\ |}}{H-C^+}} < & \underset{\substack{| \\ H}}{\overset{\substack{H \\ |}}{R-C^+}} < & \underset{\substack{| \\ H}}{\overset{\substack{R \\ |}}{R-C^+}} < & \underset{\substack{| \\ R}}{\overset{\substack{R \\ |}}{R-C^+}}
\end{array}
$$

Methyl carbocation   1° carbocation   2° carbocation   3° carbocation
(least stable)

Q: How could electron-releasing groups help to stabilize a carbocation?

A: The stability of charge particles depends on the extent of charge dispersal. For a carbocation, if the positive charge can be made less concentrated on the carbon bearing it, then there is less tendency for it to be attacked by nucleophiles, rendering it more stable. The way to disperse the positive charge is to "neutralize it" by distributing it over the entire cation rather than residing on a single atom.

The greater the number of electron-releasing alkyl groups, the more stable is the carbocation since there is greater charge dispersal (or less electron deficiency) on the positively charged carbon atom.

$$
\begin{array}{ccc}
\underset{\substack{| \\ H}}{\overset{\substack{H \\ |}}{H-C+}} & \underset{\substack{| \\ H}}{\overset{\substack{CH_2CH_3 \\ \downarrow}}{H-C+}} & \underset{\substack{| \\ H}}{\overset{\substack{CH_3 \\ \downarrow}}{CH_3 \rightarrow C+}}
\end{array}
$$

On the other hand, substituents such as halogens (e.g. $-Cl$) or $-COOH$ groups are electron-withdrawing groups that intensify the positive charge on the carbon and destabilize the carbocation:

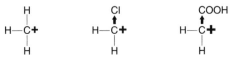

This brings us to the relative reactivities of alkenes, which mirror the relative stabilities of the carbocations. In order of increasing reactivity, we have:

least reactive

Q: In the earlier section, it was mentioned that a tetra-substituted alkene is the most stable, but here, it is stated to be the most reactive. Isn't this contradictory?

A: There is no contradiction here. The more highly substituted alkene would more likely be formed than the less substituted alkene from an elimination reaction involving an alcohol due to the specificity of that elimination reaction mechanism. In electrophilic addition, the more highly substituted alkene would be more likely to react due to the more stable carbocation that would be formed.

Q: Would a more highly substituted alkene be more sterically hindered from the attack of an electrophile, thus less reactive as compared to a less substituted alkene?

A: Yes, certainly it would. Thus, if you compare two different alkenes and find that the more substituted one is actually less reactive, then one reason that you could use to account for the observed difference in reactivities would be the steric factor.

Q: Why would a more stable carbocation be formed faster?

A: Before the formation of the carbocation, the activated complex of the transition state would look like the following for different types of carbocation:

Activated complex of a 1° carbocation   Activated complex of a 2° carbocation

It is clear that the activated complex of the 2° carbocation is more stable than that for the 1° carbocation due to the presence of more electron-releasing alkyl groups. As a result, the activation energy to form the 2° carbocation is lower than that to form the 1° carbocation. Thus, at the same temperature, the rate to form the 2° carbocation would be higher than that to form the 1° carbocation.

A quick way to decide the major product of an electrophilic addition reaction is to make use of **Markovnikov's rule**. The rule states that with the addition of a protic acid HX (e.g. HCl, HBr, HI, H_2SO_4, etc.) to an alkene, the acidic hydrogen (H) would become attached to the carbon with *fewer* alkyl substituents, and the halide (X) group would become attached to the carbon with more alkyl substituents.

Q: How is Markovnikov'srule related to what has been discussed on the carbocations?

A: This rule is purely based on experimental observations that for an unsymmetrical alkene, the major product formed from the addition reaction with an acidic molecule (a proton donor) consists of the $^{\delta+}$H atom attaching itself to the doubly bonded C atom that has more H atoms bonded to it. Later, this phenomenon can actually be accounted for by looking into the mechanism and the relative stabilities of the carbocations involved.

5.5.3 *Formation of Dihalide — Electrophilic Addition*

Reagents:	Br_2 in CCl_4
Conditions:	Room temperature
Options:	Cl_2 in CCl_4

In this electrophilic addition reaction, decolorization of reddish-brown Br_2 will be observed.

This reaction serves as a useful distinguishing test to determine the presence of an alkene in an unknown organic sample. If you are provided with two test-tubes, one of which contains an alkene and the other an alkane, the distinguishing test using Br_2 must be carried out in the dark. This is done so as to prevent free radical substitution from occurring. There will then be one positive result — the decolorization of Br_2 for the test–tube that contains the alkene; and a negative result — no decolorization of Br_2 for the test–tube that contains the alkane. Actually, if the differentiation test is not being performed in the dark, the alkene can still be differentiated from the alkane because the decolorization for alkene is much more rapid than for the alkane.

5.5.4 *Formation of Halohydrin — Electrophilic Addition*

major product

Reagents:	$Br_2(aq)$
Conditions:	Room temperature
Options:	$Cl_2(aq)$

Similar to the reaction with Br_2 in CCl_4, addition of aqueous Br_2 to an alkene serves as a distinguishing test. Decolorization of yellowish-brown $Br_2(aq)$ will be observed.

Due to the type of product that is formed, this reaction clearly testifies for the presence of a carbocation as an intermediate during the electrophilic addition reaction. The possibility of two carbocations being formed in the electrophilic addition reaction accounts for the products, 2-bromopropan-1-ol and 1-bromopropan-2-ol, obtained. However, 1-bromopropan-2-ol is the major product, since it is formed from the more stable $2°$ carbocation, $CH_3CH^+CH_2Br$.

Q: But why did you say that "this reaction clearly testifies for the presence of a carbocation as an intermediate during the electrophilic addition reaction?" Do you mean that the previous few reactions do not?

A: For the reaction of an alkene with HBr or Br_2, the attacking molecule can be seen as just "added across" to the double bond. That means the breaking of the pi and Br–Br bonds and the formation of two sigma bonds with the two Br atoms can be perceived as a concerted step. This is not the case for the formation of the halohydrin. The fact that there is a OH–group in the product, where this OH–group must come from a H_2O molecule, is a clear indication that a carbocation intermediate exists, "awaiting" the attack by a water molecule.

1,2-dibromopropane (least yield)	2-bromopropan-1-ol	1-bromopropan-2-ol (most yield)

In addition, 1,2-dibromopropane is also formed, but as a minor product. The explanation for its relatively lower yield centers on the fact that there are two nucleophiles (or Lewis base) — H_2O and Br^- — vying to react with the carbocation intermediate (a Lewis acid). As H_2O molecules are present in much greater quantity (since water is the solvent), the chances of a carbocation reacting with H_2O are much higher, and this leads to a greater proportion of the halohydrin products formed.

Q: How does H_2O behave as a nucleophile?

A: Considering the 2nd step of the electrophilic addition mechanism that involves the more stable carbocation, we have the O atom of the water molecule donating its lone pair of electrons to form a dative covalent bond with the electron-deficient C atom.

However, in so doing, the O atom now acquires the positive charge. Since it is very electronegative, the O atom does not stay electron-deficient for long, and it subsequently loses a H^+ through the heterolytic cleavage of an $O-H$ bond.

$$
\begin{array}{c}
\text{CH}_3 \\
| \\
\text{H}-\text{C}-\overset{+}{\text{O}}\underset{\text{H}}{\overset{\text{H}}{\diagup}} \\
| \\
\text{H}-\text{C}-\text{Br} \\
| \\
\text{H}
\end{array}
\longrightarrow
\begin{array}{c}
\text{CH}_3 \\
| \\
\text{H}-\text{C}-\text{O}-\text{H} \\
| \\
\text{H}-\text{C}-\text{Br} \\
| \\
\text{H}
\end{array}
\ + \ \text{H}^+
$$

Q: Can the water molecule in aqueous Br$_2$ act as an electrophile?

A: Although the hydrogen atom of a water molecule is electron-deficient, the water molecule is not really a good electrophile because it is energetically demanding to cleave the O$-$H bond of the water molecule. Substantiating this point are the drastic reaction conditions and the need of an acid catalyst when trying to convert an alkene to an alcohol in the presence of water.

Example 5.1: What are the possible products formed when propene reacts with Br$_2$(aq) in the presence of NaCl(aq)?

Approach: Essentially, most addition reactions of alkenes, except for catalytic hydrogenation, proceed via the electrophilic addition mechanism. Did you notice that the electrophilic addition reactions of propene that were discussed so far all involve the same carbocation intermediate that gives the major product?

$$
\begin{array}{c}
\text{CH}_3 \quad\quad \text{H} \\
\diagdown \quad\diagup \\
\text{C} \\
\| \\
\text{C} \\
\diagup \quad\diagdown \\
\text{H} \quad\quad \text{H}
\end{array}
\xrightarrow{\text{Step 1}}
\begin{array}{c}
\text{CH}_3 \\
| \\
\text{H}-\text{C}^+ \\
| \\
\text{H}-\text{C}-\text{E} \\
| \\
\text{H}
\end{array}
$$

2° carbocation   where E = electrophile

Solution: Considering the reaction of the more stable 2° carbocation with the nucleophiles, the following products would be formed:

$$
\begin{array}{ccc}
\text{H} \quad \text{Cl} \quad \text{Br} & \text{H} \quad \text{Br} \quad \text{Br} & \text{H} \quad \text{OH} \quad \text{Br} \\
| \quad\ | \quad\ | & | \quad\ | \quad\ | & | \quad\ \ | \quad\ | \\
\text{H}-\text{C}-\text{C}-\text{C}-\text{H} & \text{H}-\text{C}-\text{C}-\text{C}-\text{H} & \text{H}-\text{C}-\text{C}-\text{C}-\text{H} \\
| \quad\ | \quad\ | & | \quad\ | \quad\ | & | \quad\ \ | \quad\ | \\
\text{H} \quad \text{H} \quad \text{H} & \text{H} \quad \text{H} \quad \text{H} & \text{H} \quad \text{H} \quad \text{H}
\end{array}
$$

1-bromo-2-chloropropane   1,2-dibromopropane   1-bromopropan-2-ol
                                                   (major product)

1-bromo-2-chloropropane is formed since Cl^- can also function as a nucleophile drawn towards the carbocation. However, 1,2-dichloropropane is not formed because the Cl^- ion cannot act as an electrophile.

$$H—\overset{\overset{\displaystyle H}{|}}{\underset{\underset{\displaystyle H}{|}}{C}}—\overset{\overset{\displaystyle Cl}{|}}{\underset{\underset{\displaystyle H}{|}}{C}}—\overset{\overset{\displaystyle Cl}{|}}{\underset{\underset{\displaystyle H}{|}}{C}}—H$$

1,2-dichloropropane ✗

5.5.5 *Formation of Alcohol — Electrophilic Addition*

$$\underset{H}{\overset{CH_3}{}}{\Large\diagup}C{=}C{\diagup}\underset{H}{\overset{H}{}} + H_2O \xrightarrow[\text{(ii) H}_2\text{O, heat}]{\text{(i) cold concentrated H}_2\text{SO}_4} H—\overset{\overset{\displaystyle H}{|}}{\underset{\underset{\displaystyle H}{|}}{C}}—\overset{\overset{\displaystyle OH}{|}}{\underset{\underset{\displaystyle H}{|}}{C}}—\overset{\overset{\displaystyle H}{|}}{\underset{\underset{\displaystyle H}{|}}{C}}—H$$

major product

	For Step (i),		For Step (ii),
Reagents:	Concentrated H_2SO_4	Reagents:	Water
Conditions:	Cold temperature	Conditions:	Heat
Options:	—	Options:	—

There are two steps in this reaction to produce alcohol:

- In the first step, we have the electrophilic addition of the acid across the double bond, (considering the formation of the more stable 2° carbocation):

alkyl hydrogensulfate(VI)

- In the second step, the alkyl hydrogensulfate(VI) intermediate then undergoes hydrolysis to generate the final alcohol product:

$$
\begin{array}{c}
\underset{H}{\overset{CH_3}{\underset{|}{\overset{|}{C}}}}\text{—O—}\overset{O}{\underset{OH}{\overset{||}{S}}}=O \\
\end{array}
+ H_2O \xrightarrow{\text{heat}}
\begin{array}{c}
\overset{CH_3}{\underset{|}{\overset{|}{C}}}\text{—OH} \\
\end{array}
+ H_2SO_4
$$

The net result of the electrophilic addition reaction followed by hydrolysis is the addition of water to the alkene to form the alcohol product.

An industrial method to produce alcohol that makes use of steam requires the following reagents and conditions:

Reagents:	$H_2O(g)$, H_3PO_4 on celite
Conditions:	$330°C$, $60\,atm$

Q: Why do we need to use the concentrated H_2SO_4 first? Can we not just use water straight away?

A: If you look closely at the structure of a sulfuric acid molecule, the hydrogen atom is very much electron-deficient, that is, it has a greater $\delta+$ than the hydrogen atom of a water molecule due to the presence of more than one highly electronegative oxygen atom. As such, the more electron-deficient hydrogen atom of the sulfuric(VI) acid molecule acts as a better electrophile. In addition, once the alkyl hydrogensulfate(VI) is formed, the carbon atom that is bonded to the hydrogensulfate(VI) group is made more electron-deficient due to the highly electron-withdrawing hydrogensulfate group. As such, it is very prone to nucleophilic attack. Thus, using water with a milder heating condition is good enough to hydrolyse the alkyl hydrogensulfate(VI) into forming the alcohol.

Q: Concentrated H_2SO_4 is an oxidizing agent, why didn't it oxidize the alkene?

A: Concentrated H_2SO_4 is just not strong enough an oxidizing agent to oxidize the alkene here. Note that an oxidizing agent need not necessary be always oxidizing, it depends on whether it is strong enough to oxidize or not!

Q: H_2SO_4 in the concentrated form can act as an oxidizing agent, but when it is in the diluted form, it is simply an acid. Why is this so?

A: This is because in the diluted state, the H_2SO_4 molecule dissociates to give the SO_4^{2-} ion. In order for the SO_4^{2-} to act as an oxidizing agent, it needs to take in electrons and be reduced. But as the SO_4^{2-} ion is negatively charged, it is less likely to accept electrons. On the other hand, when H_2SO_4 is in the concentrated state, most of its molecules are undissociated. The sulfur atom of the undissociated H_2SO_4 molecule is highly electron-deficient, and it accepts electrons more readily to undergo reduction. Hence, it can act as an oxidizing agent.

5.5.6 *Combustion — Oxidation*

$$CH_3 \diagdown C=C \diagup H \quad + \quad \tfrac{9}{2}O_2 \quad \longrightarrow \quad 3CO_2 \quad + \quad 3H_2O$$

Reagents:	$O_2(g)$
Conditions:	Heat at high temperature
Options:	—

Similar to alkanes, alkenes undergo complete combustion to give CO_2 and H_2O. Incomplete combustion would lead to the formation of CO and non-combusted carbon. Chemically, combustion is an oxidation process, but naming a combustion as an oxidation process would cause the reaction to lose its specificity.

5.5.7 *Formation of Diol — Oxidation*

$$CH_3 \diagdown C=C \diagup H \quad + \quad [O] \quad + \quad H_2O \xrightarrow{\text{cold KMnO}_4\text{, NaOH}} H-C-C-C-H$$

Reagents:	$KMnO_4(aq)$, $NaOH(aq)$
Conditions:	Cold temperature
Options:	$KOH(aq)$ as base

In this redox reaction, decolorization of purple $KMnO_4(aq)$ and a brown precipitate of $MnO_2(s)$ will be observed. Such observations are thus a useful characteristics test for the presence of an alkene functional group. The pi bond in the alkene cleaved to form two new sigma bonds with the oxygen atoms of the hydroxyl groups, generating a vicinal diol, which is useful for the manufacturing of polyester (see Chapter 13).

Q: What is a molecule with two -OH groups bonded to the same carbon atom known as?

A: It is known as a geminal diol.

5.5.8 *Formation of Carbon Dioxide, Ketone and Carboxylic Acid — Oxidation*

$$CH_3\diagdown_{H}\diagup C=C\diagup^{H}_{\diagdown H} + 5[O] \xrightarrow[heat]{KMnO_4, H_2SO_4} CH_3\diagdown_{HO}\diagup C=O + CO_2 + H_2O$$

Reagents: $KMnO_4(aq)$, $H_2SO_4(aq)$
Conditions: Heat
Options: —

Unlike cold alkaline $KMnO_4$, this oxidation reaction involves total cleavage of the C=C bond. The type of products formed depends on the type of substituents attached to the doubly bonded carbon atoms:

$$H\diagdown_{H}\diagup C \doteq C\diagup^{H}_{\diagdown H} + 6[O] \longrightarrow 2(CO_2 + H_2O)$$

$$R\diagdown_{H}\diagup C \doteq C\diagup^{R}_{\diagdown H} + 4[O] \longrightarrow 2 \ R\diagdown_{HO}\diagup C=O \quad \text{carboxylic acid}$$

$$R\diagdown_{R}\diagup C \doteq C\diagup^{R}_{\diagdown R} + 2[O] \longrightarrow 2 \ R\diagdown_{R}\diagup C=O \quad \text{ketone}$$

In this redox reaction, decolorization of purple $KMnO_4(aq)$ will be observed. If the alkene contains a $=CH_2$ group, carbon dioxide gas will be evolved, which forms a white precipitate with $Ca(OH)_2(aq)$. The evolution of CO_2 is a characteristics test for the presence of a terminal alkene double bond. $CO_2(g)$ is generated because the carbonic acid, H_2CO_3, formed from the oxidation process is relatively unstable in water:

$$H_2CO_3(aq) \rightleftharpoons CO_2(g) + H_2O(l)$$

$$CO_2(g) + 2OH^-(aq) \longrightarrow CO_3^{2-}(aq) + H_2O(l)$$

$$Ca^{2+}(aq) + CO_3^{2-}(aq) \longrightarrow CaCO_3(s)$$

Q: Why does the reaction with $KMnO_4$ in different mediums and conditions produce different products?

A: $KMnO_4$ is a stronger oxidizing agent in acidic solution than in alkaline solution.

$$MnO_4^-(aq) + 8H^+(aq) + 5e^- \rightleftharpoons Mn^{2+}(aq) + 4H_2O(l)$$
$$E^\ominus = +1.51\,V$$
$$MnO_4^-(aq) + 2H_2O(l) + 3e^- \rightleftharpoons MnO_2(aq) + 4OH^-(aq)$$
$$E^\ominus = +1.23\,V$$

As a milder oxidizing agent, it can only cleave the weaker carbon–carbon pi bond, but as a strong oxidizing agent, it can completely cleave the C=C bond.

The reaction between propene and MnO_4^- in an alkaline medium is represented as follows:

$$3CH_3CH = CH_2 + 2MnO_4^- + 4H_2O \longrightarrow 3CH_3CH(OH)CH_2OH$$
$$+ 2MnO_2 + 2OH^-$$

Based on the reaction, there is a decrease in oxidation state of Mn from +7 in MnO_4^- to +4 in MnO_2, a change of 3 units. When the reaction is carried out in an acidic solution, MnO_4^- is further reduced to Mn^{2+}, and there is a greater decrease in oxidation state of Mn by 5 units from +7 in MnO_4^- to +2 in Mn^{2+} (we are assuming that the alkene is converted to a diol here for easy comparison, which in fact the alkene should undergo oxidative cleavage):

$$5CH_3CH = CH_2 + 2MnO_4^- + 2H_2O + 6H^+$$
$$\longrightarrow 5CH_3CH(OH)CH_2OH + 2Mn^{2+}$$

Q: What happen if we use hot alkaline $KMnO_4$ as the oxidizing agent?

A: Hot alkaline $KMnO_4$ is a stronger oxidizing agent than cold alkaline $KMnO_4$. It would cause the cleavage of the C=C bond, in contrast to just the pi bond cleavage when cold alkaline $KMnO_4$ is used. But one has to take note that a carboxylate salt would form in place of a carboxylic acid, if any, as the latter undergoes an acid−base reaction in the alkaline medium:

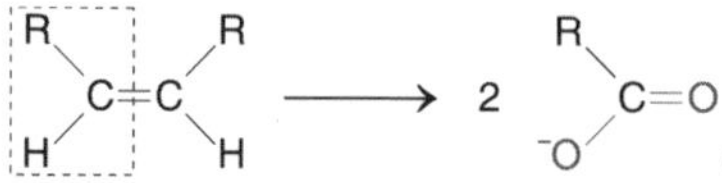

Example 5.2: What are the products formed when (i) $CH_2=CHCH_2CH=C(CH_3)_2$ and (ii) $CH_2=CHCH=C(CH_3)_2$, each react with hot, acidified $KMnO_4(aq)$?

Solution:

(i) The products formed are $HOOCCH_2COOH$, CH_3COCH_3 and CO_2 as shown:

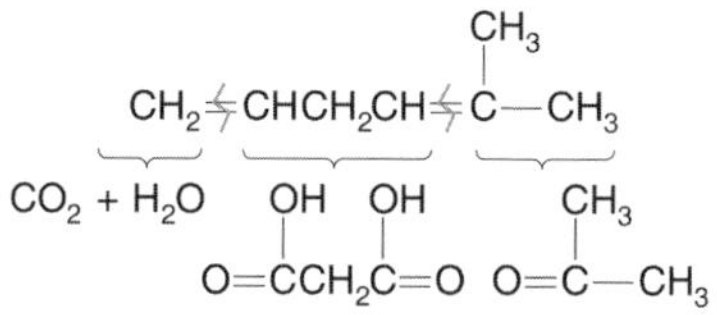

(ii) The products formed are CH_3COCH_3 and CO_2. In this case, ethanedioic acid, $(COOH)_2$, is not obtained. The C−C bond in ethanedioic acid is relatively weak due to the presence of the highly electronegative O atoms attached to these C atoms. As such, it is readily oxidized to form CO_2 and H_2O:

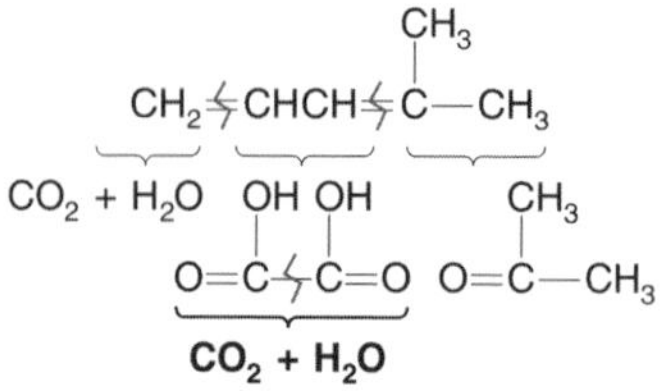

The formation of ethanedioic acid and subsequently CO_2 and H_2O are a characteristics test of the presence of a conjugated alkene functional group, i.e. C=C–C=C. In addition, by analyzing the composition of the oxidative products, coupled with other analytical techniques, one can determine the structure of an alkene molecule (refer to Q2 in My Tutorial).

5.5.9 *Formation of Aldehyde and Ketone — Oxidation*

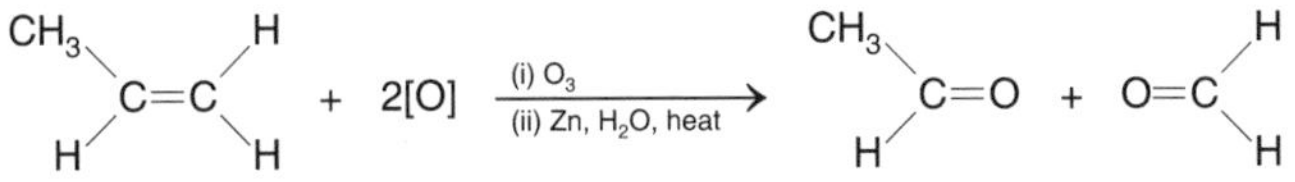

$$\underset{H}{\overset{CH_3}{>}}C=C\underset{H}{\overset{H}{<}} + 2[O] \xrightarrow[\text{(ii) Zn, H}_2\text{O, heat}]{\text{(i) O}_3} \underset{H}{\overset{CH_3}{>}}C=O + O=C\underset{H}{\overset{H}{<}}$$

For Step (i),		For Step (ii),	
Reagents:	$O_3(g)$	Reagents:	$Zn(s)$, $H_2O(l)$
Conditions:	—	Conditions:	Heat
Options:	—	Options:	—

The above reaction is known as ozonolysis, and it involves the total cleavage of the C=C bond. As H_2O_2 is formed, zinc serves as the reducing agent to remove it so that it will not further react with the products, i.e. the aldehydes or ketones that are formed. The type of product formed depends on the type of substituent attached to the doubly bonded carbon atoms:

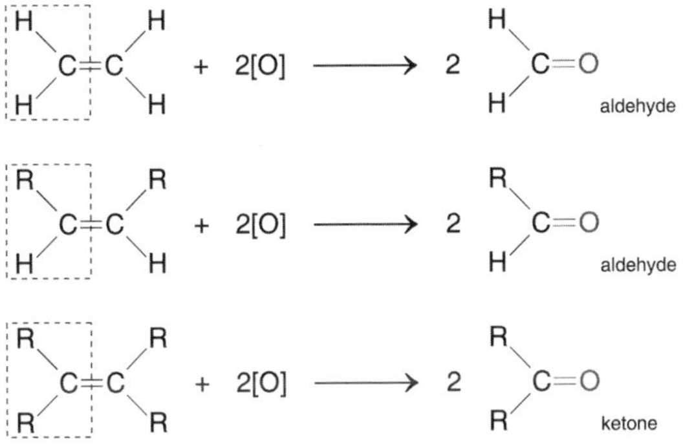

$$\underset{H}{\overset{H}{>}}C\dotplus C\underset{H}{\overset{H}{<}} + 2[O] \longrightarrow 2\ \underset{H}{\overset{H}{>}}C=O \quad \text{aldehyde}$$

$$\underset{H}{\overset{R}{>}}C\dotplus C\underset{H}{\overset{R}{<}} + 2[O] \longrightarrow 2\ \underset{H}{\overset{R}{>}}C=O \quad \text{aldehyde}$$

$$\underset{R}{\overset{R}{>}}C\dotplus C\underset{R}{\overset{R}{<}} + 2[O] \longrightarrow 2\ \underset{R}{\overset{R}{>}}C=O \quad \text{ketone}$$

Ozonolysis is sometimes preferred over oxidation using acidified $KMnO_4$ if one's intention is to obtain the aldehyde product from the oxidation of an alkene rather than the carboxylic acid.

Q: If O_3 can oxidize alkene to aldehyde and ketone, why not O_2?
A: O_2 is not as reactive as O_3, from the following equation:

$$O_3 \longrightarrow 3/2 O_2 \ \Delta H = -142\,\text{kJ mol}^{-1}$$

Ozone has 142 kJ of energy over oxygen, which thus makes it more reactive.

5.6 Summary

The following summary depicts both the preparation methods and the reactions of alkenes as exemplified by propene:

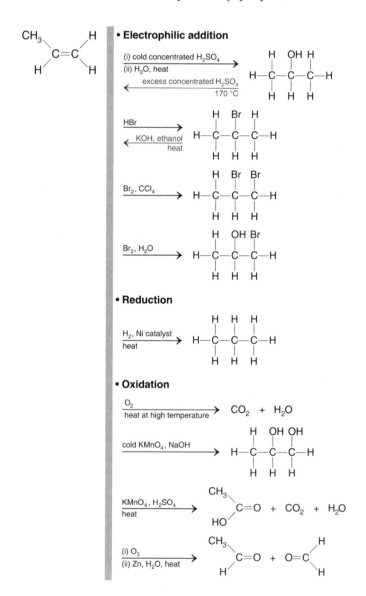

My Tutorial

1. The following compound **X**, 2-ethylhexyl 4-methoxycinnamate, is capable of UV absorption, hence it is commonly used in sunscreens lotion. The structural formula of **X** is:

$$CH_3O - \langle\bigcirc\rangle - CHCHCOOCH_2CH(CH_2CH_3)CH_2CH_2CH_2CH_3$$

(a) (i) Name three functional groups present in **X**.

(ii) Identify the sp^2 hybridized carbon atoms in **X**.

(iii) Compound **X** is an unsaturated compound. Describe a chemical test to demonstrate the presence of unsaturation in this molecule. Give the reagents, conditions, balanced equation, as well as observations.

(iv) Compound **X** exhibits stereoisomerism. Name the types of stereoisomerism that is/are present in compound **X** and draw all the possible stereoisomers.

(b) Give the structures of the organic products formed when compound **X** undergoes the following reactions:

(i) Addition of cold alkaline potassium manganate (VII).

(ii) Heating under reflux with alkaline potassium manganate (VII).

(iii) Heating under reflux with alkaline potassium manganate (VII), followed by acidification.

(iv) Heating under reflux with acidified potassium manganate (VII).

(v) Addition of LiAlH₄ in dry ether, followed by warming with water.

(vi) Heating with hydrogen in the presence of a nickel catalyst.

(vii) Heating under reflux with aqueous hydrochloric acid.

(viii) Heating under reflux with aqueous sodium hydroxide.

(c) Compound **X** reacts with iodine chloride (ICl) to give two different products. Give the mechanism for the formation of the major product formed and reasons for its formation.

2. Alkene can be selectively oxidized to give a range of products such as diol, aldehyde, ketone, carboxylic acid, and carbon dioxide. In

such an experiment, an alkene C_6H_{12}, gave an aldehyde **P** and a ketone **Q**, both with the molecular formula C_3H_6O.

(a) Draw all the possible constitutional/structural formulae corresponding to the molecular formula C_6H_{12}.

(b) Suggest a simple chemical test to differentiate **P** from **Q**. (Refer to Chapter 9.)

(c) (i) Draw two constitutional/structural isomers of an alkene, C_6H_{12}, which exhibit *cis-trans* (geometrical) isomerism.

(ii) How can you use ozonolysis to differentiate between these two constitutional/structural isomers of C_6H_{12}?

3. Account for why $CH_3CH_2CH_2Cl$ reacts rapidly with aqueous potassium hydroxide, whereas $CH_3CH=CHCl$ reacts slowly.

CHAPTER 6

Arenes

6.1 Introduction

Arenes are a family of aromatic compounds known for their characteristic feature — the benzene ring. Benzene, C_6H_6, is made up of six sp^2 hybridized carbon atoms, and its molecular structure is a resonance hybrid described by "an average" of two equivalent resonance forms (see Fig. 6.1). The six electrons are delocalized throughout the six-membered ring structure. As a result, the whole benzene molecule is planar in shape, with the 12 atoms (six carbon and six hydrogen) lying on the same plane.

Q: Since arenes are aromatic compounds, do they all have nice fragrance?

A: As a matter of fact, not all arenes have a nice fragrance, although the term "aromatic" literally means nice fragrance. In fact, the term "aromatic" refers to compounds that have a *resonance stabilized pi network of electrons*, consisting of a system of conjugated pi bonds. Based on molecular orbital theory (refer to Chapter 15), the German physicist Erich Hükel considered a molecule to be aromatic if the molecule possessed $(4n + 2)$ number of conjugated pi electrons. Note that this rule is good enough to help us predict whether a molecule is aromatic, though there are some molecules that are aromatic in nature but do not follow the Hükel's $(4n + 2)$ rule. This would mean that such molecules are still resonance stabilized but do not have $(4n + 2)$ number of conjugated pi electrons.

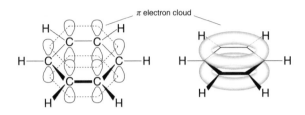

Fig. 6.1.

Q: What is resonance?

A: Resonance refers to the delocalization of pi electrons as a result of the side-on overlapping of a few *p* orbitals that are parallel to each other. Note that the way the electrons are delocalized here is different from that in metal because of the need for *p* orbitals to be involved, whereas in metal it is not necessary.

Q: Is there a limit to the number of atoms that can be involved in the delocalization?

A: There must be a minimum of three atoms, but there is no maximum limit. Look at the porphyrin ring of chlorophyll — it is highly conjugated!

This resonance phenomenon accounts for the observed partial double bond length (i.e. a bond that is intermediate between a single and a double bond) between adjacent carbons in benzene. What's more important is that it is able to account for a higher than expected stability of benzene as concluded from the thermochemical data.

Thermochemical data are basically data obtained from experimental studies focusing on energy changes and the flow of energy from one substance to another. Such studies are useful for predicting the behaviors of chemical systems; to determine the thermodynamic feasibility of a change or reaction occurring.

A general rule of thumb indicates that if the product that is formed has a lower energy level than the reactant, then the reaction is energetically feasible. A lower energy level indicates greater stability, which results from greater bond strength. As a result, when such a reaction proceeds, energy is released, and such reaction is deemed to be an exothermic reaction. The quantity of heat released or absorbed (in endothermic reactions) is calculated as the enthalpy change of reaction.

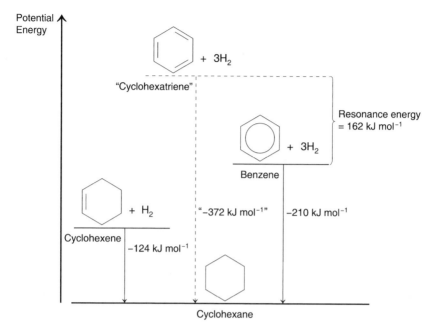

Fig. 6.2.

The hydrogenation of alkene is an example of an exothermic reaction that results in the formation of a more stable compound, the alkane. Cyclohexene can be hydrogenated to form cyclohexane. The enthalpy change of hydrogenation (denoted as $\Delta H_{\text{hydrogenation}}$) is illustrated in the energy level diagram shown in Fig. 6.2.

Q: Why is $\Delta H_{\text{hydrogenation}}$ exothermic in nature?

A: Consider the hydrogenation of cyclohexene; a carbon–carbon pi bond is broken {BE(C=C) − BE(C−C) = 610 − 350 = +260 kJ mol^{-1}} and a H–H sigma bond {BE(H−H) = +436 kJ mol^{-1}} is also broken. After hydrogenation, two C–H sigma bonds are formed, which releases a total of 820 kJ mol^{-1} of energy. Thus, the discrepancy in energy absorbed and released accounts for the exothermic nature of hydrogenation. Basically, if overall the bonds in the products are stronger than those in the reactants, the enthalpy change of reaction would be exothermic in nature!

Now, if benzene is depicted as a structure with alternating C–C and C=C double bonds, as in "cyclohexa-1,3,5-triene," we would expect its hydrogenation to release thrice the amount of energy as that for the hydrogenation of cyclohexene, since thrice the number of C=C bonds are subjected to the same hydrogenation reaction.

But experimental data has proven otherwise. The amount of energy released from the hydrogenation of benzene is only 210 kJ mol^{-1}, and this is much less than what was expected. Energy cannot be created or destroyed, but is transferred. So where did the energy, amounting to the difference between the expected energy released and that which is actually released, go?

If you compare the number of H–H sigma bonds to be broken during the hydrogenation of "cyclohexa-1,3,5-triene" and benzene, it is the same (three of them). If you also compare the number of C–H sigma bonds that are formed after hydrogenation for both "cyclohexa-1,3,5-triene" and benzene, it is also the same (six of them). Thus, the main difference must lie in the breaking of the three pi bonds in "cyclohexa-1,3,5-triene" and partially breaking the carbon–carbon bonds in benzene during hydrogenation. Since the energy that is released during the hydrogenation of benzene is less than expected if the benzene has the "cyclohexa-1,3,5-triene" structure, it goes to say that the bonds in benzene are stronger than in "cyclohexa-1,3,5-triene," thus more energy is needed to break them. This "extra" energy that goes into breaking the bonds in the resonance structure is thus known as "resonance stabilization energy."

To conclude, benzene actually does not have the structure of "cyclohexa-1,3,5-triene," but instead, a more stable structure — the resonance structure. In fact, "cyclohexa-1,3,5-triene" is just a hypothetical compound that does not exist because its structure cannot account for the various experimental observations and data.

Q: How does delocalization of electrons over a greater network of atoms result in greater stability?

A: The greater charge dispersal, brought about by the ability of the π electrons to delocalize over the entire network of atoms, results in stabilization as inter-electronic repulsion is decreased.

Q: When the electrons in the three C=C double bonds of "cyclohexa-1,3,5-triene" are delocalized, would it not have also weakened the carbon–carbon bond? How can the overall bond strength increase for benzene?

A: If you compare the carbon–carbon double and single bonds, you would know that (i) a pi bond is weaker than a sigma bond, and (ii) the sigma bond in a C=C double bond is not of the same strength as that in a C–C single bond. In fact, the presence of an extra pair of pi electrons in the C=C double bond actually weakens the sigma bond in it due to an increase in inter-electronic repulsion. Now, comparing the theoretical "cyclohexa-1,3,5-triene" and benzene, through delocalization, when the pi electrons of "cyclohexa-1,3,5-triene" are "spread" out, the strength of the sigma bond in the original C=C double bond is enhanced while the pi bond in it is weakened. All this is a result of weaker inter-electronic repulsion. These two opposing effects (i.e. the weakening and strengthening) do not actually cancel out each other. How do we know? The carbon–carbon bond length in benzene is about 139 pm, which is slightly longer than the 134 pm for the C=C double bond in ethene. This shows that the weakening of the pi bond overshadows the strengthening of the sigma bond, resulting in an *increase* of about 5 pm in bond length. If we now look at the original C–C sigma bond in "cyclohexa-1,3,5-triene," the delocalization of the pi electrons takes place at the expense of the original C–C sigma bond, which is now weakened. But at the same time, the formation of the pi bond due to the delocalization strengthens the carbon–carbon bond. Overall, the formation of the pi bond overshadows the weakening of the sigma bond, leading to a stronger carbon–carbon bond. The evidence can be observed if one compares the carbon–carbon bond length of benzene (139 pm) to the C–C single bond (147 pm) of buta-1,3-diene. The strengthening effect creates a *decrease* of about 8 pm in bond length. Hence, one would see that the overall effect of delocalization would lead to a net strengthening of the carbon–carbon bonds (+5 pm versus −8 pm). As a result, during hydrogenation, more energy is required to break the six carbon–carbon bonds in benzene as compared to breaking the three pi bonds in "cyclohexa-1,3,5-triene," thus resulting in less energy being released.

Q: Why did you use the C–C single bond (147 pm) of buta-1,3-diene to do the comparison and not the C–C single bond (154 pm) of ethane?

A: The C–C single bond of buta-1,3-diene is due to the sp^2–sp^2 overlap, whereas for ethane, it is due to sp^3–sp^3 overlap. The

C–C single bond of "cyclohexa-1,3,5-triene" is also of the sp^2–sp^2 overlap, hence it is more appropriate to use the C–C single bond of buta-1,3-diene.

Q: So, with the six pi electrons moving among the six carbon atoms, does it mean that there are six carbon–carbon double bonds at any point in time?

A: Two electrons are needed to form a pi bond. There are only six pi electrons, so there is no way to form six pi bonds. Thus, in actual fact, the six carbon–carbon bonds are neither double bond nor single bond. Each carbon–carbon bond is intermediate between a single and a double bond. You can consider it to have a bond order of 1.5.

Q: What is bond order?

A: Refer to Chapter 15 on molecular orbital theory.

6.2 Nomenclature

Unlike the alkanes and alkenes that we discussed earlier, the nomenclature for substituted benzene ring compounds is less systematic. Some of the monosubstituted benzenes listed below have unique common names which they are associated with:

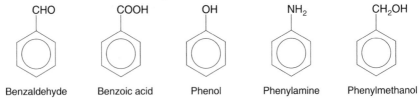

It is not an easy task to remember these names. However, for many of benzene derivatives, the substituents are just added as prefixes to the root word *–benzene*, such as in the following examples:

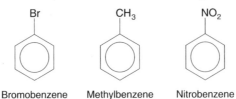

For polysubstituted benzene derivatives, substituents, other than those included in the common names are stated as prefixes, and in alphabetical order.

OH

Br

Br

Br

2,4,6-Tribromophenol

Numbering of positional number starts from the carbon atom bearing the principal functional group and in the direction that utilizes the lowest possible numbers for these substituents.

COOH

CH_3

2-Methylbenzoic acid ✔

COOH

CH_3

6-Methylbenzoic acid ✘

6.3 Physical Properties

6.3.1 *Melting and Boiling Points*

Benzene is a colorless and highly flammable liquid with a characteristic pleasant smell that gives it the title of "aromatic" compound. As it is non-polar, there exist only instantaneous dipole–induced dipole (id–id) attractive forces among the benzene molecules.

As with the trend for homologues of each homologous series, the melting and boiling points of arenes increase with the increase in molecular mass. This is because an increase in molecular mass is associated with an increase in the number of atoms in the molecule, thus resulting in an increase in the total number of electrons. The greater number of electrons contributes to a more polarizable electron cloud, leading to stronger id–id interactions. The boiling point of benzene is about 80°C.

6.3.2 *Solubility*

As with all hydrocarbons, arenes are soluble in non-polar organic solvents and insoluble in polar solvents. Benzene used to be an excellent solvent for organic compounds, but things have changed since it was found that benzene is toxic and carcinogenic. With similar solvent properties and being less toxic, methylbenzene has replaced benzene as a common solvent in the laboratory.

6.4 Chemical Properties of Benzene

In order to "retain" the added stability brought about by the delocalization of pi electrons over the network of six carbon atoms, benzene undergoes mostly substitution reactions, in contrast with an alkene molecule. As the more "loosely" bounded pi electrons are available, benzene is an attractive site for strong electrophiles to attack. Benzene does not behave chemically the same as alkenes. This difference is clearly seen in their behavior towards the same reagents:

- When Br_2 in CCl_4 is added separately to cyclohexene and benzene, there is instant decolorization of reddish-brown Br_2 observed when it is mixed with cyclohexene. No decolorization of Br_2 is observed when mixed with benzene.
- When cold alkaline $KMnO_4(aq)$ is added separately to both hydrocarbons, decolorization of purple $KMnO_4(aq)$ and a brown precipitate of $MnO_2(s)$ will be observed for cyclohexene but not benzene.

Benzene does undergo addition reactions only when reaction conditions are harsh enough. For instance, it can undergo hydrogenation, using a nickel catalyst, and be converted to cyclohexane. But this reaction requires a much higher temperature (150–250°C) and pressure (25 atm) compared to the hydrogenation of cyclohexene.

In the following section, we will look into the reactions of benzene, which is typical of all arenes.

6.4.1 *Formation of Halogenoarene — Electrophilic Substitution*

Reagents:	X_2 with FeX_3 catalyst (where X = Br or Cl)	
Conditions:	Room temperature; Anhydrous	
Options:	AlX_3 or *Fe as catalyst	

Q: How do Fe filings act as a catalyst in the above reaction?

A: When Fe(s) comes in contact with Br_2 or Cl_2, the FeX_3 is generated *in situ*.

Q: If Br_2 can be polarized by an alkene and serve as an electrophile, why can't the same reaction occur between benzene and Br_2? Why is a catalyst needed? Since there are six pi electrons in the benzene ring, shouldn't the ring be electron-rich enough to bring about the polarization?

A: On the surface, it seems that a benzene molecule is more electron rich than an alkene molecule as there are six pi electrons. But on average, there is only one pi electron between two carbon atoms. Thus, the pi electron cloud is not rich enough to polarize the Br_2 molecule. To create a stronger electrophile, Lewis acids such as $FeBr_3$ and $AlBr_3$ are used, i.e.

$$\text{Br---Br} + FeBr_3 \ \rightleftharpoons \ ^{\delta+}\text{Br}\overset{\delta-}{\text{---}}\text{Br}\cdots\cdots FeBr_3 \ \rightleftharpoons \ Br^+ + [FeBr_4]^-$$
$$\text{polarised}$$

Thus, a Br^+ is certainly more electrophilic than a partially polarized Br_2. So you can see that if the molecule involved is less reactive — in this case, benzene is less reactive than an alkene — one can simply make the reaction go by increasing the reactivity of the attacker, which is an electrophile here.

Q: Is it alright to use either $FeCl_3$ or $AlCl_3$ as a catalyst if the purpose is to incorporate a Br atom onto the benzene ring?

A: Absolutely not. If you want to put a Br atom onto the benzene ring, then use either $FeBr_3$ or $AlBr_3$. If you use $FeCl_3$ instead, then during the generation of the Br^+ electrophile, the Cl^+ electrophile would also be generated due to halogen exchange. Hence, you would get some unnecessary side products — it's a waste of reactants!

6.4.2 *Formation of Alkylbenzene — Electrophilic Substitution*

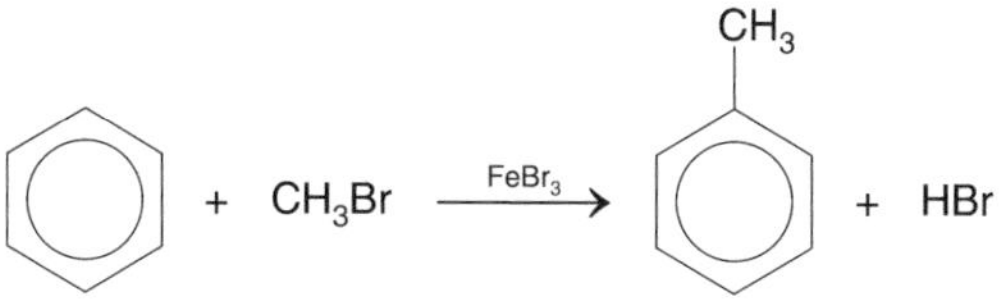

Reagents:	CH_3X with FeX_3 catalyst (where $X = Br$ or Cl)
Conditions:	Room temperature; Anhydrous
Options:	Use different RX if the purpose is to incorporate an R-group
	AlX_3 as catalyst

This reaction is known as Friedel–Crafts alkylation, in which an alkyl group is introduced onto the benzene ring. It is similar to the halogenation reaction in terms of reaction mechanism, and the difference only lies in the type of electrophile. The Lewis acid, $FeBr_3$ or $AlBr_3$, is used to generate the electrophile, CH_3^+, as follows:

Alternatively, $AlBr_3$ can also be used to generate the same electrophile:

The two-step electrophilic substitution mechanism is as follows:

Step 1: CH_3^+ electrophile attacks benzene to form a carbocation intermediate.

carbocation

Step 2: Deprotonation of carbocation.

The regeneration of $AlBr_3$ makes it a catalyst, as well as a Lewis acid.

Q: Why is it important to know the Friedel-Crafts alkylation process?

A: The reaction is important as it allows us to form a carbon-carbon bond which allows us a process to extend the carbon chain length.

6.4.3 *Formation of Nitrobenzene — Electrophilic Substitution*

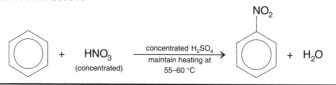

Reagents:	Concentrated $HNO_3(l)$, concentrated $H_2SO_4(l)$
Conditions:	Maintain heating at 55–60°C
Options:	—

Nitration is the process of introducing the nitro group onto the benzene ring. Later in the book, we will see that nitrobenzene is an important precursor for phenylamine (Chapter 11). In addition, nitration is crucial for converting methylbenzene (or toluene) to trinitrotoluene (TNT), an important explosive.

Q: So, we do not have to use reflux when preparing nitrobenzene?

A: No! At the temperature range of 55-60°C, vaporization is not very serious.

Q: What does "reflux" actually mean? Can we use the terms "heat" and "reflux" interchangeably?

A: Reflux simply refers to continuous heating with condensation of the vapor back into the reaction mixture. This is done by connecting a Liebig condenser, as shown in the following set-up. We cannot use heat and reflux interchangeably because heating does not imply there is a condensation process taking place. The main purpose of reflux is to have continuous heating without causing the contents to dry up. When heating is required in organic synthesis, it is usually carried out under reflux, especially if the content is relatively volatile at the temperature of heating.

Q: Then why did you not use the term "reflux" throughout the text so far? Most of the time, the term "heat" was used instead.

A: I have used the term "heat" thus far just to indicate that a higher temperature is needed for the reaction to happen. But this reaction may simply be a distinguishing chemical test rather than for a synthetic one. If it is a chemical test, which is usually done in a test-tube, then it is inappropriate to use the term "reflux," as it is impossible to connect a Liebig condenser to a test-tube. So to prevent misconception and inappropriate usage, the term "heat" was commonly used. But one should note that if it is a chemical reaction in a synthetic pathway, where prolonged heating is required, then the term "reflux" is preferred, unless stated otherwise. Some examples in which reflux is not needed would be: the dehydration of alcohol to alkene using concentrated H_2SO_4, the preparation of nitrobenzene and nitration of methylbenzene.

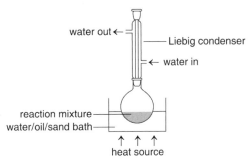

The electrophile NO_2^+ is generated in the acid-base reaction below:

$$HNO_3 + 2H_2SO_4 \rightleftharpoons NO_2^+ + H_3O^+ + 2HSO_4^-$$
$$\text{base} \qquad \text{acid}$$

Q: Why do you say that the above reaction is an acid-base reaction?
A: If you look closely at the above reaction, you would know that when a HNO_3 molecule becomes NO_2^+, effectively an OH^- species has been "extracted" from HNO_3. Thus, from the acid-base concept, the HNO_3 is acting as a base here as it provides an OH^- ion. Sulfuric acid is acting as an acid since it is a proton provider here, as HSO_4^- is being generated. Hence, the above reaction is a good demonstration that an acid can play the role of a base in the presence of another acid that is stronger than itself.

Q: Why must concentrated H_2SO_4 be introduced? Isn't concentrated HNO_3 capable of "self-generating" the NO_2^+ electrophile?

A: Concentrated nitric acid contains very little of the nitronium ion, NO_2^+, and hence it has hardly any effect on benzene in the absence of sulfuric acid. In the presence of H_2SO_4, more NO_2^+ can be generated as H_2SO_4 is a stronger acid than HNO_3. But later on in Chapter 8, we will see that if the benzene-substituted compound is very reactive, even diluted HNO_3 is good enough to generate the required electrophile.

The mechanism involved in this electrophilic substitution reaction is similar to that for the halogenation reaction that benzene undergoes:

Step 1: NO_2^+ electrophile attacks benzene to form a carbocation intermediate

carbocation

Step 2: Deprotonation of carbocation

Q: Why is Step 1 a slow step, or the rate-determining step?

A: Well, it is not difficult to understand this if one remembers that a benzene molecule is resonance stabilized. Thus, any attempt to break the resonance stability, in this case to form the carbocation, would be energetically demanding. Such understanding is also useful to explain why Step 2 is a fast step as compared to Step 1. This is because in Step 2, the resonance stability is regenerated.

6.4.4 *Catalytic Hydrogenation — Reduction*

Like all unsaturated organic compounds, benzene is capable of undergoing hydrogenation, forming the saturated cyclohexane. Some may wonder, what is the purpose of hydrogenating aromatic compounds?

Aromatic compounds can be carcinogenic in nature, and if aromatic compounds are involved in combustion, one likely product is non-combusted carbon, which results in undesired particle emissions in exhaust gases. Thus, converting aromatic compounds to saturated ones not only decreases health-related risks but also improves fuel efficiency.

Reagents:	$H_2(g)$ with $Ni(s)$ catalyst
Conditions:	Heat under high temperature and pressure
Options:	$Pt(s)$ or $Pd(s)$ as catalyst

6.4.5 *Combustion — Oxidation*

$$\text{benzene} + \tfrac{15}{2}O_2 \longrightarrow 6CO_2 + 3H_2O$$

Reagents:	$O_2(g)$
Conditions:	Heat at high temperature
Options:	—

With a high C to H ratio, benzene burns with a smoky luminous flame. Combustion can be used to differentiate between saturated hydrocarbons and highly unsaturated hydrocarbons. In the presence of excess oxygen, the former will produce non-sooty flame whereas the latter will produce sooty flame. The production of soot is a result of the different combustion routes taken by both the saturated and highly unsaturated hydrocarbons.

6.5 Benzene Derivatives

Similarly to benzene, substituted benzenes also undergo electrophilic substitution reactions, but the nature of the substituent on the ring affects both the reaction rate and the orientation of electrophilic attack. Not just any of the remaining aromatic hydrogen atoms on the substituted benzene ring will be replaced by the attacking electrophile. For purposes of discussion, we will focus on the reactivity of monosubstituted benzene derivatives.

6.5.1 *The Influence of the Substituent on the Rate of Electrophilic Substitution*

Substituents that are electron-releasing enhance the electron density of the benzene ring as compared to an unsubstituted benzene molecule, making it more susceptible to attack by electrophiles. Hence, the rate of electrophilic substitution for such benzene derivatives is considerably faster than that for unsubstituted benzene. Such substituents are termed **activating groups.**

Deactivating groups are substituents that withdraw electron density from the benzene ring. This causes the electron density of the ring to decrease, and electrophiles are less attracted to it. As such, electrophilic substitution occurs at a slower rate for these benzene derivatives compared to unsubstituted benzene. To effect a considerable rate, prolonged heating at a higher temperature may be required.

Electrons can either be donated to the benzene ring or withdrawn from it in two ways: (1) through the sigma bond or (2) through the pi bond. When electron flow occurs through the sigma bond, it is termed the **inductive effect.** This mode of "transmission" is viable when there is a difference in electronegativity between the bonding atoms. Alkyl groups attached to the benzene ring inductively donate electrons through the sigma bond. This is because the tetrahedral configuration about the carbon atom of an alkyl group results in a net dipole moment pointing away from the alkyl group, towards the benzene ring, as shown via a methyl group:

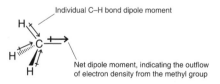

In addition, like the carbon atom of the C=C in alkene, the carbon atom of the benzene ring is sp^2 hybridized whereas that of the alkyl group is sp^3 hybridized. This make the carbon atoms of the benzene ring to have the electron withdrawing effect. But from the perspective of the alkyl group, we can say that the alkyl group is electron donating. In contrast, substituents such as halogens, aldehyde ($-CHO$), hydroxyl ($-OH$), amino ($-NH_2$), and $-NO_2$ groups inductively withdraw electrons from the benzene ring via the sigma bond. Take, for instance, benzaldehyde; the carbonyl C atom is electron deficient and it seeks to enhance its electron density by withdrawing electrons from the more electron-rich benzene ring. This causes an "outflow" of electron density from the ring, making it less prone to electrophilic attack. The aldehyde group is said to deactivate the ring.

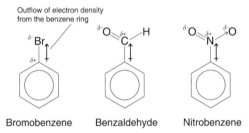

Bromobenzene   Benzaldehyde   Nitrobenzene

When electron flow occurs through the pi bond, it is termed the **resonance effect** or **mesomeric effect**. This mode of "transmission" is possible when there is overlapping of p orbitals — between the sp^2 hybridized carbon atom of the benzene ring and the substituent that it is bonded to.

Substituents that donate electrons by resonance include halogens, hydroxyl and amino groups. A common feature among these groups is that they contain at least a lone pair of electrons residing in the p orbital.

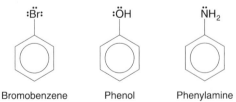

Bromobenzene   Phenol   Phenylamine

With this filled p orbital overlapping with that of the sp^2 hybridized-carbon atom, an "inflow" of electron density into the ring makes it more prone to electrophilic attack.

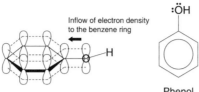

Phenol

Such electron-releasing groups are said to activate the ring, as one can see the delocalization of the negative charge, which represents an electron, into the benzene ring in the following diagram. Note that there are only three specific carbon atoms of the benzene ring, i.e. carbons 2, 4 and 6, that possess this negative charge. Later on, we will see that such electron-rich carbon atoms are more susceptible to electrophilic attack.

Q: Since there are four resonance structures above, are they all equivalent to each other?

A: No, absolutely not. The carbon atom is not a very electronegative atom. For it to be negatively charged would be highly unstable. Thus, we would expect that the contribution of the last three resonance structures to the overall resonance hybrid not to be as prominent as compared to the first structure. In addition, an oxygen atom, being highly electronegative, would not "harbor" the positive charge "comfortably." This means that if we were to map out the electron distribution profile for phenol above, the OH-group would only be slightly electron deficient, and the benzene ring, slightly electron-rich. But note that this is good enough to make the molecule more susceptible to electrophilic attack.

Q: Does it mean that the carbons 3 and 5 would have the same electron density as the carbon atoms in an unsubstituted benzene?

A: No. Comparatively, carbons 3 and 5 would not be as electron-rich as carbons 2, 4 and 6. But if compared to an unsubstituted benzene, carbons 3 and 5 are still relatively more electron-rich.

Substituents that withdraw electrons by resonance include the carbonyl and nitro groups. The aldehyde carbon atom is sp^2 hybridized, and its p orbital overlaps with the p orbital of the carbon atom of the benzene ring it is bonded to. The aldehyde oxygen atom also has its p orbital as part of this extensive network of p orbitals, i.e.

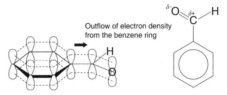

Benzaldehyde

Due to the oxygen atom being more electronegative than the carbon atom, it withdraws electron density towards itself and away from the benzene ring. The "outflow" of electron density from the ring deactivates the ring towards electrophilic attack. A similar picture is painted in the case of nitrobenzene. Note that there are only three specific carbon atoms of the benzene ring, i.e. carbons 2, 4 and 6, that possess this positive charge.

Take note that all six carbon atoms are actually more electron-deficient than those in an unsubstituted benzene ring, but with carbons 2, 4 and 6 being much more electron-deficient than carbons 3 and 5. Later on, we will see that such electron-deficient carbon atoms are less susceptible to electrophilic attack.

Q: If a substituent can be both electron-withdrawing via inductive effect and electron-releasing by resonance, how would the electron density of the benzene be affected by these two effects? Who would "win" out, and why?

A: The substituents that are both electron-withdrawing via inductive effect and electron releasing via resonance effect are halogens,

−OH, −OR and −NH$_2$ groups. The electron-withdrawing effect arises because of the greater electronegativity of the Cl, Br, O, or N atom that is bonded adjacent to the carbon atom. The electron-releasing via resonance has different impact on the electron density of the benzene ring depending on the adjacent heteroatom. What do I mean by that? The electron-releasing via resonance effect of Cl and Br groups is not as effective as that of an O or N atom. This arises because the electron-releasing via resonance effect of Cl and Br groups occurs via 3p (for the Cl group) or 4p (for the Br group) overlap with the 2p orbital of the carbon atom. Since the orbitals used for overlapping from the Cl or Br atom are more diffuse, the overlapping with the 2p orbital of a carbon atom is less effective. As a result, the electron-releasing via resonance effect is overshadowed by the electron-withdrawing effect via inductive effect. The benzene ring of chlorobenzene and bromobenzene are less electron rich than an unsubstituted benzene ring. Thus, more drastic conditions, such as higher reaction temperature, are required to bring about the same electrophilic reaction for halogenoarene as compared to the unsubstituted benzene. As for the −OH and −NH$_2$ groups, the electron-releasing via resonance effect is much more effective as it involves the 2p–2p overlap between an O or N atom with the carbon atom. As a result, the benzene ring is much more electron rich than an unsubstituted benzene ring because the electron-withdrawing via inductive effect is overshadowed by the electron-releasing via resonance effect. Thus, a benzene ring containing the −OH or −NH$_2$ substituent requires milder reacting condition than an unsubstituted benzene molecule. We will see more of these later under the chemistry of phenol and phenylamine.

There are no contradicting effects for substituents that consist of the carbonyl, cyano (−CN) and nitro groups. This is because these substituents are electron-withdrawing via both inductive and resonance effects. The benzene ring containing these substituents is being deactivated as compared to an unsubstituted benzene molecule, thus more drastic reacting conditions would be needed to bring about an electrophilic reaction.

6.5.2 *The Influence of the Substituent on the Orientation of Electrophilic Attack*

Apart from their influence on the reactivity of the ring towards electrophilic substitution and the rate of such reaction, these substituents also exert different influences on the position of the electrophilic attack. The deciding factor lies with the relative stability of the carbocation intermediate that is formed in the first step of the mechanism.

As was covered in Chapter 5 on Alkenes, the stability of charge particles depends on the extent of charge dispersal. The less concentrated the positive charge residing on the carbon atom bearing it, the more stable the carbocation will be.

It can be generalized that the stability of a carbocation increases in the order:

$$CH_3^+ < \qquad RCH_2^+ < \qquad R_2CH^+ < \qquad R_3C^+$$
methyl carbocation   1° carbocation   2° carbocation   3° carbocation

Referring to the diagram below, there are three different sites that an electrophile, E^+, can attack:

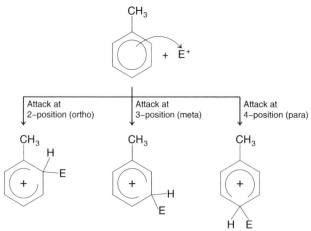

For purposes of discussion, we will use one of the localized structures of methylbenzene (see Fig. 6.3). If the electrophile attacks at the 2-position, the positive charge resides on a 3° carbocation — which is most stable. If the attack occurs at the 4-position, the positive charge

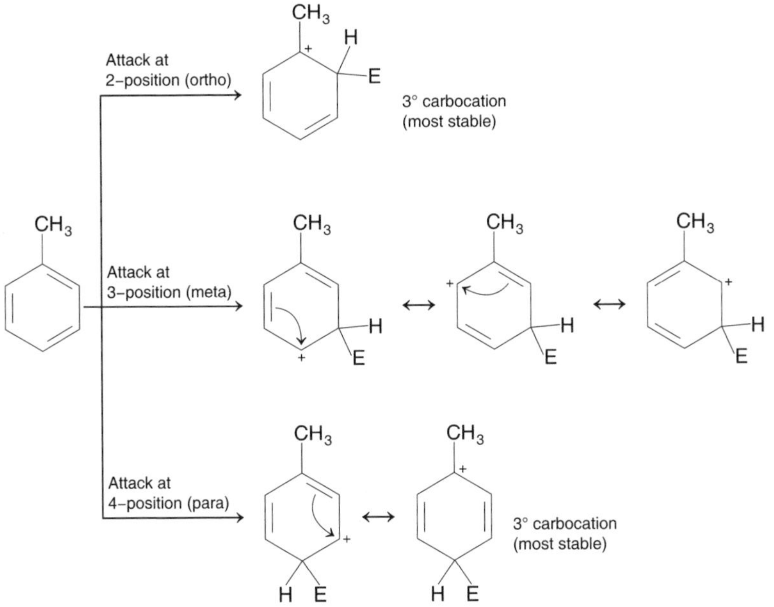

Fig. 6.3.

(through a shift in electron density), can also "be made to" reside on the 3° carbocation. However, if the attack is on the 3-position, there is no way to get the positive charge on the 3° carbocation; it always rest on a 2° carbocation.

Hence, for methylbenzene, electrophilic attacks directed at the 2- and 4-positions are preferred since these result in the formation of a more stable carbocation. Thus, we say that methylbenzene is 2- and 4-directing or ortho- and para-directing.

Therefore, the major products that would be formed would be the 2- and 4-substituted products. The 3-substituted product would still be formed, but in a more minute quantity.

Q: Is the amount of 2-substituted and 4-substituted product the same?

A: There are two 2-positions available for electrophilic attack as compared to only one 4-position available. Thus, from a statistical point of view, twice as much of the 2-substituted product would be formed compared to the 4-substituted product. But when an electrophile attacks the 2-position, it would face greater steric hindrance than when it attacks at the 4-position. Thus, this

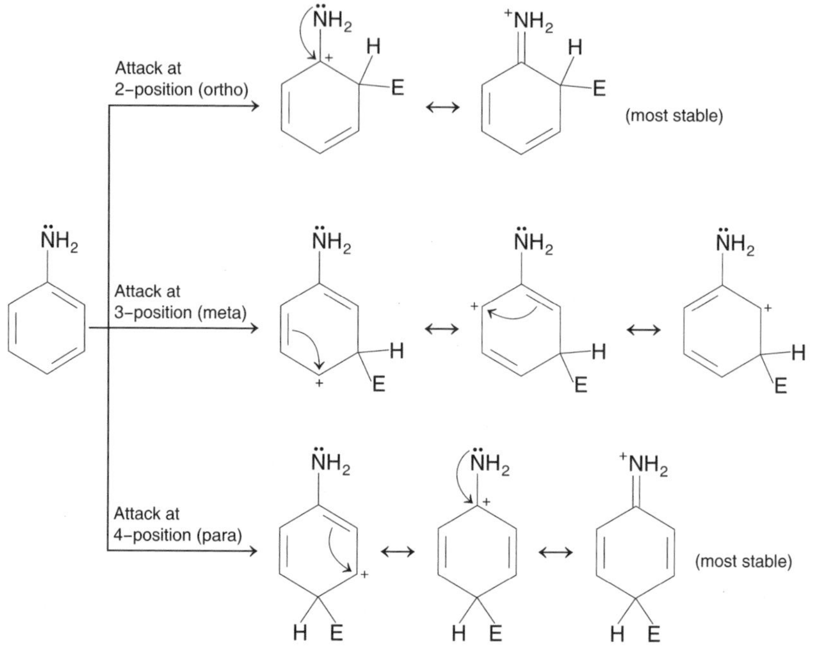

Fig. 6.4.

steric effect would diminish the amount of 2-substituted product being formed. So, whether you would get more 2-substituted
or 4-substituted product would really depend on which factor wins,
i.e. statistical versus steric factors. How would we know? Well, by
making a synthesis and analyzing the composition of the products.

Let's now discuss phenylamine (see Fig. 6.4). If the electrophile
attacks at the 2-position, it is possible to have the N atom donate
its lone pair of electrons by resonance to lower the concentration
of the positive charge on the adjacent carbocation. The same scenario can also be found if the attack is at the 4-position. However,
such a stabilizing effect is not possible if the attack occurs at the
3-position. Hence, phenylamine is also 2- and 4-directing since electrophilic attack directed at these positions results in a more stable
carbocation formed.

Q: Since the nitrogen atom of the NH_2-group is quite electronegative, can't it prevent its lone pair from being involved in the
delocalization?

A: It can't. This is because the lone pair "sits" in a *p* orbital which is parallel to the other *p* orbitals. You can't expect the nitrogen atom to bend its *p* orbital in such a way that it is not parallel to the rest of the *p* orbitals, right? Hence, once you have a series of *p* orbitals that are next to one another, delocalization is "automatic." At most, the delocalization of electron density can be diminished through electron-withdrawing via the inductive effect.

We know that the aldehyde substituent is a deactivating group. So let's see how it affects the relative stability of the carbocation intermediate formed (see Fig. 6.5).

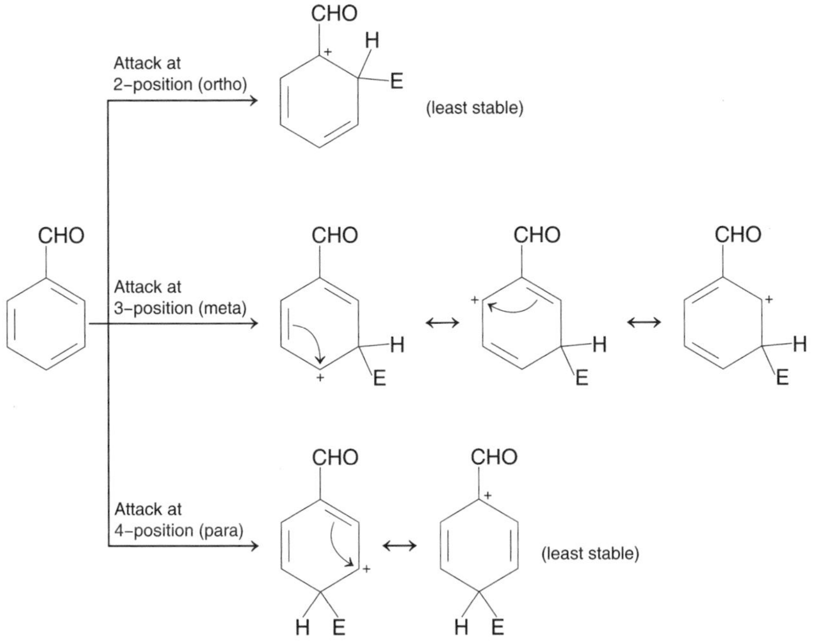

Fig. 6.5.

If the electrophile attacks at either the 2- or 4-position, the positive charge resides on the carbon atom containing the electron-withdrawing group. However, this now presents a least stable structure since it is attached to the electron-withdrawing aldehyde group, which intensifies the positive charge on this carbocation.

If the attack occurs at the 3-position, no such destabilizing effect is observed. Hence, for benzaldehyde, it is meta-directing or 3-directing.

In general, functional groups with no unsaturated bonds are mostly 2- and 4-directing. This list includes the halogens, amino ($-NH_2$), hydroxyl ($-OH$) and alkyl groups. Among these substituents, halogens are ring deactivating but 2, 4-directing, while alkyl, amino and hydroxyl are both ring activating and 2, 4-directing.

Functional groups with unsaturated bonds are mainly 3-directing. Examples include the nitro ($-NO_2$), cyano ($-CN$) and carboxyl ($-COOH$) groups, which are also ring deactivating.

In the following section, we will discuss the chemistry of methylbenzene, which is 2- and 4-directing.

6.6 Chemical Properties of Methylbenzene

Methylbenzene, commonly known as toluene, undergoes the following reactions:

- Electrophilic substitution of the parent benzene ring
- Free radical substitution at its alkyl side-chain
- Oxidation of alkyl side-chain

6.6.1 *Formation of Halogenoarene — Electrophilic Substitution*

major products

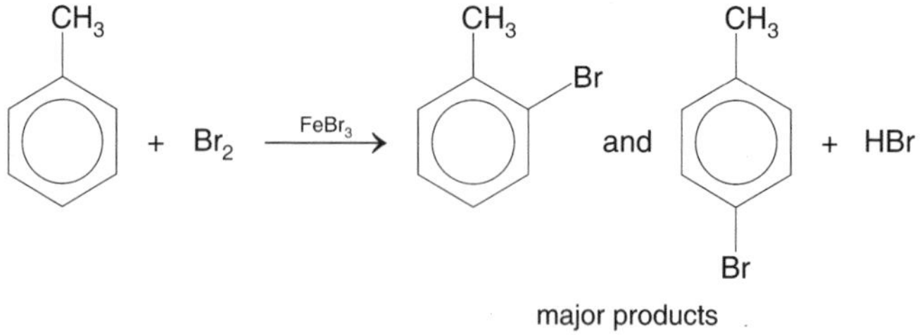

Reagents:	X_2 with FeX_3 catalyst (where $X = Br$ or Cl)
Conditions:	Room temperature; Anhydrous
Options:	AlX_3 or Fe as catalyst

Electrophilic substitution occurs at the ortho and para positions (i.e. 2- and 4-positions) with both isomers produced as major products as compared to the meta-substituted product, which we are ignoring it here although it is still being formed in minute quantity. If the reaction were to occur in the presence of sunlight, there would also be free-radical substitution of the alkyl side-chain.

Q: How do you separate the mixture of products formed?

A: Well, if the products are in the liquid form, you can separate them by fractional distillation, since these products have different boiling points. If they are in the solid form, you can use chromatographic method to separate them.

6.6.2 *Formation of Alkyl(Methylbenzene) — Electrophilic Substitution*

major products

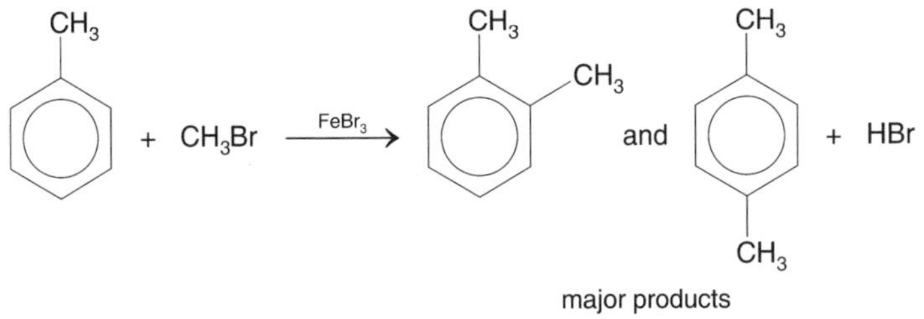

Reagents:	CH_3X with FeX_3 catalyst (where X = Br or Cl)
Conditions:	Room temperature; Anhydrous
Options:	Use different RX if the purpose is to incorporate an R-group
	AlX_3 as catalyst

6.6.3 *Formation of Nitroarene — Electrophilic Substitution*

major products

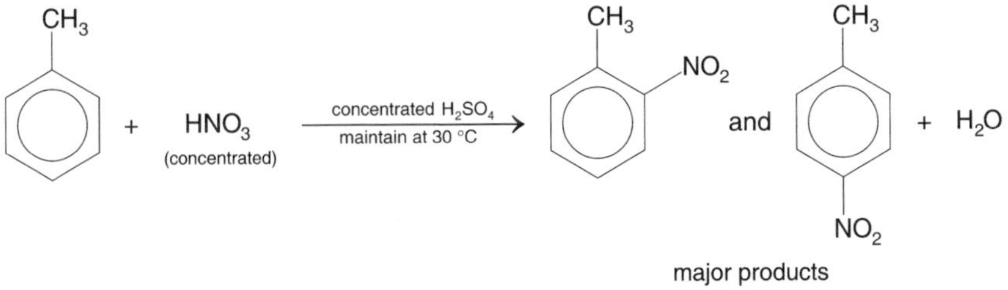

Reagents:	Concentrated HNO_3(l), concentrated H_2SO_4(l)
Conditions:	Maintain at 30°C
Options:	—

Not only is the nitration of methylbenzene considerably faster than the nitration of benzene, a lower working temperature is required — simply due to the electron-releasing effect of the methyl substituent that makes the benzene ring more susceptible to electrophilic attack.

Q: If 4-nitromethylbenzene is subjected to nitration, which position would electrophilic substitution occur at given that it contains both an activating and deactivating group?

A: The methyl substituent is 2- and 4-directing, whereas the nitro-group is 3-directing. If we consider the directing effect of each group in turn, the respective sites of attack are shown below:

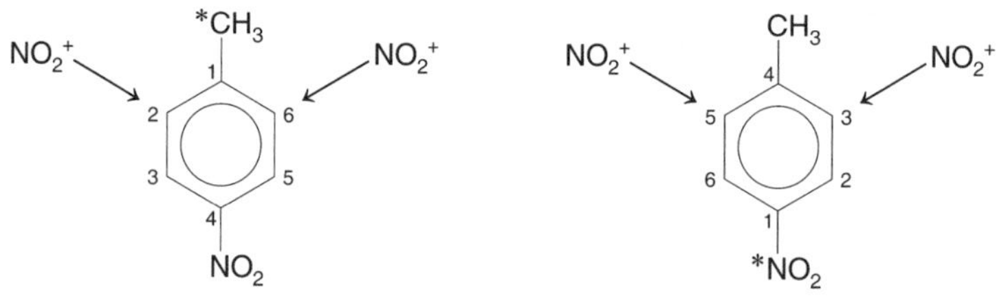

In this case, both groups' directing properties actually complement each other with the sites of attack occurring at positions 2 and 6. Substitution at these positions will result in the identical product, 2,4-dinitromethylbenzene, being formed:

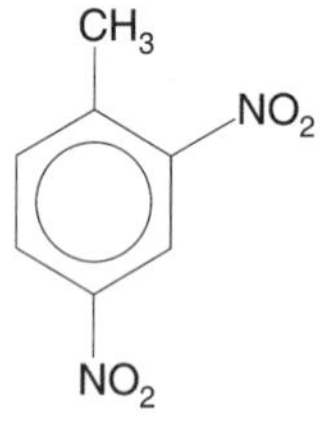

Substitution does not occur at position 4 since it is taken up by the nitro group.

Q: So, does that mean the above molecule is the "only" product formed?

A: No! Not at all. In the actual organic synthesis, you would still get a mixture of products. What do we mean? You can still get minute amount of side-products, for example, one of the side-product formed is one in which all the remaining unsubstituted positions of 4-nitromethylbenzene got substituted by the nitro group. Remember, "particles have no feeling" and they actually do not know where to attack. Unlike human, the particles do not have "goal" or direction. Chemical reaction happened simply through molecular collision with the "guide" of electrostatic attraction. So, as long the particles possess the minimum amount of kinetic energy to overcome the activation barrier and they collide with the correct orientation, chemical reaction will occur.

Q: Given that the bond energies of the C–C bond ($350 \, \text{kJ} \, \text{mol}^{-1}$) and C–N bond ($305 \, \text{kJ} \, \text{mol}^{-1}$) are lower than that of the C–H bond ($410 \, \text{kJ} \, \text{mol}^{-1}$), wouldn't the electrophilic substitution result in the weaker bonds being preferentially cleaved?

A: That is a good question to think about because it enables us to think about the complexity that is involved in a chemical reaction. Does the attack occur before the bond breakage? The electrophile would be less likely to attack the carbon atom bearing the methyl group because it is more sterically hindered than the one that is bonded to a hydrogen atom. Now, the electrophile is an electron-deficient species, it would be less likely to attack the carbon atom bearing the $-NO_2$ group as it is electron deficient due to the more electronegative N atom that is bonded to it.

Q: If 2-nitromethylbenzene undergoes nitration, at which position would electrophilic substitution occur?

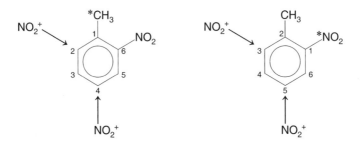

A: Yet again, both groups' directing properties actually complement each other with the sites of attack occurring at positions 2

and 4. Substitution at these positions will then result in two major products being formed:

CH₃ NO₂ NO₂ 2,6-dinitromethylbenzene   CH₃ NO₂ NO₂ 2,4-dinitromethylbenzene

Q: If 3-nitromethylbenzene undergoes nitration, at which position would electrophilic substitution occur?

A: In this situation, wherein the directing properties of two groups do not complement each other, the more dominating group would dictate the preferential site of attack. In this case, a mixture of products of varying quantities would be formed. Between the $-CH_3$ and $-NO_2$ groups, the former is in dominance, and electrophilic substitution takes place at positions 2, 4 and 6 rather than at position 3. Figure 6.6 shows the relative strengths of some activating and deactivating substituents.

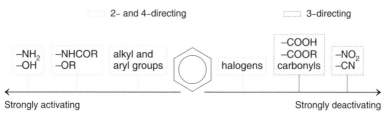

Fig. 6.6. Substituents and their effect on rate and orientation of electrophilic attack.

6.6.4 *Halogenation — Free Radical Substitution (at Alkyl Side-Chain)*

CH₃   CH₂Br

+ Br₂ $\xrightarrow{\text{UV light}}$ + HBr

Reagents:	Halogens such as $F_2(g)$, $Cl_2(g)$ or $Br_2(l)$
Conditions:	UV light or heat
Options:	—

In methylbenzene, the methyl substituent undergoes free radical substitution via the same mechanism as we have seen for alkanes.

Q: Considering only the monochlorination of ethylbenzene, which one of the following pair is the major product?

CH₂CH₃   CHClCH₃   CH₂CH₂Cl

+ Cl₂ $\xrightarrow{\text{UV light}}$ and

(1-chloroethyl)benzene   (2-chloroethyl)benzene

A: Based on the statistical factor (the number of H atoms bonded to carbon atoms), the ratio wouldbe 2:3 for (1-chloroethyl)benzene and (2-chloroethyl)benzene, respectively. On the other hand, with regards to the electronic factor, (1-chloroethyl)benzene will be the major product since it is formed from the more stable Radical 1 compared to Radical 2 below.

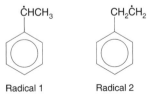

ĊHCH₃   CH₂ĊH₂ .

Radical 1   Radical 2

Radical 1 is more stable than Radical 2. The electron deficiency on the former's carbon atom possessing the lone electron is diminished by a greater extent since it is bonded to a methyl group that is inductively electron-releasing, and the benzyl

($-C_6H_5$) group, which allows the lone electron to delocalize into the benzene ring, as shown below:

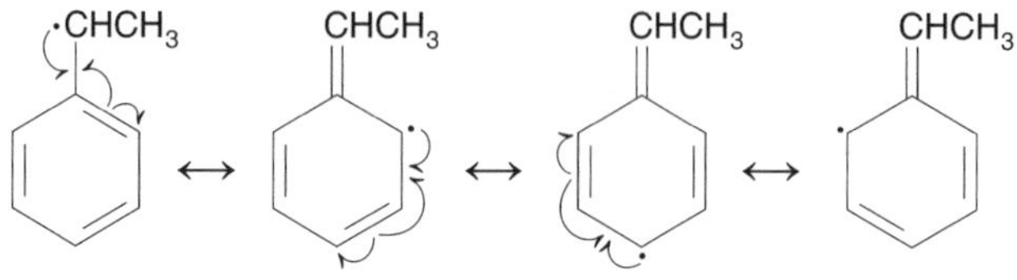

Q: So, through resonance, the electron deficiency is actually "shared" by the other carbon atoms of the benzene ring?

A: You are right! The electron deficiency is not "painfully shouldered" by a single carbon alone.

6.6.5 *Formation of Benzoic Acid — Oxidation (at Alkyl Side-Chain)*

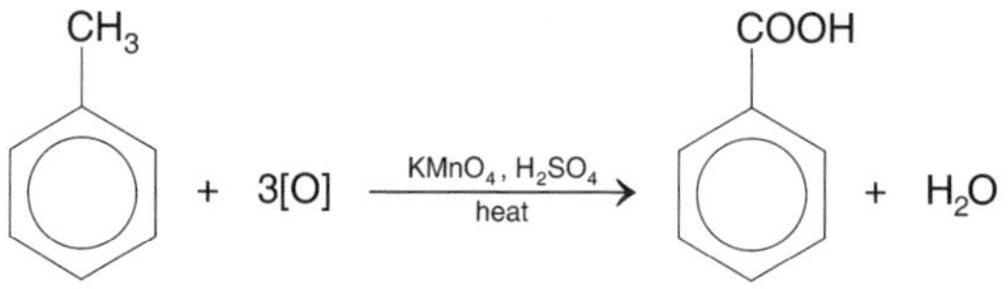

Reagents:	$KMnO_4(aq)$, $H_2SO_4(aq)$
Conditions:	Heat
Options:	Alkaline medium using $NaOH(aq)$ followed by acidification

When refluxed with a strong oxidizing agent such as acidified potassium manganate, methylbenzene can be oxidized to benzoic acid. This reaction can be used to detect the presence of alkylbenzenes. A positive result is obtained when there is decolorization of purple $KMnO_4$ and the formation of benzoic acid — a white solid which is insoluble in cold water, becomes soluble on warming (solubility increases with temperature here) and reappears as a white solid on cooling.

In fact, the vigorous oxidation of any alkylbenzene will result in the formation of benzoic acid and the corresponding amount of carbon dioxide (depending on the number of carbon atoms in the alkyl chain), as shown in the following examples:

$$CH_3CH_2\text{—}C_6H_5 \quad + \quad 6[O] \quad \longrightarrow \quad COOH\text{—}C_6H_5 \quad + \quad CO_2 \quad + \quad 2H_2O$$

$$CH_3CH_2CH_2\text{—}C_6H_5 \quad + \quad 9[O] \quad \longrightarrow \quad COOH\text{—}C_6H_5 \quad + \quad 2CO_2 \quad + \quad 3H_2O$$

When two alkyl substituents are present, both will be oxidized to form a benzenedicarboxylic acid product:

$$(CH_3)_2C_6H_4 \quad + \quad 6[O] \quad \longrightarrow \quad (COOH)_2C_6H_4 \quad + \quad 2H_2O$$

If alkaline $KMnO_4$ is used, the carboxylate salt is formed instead as the carboxylic acid is neutralized by the basic medium presence. The only alkylbenzene that would not be oxidized is the following compound:

$$R\text{—}C(R)(R)\text{—}C_6H_5$$

The absence of a hydrogen atom at the quarternary carbon atom prevents oxidation from taking place.

Q: Why use an alkaline medium to carry out the oxidation, then follow that with acidification? What is the purpose of this extra step?

A: In an alkaline medium, the strength of the oxidizing agent is weaker than if it is in an acidic medium. Using a weaker oxidizing medium would help to preserve other side-chains from being oxidized.

6.7 Summary

The following summary depicts the reactions of arenes as exemplified by benzene and methylbenzene (note that only major products are shown):

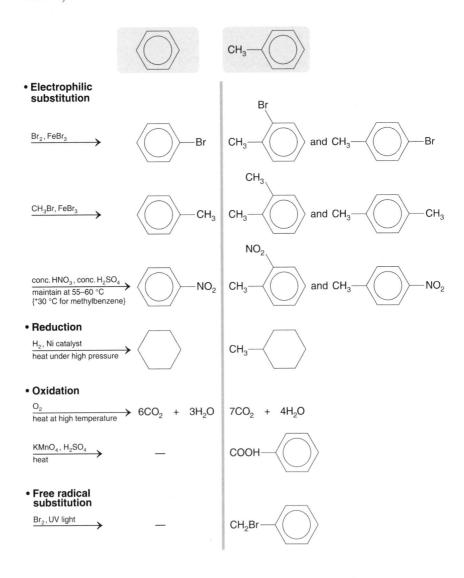

My Tutorial

1. Benzene, C_6H_6, is an aromatic hydrocarbon with six delocalized pi electrons. On treatment with a mixture of concentrated nitric acid and sulfuric acid, benzene undergoes electrophilic substitution reaction to give nitrobenzene. Pyridine, C_5H_5N, exhibits chemical properties very similar to that of benzene.

Pyridine

(a) (i) Suggest one substitution reaction of pyridine, and give the structure of a possible product.

    (ii) Pyridine does not undergo the reaction that is mentioned in (a)(i) as readily as benzene. Suggest a possible reason for this observation.

    (iii) Benzene does not react with sulfuric acid, but pyridine does. Explain this observation and give a balanced equation for its reaction.

(b) Pyridine has a higher boiling point than benzene. Explain.

(c) Benzene also undergoes the same reaction when treated with ethanoyl chloride (CH_3COCl), in the presence of aluminum chloride under anhydrous conditions, to give phenylethanone ($C_6H_5COCH_3$).

    (i) Give the formula of the electrophile that is involved in the reaction.

    (ii) Suggest a reason for the requirement of anhydrous conditions.

    (iii) If ethanoyl chloride is replaced by chloroethane under the same conditions, ethylbenzene is formed. Give a mechanistic account of the formation of ethylbenzene.

(d) Account for the differences in the rate of reactions nitrobenzene, phenol and methylbenzene have with concentrated nitric acid.

(e) A catalyst is required for the reaction of benzene with bromine but not for the reaction of ethene with bromine. Explain.

2. To prepare methyl 3-nitrobenzoate, methyl benzoate is dissolved in concentrated sulfuric acid and the mixture is cooled in an ice

bath. The nitrating agent is then added to the cooled methyl benzoate, maintaining the temperature of the mixture at about 8°C.

(a) (i) Give the reagents that are required to make the nitrating agent.

(ii) Suggest reasons why dilute sulfuric acid is not used as the medium to dissolve methyl benzoate.

(iii) Give the mechanism showing the nitration of methyl benzoate.

(iv) The amount of methyl 3-nitrobenzoate formed would decrease once the temperature is above 8°C. Suggest a reason for the observation.

(b) Write balanced equations for the conversion of methyl 3-nitrobenzoate to each of the following compounds:

(i) 3-aminobenzoic acid

(ii) 3-nitrobenzoyl chloride

(iii) *N*-methyl-3-nitrobenzamide

(iv) 3-aminophenyl methanol

(c) With appropriate reagents and conditions, describe a simple chemical test to differentiate the following pairs of compounds:

(i) methyl 3-nitrobenzoate and methyl benzoate

(ii) methyl 3-nitrobenzoate and 3-aminobenzoic acid

(iii) 3-aminobenzoic acid and 3-nitrobenzoyl chloride

(iv) 3-aminophenyl methanol and 3-nitrobenzoyl chloride

(v) *N*-methyl-3-nitrobenzoyl amide and 3-nitrobenzoyl chloride

(d) The melting point of 2-aminobenzoic acid is lower than that of 4-aminobenzoic acid. State the type of isomerism that both compounds exhibit and give reasons for the difference in the observed melting point.

(e) The solubility of 2-nitrobenzoic acid is lower than 4-nitrobenzoic acid. Give reasons for the observation.

3. With reference to the reaction of methylbenzene with bromine, show how the experimental conditions are important in determining the nature of the products of an organic reaction.

CHAPTER 7

Halogen Derivatives

7.1 Introduction

Halogenoalkanes, also known as alkyl halides, are saturated organic compounds that contain the $-C-X$ functional group (X = F, Cl, Br or I). They are important derivatives of alkanes and have the general formula $C_nH_{2n+1}X$. An example is bromoethane:

$$
\begin{array}{c}
\overset{\displaystyle H}{\underset{\displaystyle H}{\vert\vert}} \overset{\delta^-}{\,}Br \\
H-\overset{\vert}{C}\overset{\delta+}{-}\overset{\vert}{C}-H \\
\overset{\vert}{H}\ \overset{\vert}{H}
\end{array}
$$

Halogenoalkanes do not occur naturally. In fact, they are the by-products of the reaction of alkanes or alkenes with halogen, as these hydrocarbons are commonly found in petroleum. Halogenoalkanes are generally known as the "workhorse" in organic chemistry as they are very useful intermediates to be converted to other more important specialty chemicals of greater economic value. Some halogenoalkanes, such as chlorofluorocarbon, can also be harmful to the environment.

Halogenoarenes (or aryl halides) are aromatic compounds with a halogen atom directly attached to the benzene ring. Similar to halogenoalkanes, halogenoarenes do not occur naturally and are in fact synthesized by reacting aromatic compounds isolated from petroleum with halogens.

7.2 Nomenclature

A halogenolkane is obtained when one or more hydrogen atoms of an alkane molecule have been replaced by halogen atoms via the free radical substitution reaction. Other than this, halogenoalkanes can also be obtained when hydrogen halide (HX) or the diatomic halogen molecules add across an alkene double bond through the electrophilic addition mechanism. Thus, one can simply perceive halogenoalkanes as substituted alkanes. Therefore, halogenoalkanes are named in a similar manner to alkanes — the suffix ends in — *ane* and the halogens, as with other substituents, are named as prefixes accompanied by the appropriate positional numbers.

Examples of halogenoalkanes, classified as being primary (1°), secondary (2°) or tertiary (3°), are shown below:

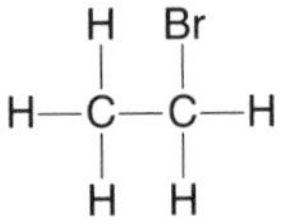

| Bromoethane | 2-Bromopropane | 2-Bromomethylpropane |
| (1° halogenoalkane) | (2° halogenoalkane) | (3° halogenoalkane) |

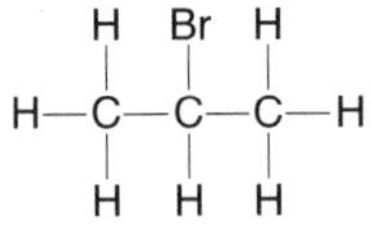

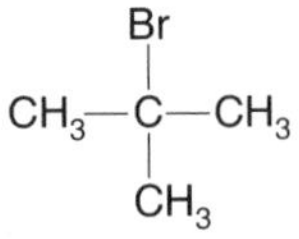

Q: Why is 2-bromomethylpropane not named as 2-bromo-2-methylpropane?

A: It is alright to exclude the positional number "2" for the methyl group here as it does not cause any ambiguity. But the positional number for the Br-group must be stated as there are actually two possible positions to attach the Br-group.

Since halogenoarenes are derivatives of benzene, the approach to the naming of benzenes also applies to them, i.e. the halogen substituents are just added as prefixes to the root word — *benzene*. For polysubstituted benzene derivatives, substituents, other than those included in the common names, are stated as prefixes and in alphabetical order. Numbering of the positional number starts from the carbon atom bearing the principal functional group and in the direction that utilizes the lowest possible numbers for these substituents. Some examples are:

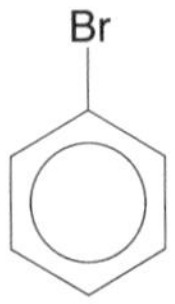

Bromobenzene 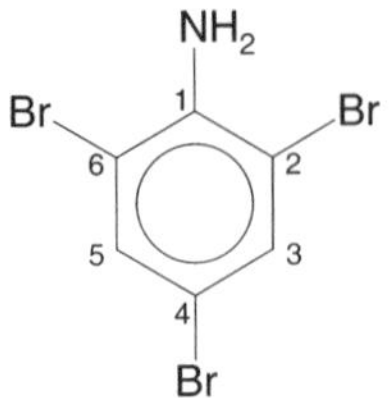   2,4,6-Tribromophenylamine

7.3 Physical Properties

7.3.1 *Melting and Boiling Points*

Similar to the trends for hydrocarbons, both the melting and boiling points of halogenoalkanes containing the same halogen atom together with the same number of carbon atoms, decrease with increasing degree of branching. While for halogenoalkanes with the same halogen atom, both the melting and boiling points increase with increasing number of carbon atoms.

In contrast, when comparing halogenoalkanes of the same chain length but with different halogens, the melting and boiling points increase from F to Cl to Br and to I.

Table 7.1

Halogenoalkane	Fluoroethane	Chloroethane	Bromoethane	Iodoethane
Structural formula	$\begin{array}{cc} \text{H} & \text{F} \\ | & | \\ \text{H—C—C—H} \\ | & | \\ \text{H} & \text{H} \end{array}$	$\begin{array}{cc} \text{H} & \text{Cl} \\ | & | \\ \text{H—C—C—H} \\ | & | \\ \text{H} & \text{H} \end{array}$	$\begin{array}{cc} \text{H} & \text{Br} \\ | & | \\ \text{H—C—C—H} \\ | & | \\ \text{H} & \text{H} \end{array}$	$\begin{array}{cc} \text{H} & \text{I} \\ | & | \\ \text{H—C—C—H} \\ | & | \\ \text{H} & \text{H} \end{array}$
Boiling point (°C)	−37.1	12.3	38.4	71.0

This is basically due to the increasing number of electrons that leads to stronger instantaneous dipole–induced dipole (id–id) attractive forces between molecules that need to be overcome before the phase change occurs.

Q: Since the polarity of the bond increases in the order: C−I < C−Br < C−Cl < C−F, the permanent dipole–permanent dipole (pd–pd) interactions should also increase in the same order. Hence, for the same carbon chain length, shouldn't the boiling point increase from I, Br, Cl to F?

A: Yes, no doubt the polarity of the C−X should result in the strongest pd–pd interaction for fluoroalkane. But this factor cannot be used to account for the observed experimental observation. From the experimental data, the dominant factor that *should be used* to account for the observed phenomenon is the strength of the id–id interactions rather than the pd–pd interactions. Iodoalkane has the greatest number of electrons among the various halogenoalkanes because of the presence of the more electron-rich iodine atom.

7.3.2 *Solubility*

Though polar, both halogenoalkanes and halogenoarenes have poor solubility in water, especially for molecules with higher molecular weight. The type of interaction that the halogenoalkane molecule could possibly have with water (i.e. pd–pd interaction for lower molecular weight molecules but id-id for higher molecular weight) is much weaker than the hydrogen bonding that exists between the water molecules. This hydrogen bonding needs to be overcome first before the halogenoalkane molecule can interact with the water molecules. Such a process needs energy. Unfortunately, the inter-molecular forces between the halogenoalkane and water molecules do not release sufficient energy to overcome the strong hydrogen bond. This would be especially true with the presence of the hydrophobic benzene ring for halogenoarenes, as the id–id interaction of the halogenoarenes and water molecules are relatively weak. Similarly for the halogenoalkanes, when the alkyl group gets bigger, the molecules become more hydrophobic in nature.

However, the solubility of halogenoalkanes in non-polar solvents is good. Some simpler halogenoalkanes are even used as solvents in reactions involving non-polar substances. For instance, the halogenation of alkenes makes use of Br_2 dissolved in tetrachloromethane, CCl_4. But the usage of halogenoalkane as solvent has been curbed because of environmental problems.

Q: You mentioned that because the hydrogen bonds between the water molecule are strong, thus, the interaction between the polar halogenoalkane and water molecule cannot provide sufficient amount of energy to overcome the strong hydrogen bonds between the water molecules. But if the halogenoalkane is of a higher molecular weight, that is have a very bulky and large alkyl group, wouldn't the id-id interaction between the alkyl groups of different halogenoalkane molecules be relatively stronger and hence would also need substantial amount of energy to be overcome?

A: When the alkyl group of a halogenolkane molecule is large, the predominant intermolecular forces between the halogenolkane molecules is considered to be the id-id type. That means that the pd-pd interaction between the halogenolkane molecules can be considered as insignificant. Take note that we are not saying that it doesn't exist! Thus, the interaction between the water molecule and the higher molecular weight halogenolkane is considered to be of the id-id type (refer to pg 111 on Alkane). This interaction would not be strong enough to release sufficient amount of energy to overcome both the hydrogen bonds that are between the water molecules and the predominantly stronger id-id interactions that are between the higher molecular weight halogenoalkane molecules. In short, the energy that is released when two different molecules interact must be sufficient enough to overcome BOTH the intermolecular forces within the solute and solvent.

7.4 Preparation Methods for Halogenoalkanes

7.4.1 *Nucleophilic Substitution of Alcohols*

$$H—\underset{\underset{H}{|}}{\overset{\overset{H}{|}}{C}}—\underset{\underset{H}{|}}{\overset{\overset{OH}{|}}{C}}—H \; + \; PCl_5 \; \longrightarrow \; H—\underset{\underset{H}{|}}{\overset{\overset{H}{|}}{C}}—\underset{\underset{H}{|}}{\overset{\overset{Cl}{|}}{C}}—H \; + \; POCl_3 \; + \; HCl$$

Reagents:	$PCl_5(s)$
Conditions:	Room temperature
Options:	$PCl_3(l)$; $PBr_3(l)$;(red $P + I_2$) or $SOCl_2(l)$

Phosphorous pentachloride, phosphorous trichloride and phosphorous tribromide are stable compounds that are commercially available. Phosphorous triiodide is not sufficiently stable to be stored and thus has to be produced *in situ* (in the reaction system), using red phosphorous and iodine.

The use of different reagents produces different side products as follows:

$$3 \; H—\underset{\underset{H}{|}}{\overset{\overset{H}{|}}{C}}—\underset{\underset{H}{|}}{\overset{\overset{OH}{|}}{C}}—H \; + \; PCl_3 \; \longrightarrow \; 3 \; H—\underset{\underset{H}{|}}{\overset{\overset{H}{|}}{C}}—\underset{\underset{H}{|}}{\overset{\overset{Cl}{|}}{C}}—H \; + \; H_3PO_3$$

$$H—\underset{\underset{H}{|}}{\overset{\overset{H}{|}}{C}}—\underset{\underset{H}{|}}{\overset{\overset{OH}{|}}{C}}—H \; + \; SOCl_2 \; \longrightarrow \; H—\underset{\underset{H}{|}}{\overset{\overset{H}{|}}{C}}—\underset{\underset{H}{|}}{\overset{\overset{Cl}{|}}{C}}—H \; + \; SO_2 \; + \; HCl$$

Among the above reactions in producing the same chloroalkane, the treatment with $SOCl_2$ is the more common method used in laboratory since the by-products are gases that can be more easily separated from the organic product. This would save a lot of effort on purification. In addition, the reactions of the alcohol with $SOCl_2$ and PCl_5 are good characteristics tests for the presence of an OH-group due to the evolution of steamy acidic gases, which turn moist blue litmus red. The SO_2 gas, if present, turns orange acidified $K_2Cr_2O_7$ green, while the HCl gas gives dense white fumes of $NH_4Cl(s)$ with concentrated $NH_3(aq)$.

7.4.2 *Formation of Halogenoalkane — Electrophilic Addition*

major product

Reagents:	Dry HCl(g)
Conditions:	Room temperature
Options:	Dry HBr(g) or HI(g)

Q: Why must the reagent be dry?

A: If water is present, it would lead to the formation of the alcohol (refer to Chapter 5 on Alkenes). In addition, if too much water is present, the HCl would dissociate and get hydrated and lose its ability to act as an electrophile.

Q: Since the H^+ ion is positively charged, shouldn't it be good electrophile?

A: The H^+ ion is "heavily" hydrated in water and due to this hydration, the overall charge density is low. In addition, the H^+ ion is actually quite far away from the point of its attack due to the many layers of water molecules surrounding it. Hence, the H^+ ion is actually not a very good electrophile.

For an unsymmetrical alkene such as propene, two possible products are formed, but in unequal proportions. The major product is formed from the more stable carbocation (refer to Chapter 5 for a detailed explanation).

2-chloropropane
(major product)

1-chloropropane
(minor product)

This is an electrophilic addition reaction that involves the electron-deficient hydrogen atom of the halide molecule — HCl, HBr or HI, making an electrophilic attack. Since the H–X bonds are of different bond strength, there would be a difference in the reaction rate when a different HX reacts with the same alkene.

The order of reactivity towards an alkene mirrors the acid strength in increasing order: HCl < HBr < HI (strongest acid). The acidity of the hydrogen halides depends on the ease of cleavage of the H–X bond; it is easiest to cleave the weakest H–I bond followed by the H–Br bond and subsequently the H–Cl bond.

This difference in covalent bond strength can be accounted for by the effectiveness of orbital overlap. As the size of the halogen atom gets bigger from Cl to I, the valence p orbital used for bonding is larger and thus more diffuse. Consequently, the overlap of the valence p orbital with the $1s$ orbital of the hydrogen atom becomes less effective and hence the bond strength is weaker as we move from H–Cl, H–Br to H–I.

Q: Since the electronegativity of the Cl atom is the greatest among the three molecules, shouldn't the electron deficiency of the hydrogen atom $(\delta+)$ in HCl be the greatest? And if this is so, shouldn't the HCl molecule be more likely to make an electrophilic attack than the HBr and HI?

A: Yes, indeed. The hydrogen atom of the HCl is the most electron-deficient among the three hydrogen halide molecules. And yes, indeed, it would thus be more likely to make an electrophilic attack. But unfortunately, this factor is not dominant enough to make the reaction between HCl and an alkene the fastest. Experimentally, it has been found that the trend of the rate of the reaction between an alkene and the HX molecules coincides with the trend of the bond strength and not with the trend of the ability to make an electrophilic attack. Thus, one cannot use the trend of the ability to make an electrophilic attack to account for the observed experimental data. The important learning point is that in the progress of a chemical reaction, multiple different factors may be affecting the reaction rate at different points of the reaction. Sometimes it may be factors before the breaking of the bonds, sometimes factors at the point of bond breaking or even at times, it may be caused by factors during bond formation. Thus, one needs to make sure that one uses the right explanation or theory to account for the observed experimental data. This is Science.

7.4.3 *Formation of Dihalide — Electrophilic Addition*

Reagents:	Cl_2 in CCl_4
Conditions:	Room temperature
Options:	Br_2 in CCl_4

The usage of Br_2 in CCl_4 offers a characteristics test for the presence of the alkene functional group. In this electrophilic addition reaction, decolorization of reddish-brown Br_2 will be observed. The halogenoalkane product, containing two halogens attached to adjacent carbon atoms, is known as a vicinal dihalide.

Q: Can we use free radical substitution of alkanes to produce halogenoalkanes?

A: Yes, you can. But then a major disadvantage is that you would get multiple halogen-substituted products. Hence, it is not recommended to use this reaction if the aim is steered towards formation of a particular halogenoalkane with the highest yield possible.

Q: But then why do we have to learn the free radical substitution of alkanes to produce halogenoalkanes?

A: The basic organic molecules that we obtain from raw petroleum are high molecular weight alkanes, alkenes and aromatic compounds. Even after subjecting the high molecular weight molecules to the cracking process, the products are still alkanes and alkenes. Thus, it is important to be able to convert the alkanes into other molecules where further conversion into more economically useful products can be done. Although an alkane molecule is not susceptible to electrophilic or nucleophilic attack as there is no particular site on the alkane molecule that is highly electron-rich or electron-deficient (this is because carbon and hydrogen have relatively similar electronegativity values), it still can react with a free radical such as a Cl or Br atom to give us alkyl halides, which are important intermediates for the manufacturing of other valuable compounds.

7.5 Chemical Properties of Halogenoalkanes

$$H-\overset{\displaystyle H}{\underset{\displaystyle H}{C}}-\overset{\overset{\displaystyle \overset{\delta-}{Cl}}{|}}{\underset{\displaystyle H}{\overset{\delta+}{C}}}-H \quad :Nu$$

Halogenoalkanes are polar molecules with the C atom of the C-Cl bond bearing the partial positive charge. Nucleophiles are attracted to this electron-deficient site and, depending on the nature of the halogenoalkane, the substitution mechanism can be either S_N1 or S_N2 (refer to Chapter 3).

When reacting with the same nucleophile, the reaction is expected to proceed the fastest for iodomethane, followed by bromoethane, chloroethane and the slowest for fluoroethane. In fact, fluoroalkanes are generally unreactive towards nucleophilic substitution. This is attributed to the increasing bond strength in the given order above (see Section 7.3.1). On one end, the C–I bond is the weakest, and on the other, we have the C–F bond as the strongest. The explanation for this is similar to that used to account for the differences in H–X bond strength, i.e. in terms of effectiveness of orbital overlap in covalent bond formation.

Q: Shouldn't the C–F bond the most polar among all? And thus shouldn't the carbon atom of the C–F bond be more susceptible to a nucleophilic attack?

A: Yes, you are right about the C–F bond being the most polar among all. But unfortunately the trend for the ability to be attacked by a nucleophile: C–F > C–Cl > C–Br > C–I, cannot be used to account for the observed experimental data here.

7.5.1 *Formation of Alcohol — Nucleophilic Substitution*

$$H-\underset{\underset{H}{|}}{\overset{\overset{H}{|}}{C}}-\underset{\underset{H}{|}}{\overset{\overset{Cl}{|}}{C}}-H \; + \; OH^- \xrightarrow{\text{heat}} \; H-\underset{\underset{H}{|}}{\overset{\overset{H}{|}}{C}}-\underset{\underset{H}{|}}{\overset{\overset{OH}{|}}{C}}-H \; + \; Cl^-$$

Reagents:	NaOH(aq)
Conditions:	Heat
Options:	KOH(aq)

This reaction is also known as alkaline hydrolysis, which is carried out in an aqueous medium. And because of the possibility of reaction between halogenoalkanes and hydroxide, PVC containers, made of none other than polyvinyl chloride (or poly(chloroethene)), cannot be used to contain the alkaline solution. In addition, this alkaline hydrolysis reaction can serve as an important way to distinguish between halogenoalkanes containing Cl, Br or I.

Q: Can the water molecules in the aqueous medium act as a nucleophile?

A: Yes, the lone pair of electrons of the water molecule can make it a nucleophile. But unfortunately, it is not as good a nucleophile as compared to the hydroxide ion as the latter is negatively charged, and thus more electron-rich. Hence, PVC can be the material for a water container.

Q: Does it mean that if 1,1-dichloroethane underwent nucleophilic substitution with aqueous KOH, a diol would be obtained?

A: No, you would not get the following geminal diol:

$$H-\underset{\underset{H}{|}}{\overset{\overset{H}{|}}{C}}-\underset{\underset{H}{|}}{\overset{\overset{OH}{|}}{C}}-OH$$

The germinal diol would undergo spontaneous dehydration to give the aldehyde (CH_3CHO), expelling a water molecule in the process. This would also mean that if 1,1,1-trichloroethane reacts with aqueous NaOH, we would not get the following germinal triol, but the carboxylic acid (CH_3COOH) instead.

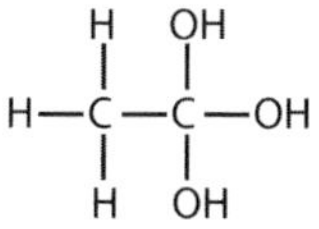

$$
\begin{array}{c}
\text{H} \quad \text{OH} \\
| \qquad | \\
\text{H—C—C—OH} \\
| \qquad | \\
\text{H} \quad \text{OH}
\end{array}
$$

Example 7.1: Describe a chemical test to distinguish between chloroethane, bromoethane and iodoethane.

Solution:

Test: Add NaOH(aq) to each of these compounds and heat. Next, cool the mixture followed by adding HNO_3(aq). Lastly, add $AgNO_3$(aq).

Rationale: When NaOH(aq) is added, the halogenoalkanes undergo hydrolysis. This reaction releases halide ions into the solution. With the silver nitrate added, the halide ions form insoluble silver halide salts that can be identified by their different color, and a conclusion can be made on the type of halogenoalkane present.

$$Ag^+(aq) + X^-(aq) \longrightarrow AgX(s)$$

Type of halide	Cl^-(aq)	Br^-(aq)	I^-(aq)
Observations	White precipitate observed.	Cream precipitate observed.	Yellow precipitate observed.
Conclusion	AgCl(s) formed. Chloroethane is present.	AgBr(s) formed. Bromoethane is present.	AgI(s) formed. Iodoethane is present.

Take note that halogenoarenes do not undergo hydrolysis unless under extreme conditions. Reasons are discussed in Section 7.6.

Q: What is the purpose of cooling the mixture and adding HNO_3(aq) before the addition of $AgNO_3$(aq)?

A: Cooling is needed to prevent $AgNO_3$ from decomposing since it is thermally unstable. The acidification using HNO_3 is to remove the excess NaOH. If not, when $AgNO_3$ is added, Ag(OH) may be precipitated out. Using HNO_3 instead of HCl or H_2SO_4 for acidification prevents introducing anions into the system that may be precipitated out by the Ag^+ ion.

7.5.2 *Formation of Nitrile — Nucleophilic Substitution*

Reagents:	NaCN(aq), ethanol
Conditions:	Heat
Options:	KCN(aq)

Q: Why is an aqueous-alcohol medium used in this reaction?

A: Being an ionic compound, NaCN is soluble in water, but an alkyl halide is relatively insoluble in water. Thus, the addition of ethanol is to help to improve miscibility. NaCN dissolves in water whereas the R–X dissolves in ethanol, and both water and ethanol mix very well.

Q: Previously, it was mentioned that halogenoalkanes are insoluble in polar solvents. But isn't ethanol a polar solvent? So, how can halogenoalkanes dissolve in ethanol?

A: Yes, ethanol is indeed a polar molecule, and it is useful for dissolving low molecular weight halogenoalkanes. It is able to do that because the ethyl group of ethanol is hydrophobic in nature, hence it can interact well with the hydrophobic part of halogenoalkanes. Water would not be able to do the job well due to the lack of this hydrophobic part. Thus, by adding ethanol to water, we are actually causing the solvent mixture to be slightly more hydrophobic and thus less hydrophilic in nature.

Q: Won't the ethanol molecule act as a nucleophile instead?

A: Ethanol molecules, like water molecules, do fulfill the criteria to act as a nucleophile. But unfortunately, they are not nucleophilic enough to attack the halogenoalkanes (see Section 7.5.4).

Nitriles are a class of useful reagents in organic synthesis. Through the formation of a nitrile, the carbon chain length can be extended by one additional C atom, hence this is a step-up reaction. In the above equation, we started with chloroethane, containing two C atoms. Upon reaction with CN^-, the organic product now contains three C atoms.

The formation of nitriles also serves as an intermediate in the synthesis of other organic compounds such as carboxylic acids and amines:

- Nitriles can undergo acidic hydrolysis to form carboxylic acids:

$$H-\underset{\underset{H}{|}}{\overset{\overset{H}{|}}{C}}-\underset{\underset{H}{|}}{\overset{\overset{CN}{|}}{C}}-H \ + \ HCl \ + \ 2H_2O \ \xrightarrow{\text{heat}} \ H-\underset{\underset{H}{|}}{\overset{\overset{H}{|}}{C}}-\underset{\underset{H}{|}}{\overset{\overset{COOH}{|}}{C}}-H \ + \ NH_4Cl$$

If basic hydrolysis is performed, the carboxylic acid will react with the base to form a carboxylate salt:

$$H-\underset{\underset{H}{|}}{\overset{\overset{H}{|}}{C}}-\underset{\underset{H}{|}}{\overset{\overset{CN}{|}}{C}}-H \ + \ NaOH \ + \ H_2O \ \xrightarrow{\text{heat}} \ H-\underset{\underset{H}{|}}{\overset{\overset{H}{|}}{C}}-\underset{\underset{H}{|}}{\overset{\overset{COO^-Na^+}{|}}{C}}-H \ + \ NH_3$$

The evolution of ammonia gas during basic hydrolysis serves as a characteristics test for the presence of a nitrile functional group, which acidic hydrolysis can't do.

- Nitriles can undergo reduction to form amines:

$$H-\underset{\underset{H}{|}}{\overset{\overset{H}{|}}{C}}-\underset{\underset{H}{|}}{\overset{\overset{CN}{|}}{C}}-H \ + \ 4[H] \ \xrightarrow[\text{dry ether}]{LiAlH_4} \ H-\underset{\underset{H}{|}}{\overset{\overset{H}{|}}{C}}-\underset{\underset{H}{|}}{\overset{\overset{CH_2NH_2}{|}}{C}}-H$$

7.5.3 *Formation of Amine — Nucleophilic Substitution*

$$H-\underset{\underset{H}{|}}{\overset{\overset{H}{|}}{C}}-\underset{\underset{H}{|}}{\overset{\overset{Cl}{|}}{C}}-H \ + \ \underset{\substack{\text{excess,}\\ \text{concentrated}}}{NH_3} \ \xrightarrow[\text{heat in sealed tube}]{\text{ethanol}} \ H-\underset{\underset{H}{|}}{\overset{\overset{H}{|}}{C}}-\underset{\underset{H}{|}}{\overset{\overset{NH_2}{|}}{C}}-H \ + \ HCl$$

Reagents:	Excess concentrated NH_3, ethanol
Conditions:	Heat in sealed tube
Options:	Use of alkylamine to form more highly substituted alkylamines

Q: Why is the reaction mixture heated in a sealed tube?

A: Well, ammonia is a gas at room temperature. The reaction is performed in a sealed tube to prevent the ammonia gas from escaping.

Q: Previously in Section 7.5.1, it was mentioned that H_2O is not a very good nucleophile to attack chloroethane to generate the alcohol. How come NH_3 molecule is able to do so?

A: Due to the higher in electronegativity of the O atom, more electron density is drawn towards the O atom as compared to the N atom. This thus makes the O atom more electron-rich, i.e. greater $\delta-$ value. But at the same time, the lone pair of electrons of the O atom is less available for donation as compared to the N atom of NH_3. Thus, all these make NH_3 a better nucleophile than H_2O.

Q: Can Br_2 act as a nucleophile since the molecule contains so many electrons?

A: Although a Br_2 molecule has many lone pair of electrons, but it is not nucleophilic enough to attack the halogenoalkane as it is a non-polar molecule. There is no net permanent dipole moment to "guide" the lone pair of electrons to make the nucleophilic attack feasible. This means that in order for a particle to be able to attack another, first there must be sufficient attractive forces to bring them together. But sometimes even if the attractive force is strong enough to bring the two particles together, the progress of the reaction may still be hindered because of the large amount of energy that is needed to break bonds. If assuming that the system has the energy to overcome the bond breaking, the reaction may still be non-fruitful because of less stable products being formed. In order to understand why some reactants can react while other combinations do not work, one needs to analyze that there are indeed many different types of reaction factors to consider in the progress of a reaction. The absence of a "one size fits all" explanation makes the learning of chemistry very frustrating but at the same time very interesting. Do you agree?

The exact chemistry behind the formation of the amine not only constitutes a nucleophilic reaction but, at the same time, an acid-base reaction as well:

weakly acidic

When the NH_3 nucleophile attaches itself to the alkyl chain, the N atom (a Lewis base) forms a dative bond to the electron-deficient C atom (a Lewis acid), and the N atom now carries a positive charge. So, what we have here is actually a salt made up of the polyatomic cation, $[CH_3CH_2NH_3]^+$, and the anion, Cl^-.

Since the desired product is the primary amine, we need to depronate the cation by treating it with a base. As NH_3 is a weak base, it can do the job as it is already present as a reactant:

$$CH_3CH_2NH_3^+ + NH_3 \rightleftharpoons CH_3CH_2NH_2 + NH_4^+$$

weak acid   weak base   1° amine

The above is a typical acid-base reaction.

Q: Why is excess ammonia needed?

A: The primary amine product that is formed can also act as a nucleophile since there is a lone pair of electrons on the nitrogen atom of the amine molecule, readily available for dative bond formation. In fact, this lone pair of electrons is more readily available than those of the N atom in NH_3 — thanks to the attached electron-releasing alkyl group that enhances the electron density on the nitrogen atom. Therefore, the excess ammonia also serves the function of converting most of the halogenoalkane to the primary amine first, based on the statistical factor. Without excess ammonia, the primary amine that is formed initially would get a chance to react with another halogenoalkane, resulting in the formation of more highly substituted amines.

Thus, as soon as the primary amine is formed, it is drawn towards the halogenoalkane and the nucleophilic substitution between these reactants results in the formation of a secondary amine:

1° amine   2° amine

And with two electron-donating alkyl groups attached to the nitrogen atom, the 2° amine is a better nucleophile than both the 1° amine and NH_3. It reacts with the halogenoalkane to form yet another product, the 3° amine:

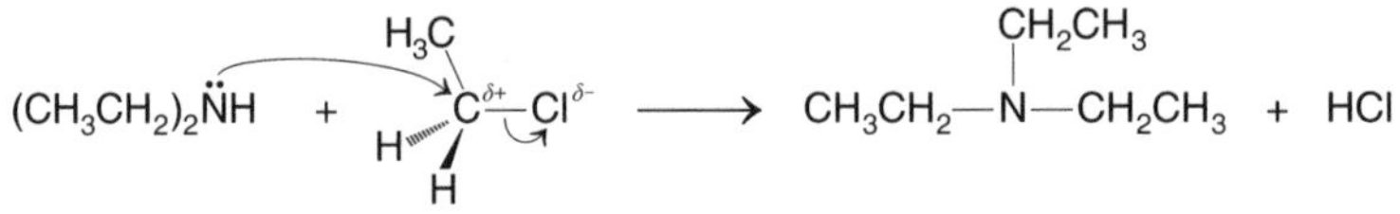

$(CH_3CH_2)_2\overset{..}{N}H$ + H_3C—$\overset{\delta+}{C}$—$Cl^{\delta-}$ $\longrightarrow$ CH_3CH_2—N—CH_2CH_3 + HCl

2° amine   3° amine

If you have notice the trend by now, the much stronger nucleophilic 3° amine will be able to react with the same halogenoalkane. This time round, a quaternary (4°) ammonium salt is formed:

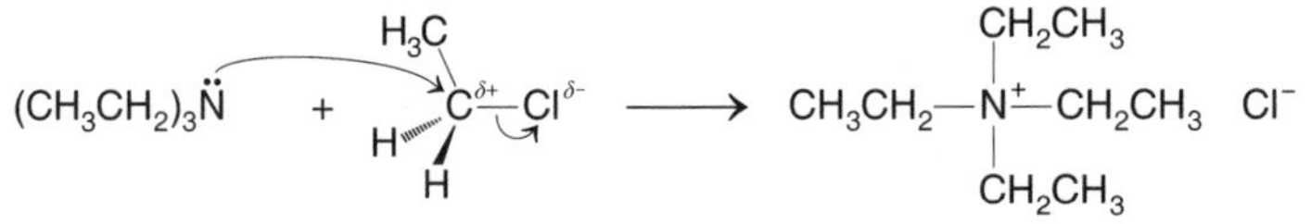

$(CH_3CH_2)_3\overset{..}{N}$ + H_3C—$\overset{\delta+}{C}$—$Cl^{\delta-}$ $\longrightarrow$ CH_3CH_2—$\overset{+}{N}$—CH_2CH_3 Cl^-

3° amine   quaternary ammonium salt

Thus, if the desired product is the 1° amine, one must ensure that the halogenoalkane is a limiting reagent and there is excess NH_3 to limit the formation of the poly substituted products. The higher concentration of NH_3 will ensure the faster depletion of the halogenoalkane and lowers the chances of the other nucleophiles attacking it.

If excess halogenoalkane is used, polyalkylation is unavoidable and a mixture of products is obtained. One would need to use fractional distillation to separate these products.

Q: The steric effect of an amine making a nucleophilic attack increases from 1° to 2° and to 3° amines. So, isn't it more difficult for polyalkylation to take place?

A: Yes, indeed it is. If the number of alkyl groups increases, the availability of the lone pair of electrons on the nitrogen atom increases. At the same time, the steric effect for the more highly substituted amines to make a nucleophilic attack also increases. But the steric factor does not totally exclude the more highly substituted amine from undergoing further nucleophilic attack. Thus, the polyalkylated products would still be formed.

7.5.4 *Formation of Ether — Nucleophilic Substitution*

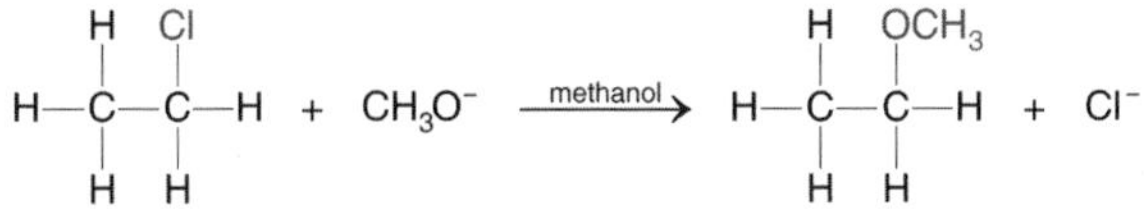

Reagents:	NaOCH$_3$, methanol
Conditions:	Room temperature
Options:	NaOR where R is an alkyl group

The alkoxide is obtained by dissolving a sodium (or potassium) metal in the corresponding alcohol, i.e. dissolve sodium in methanol to obtain sodium methoxide, $CH_3O^-Na^+$. You will learn in Chapter 8 that this is actually an acid-metal reaction, wherein the weakly acidic alcohol reacts with the reactive Na metal. The alcohol molecule is not as good a nucleophile when compared to the negatively charged more electron-rich alkoxide ion.

Q: Can ethanol be used as a solvent for the above reaction?

A: No. If ethanol is used, then ethanol would react with the methoxide ion via an acid-base reaction to generate the ethoxide nucleophile. Hence, the corresponding alcohol of the alkoxide to be substituted is used as a solvent instead.

7.5.5 *Formation of Alkene Via Dehydrohalogenation — Elimination*

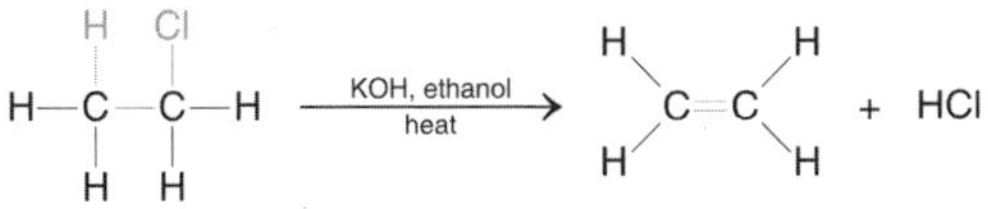

Reagents:	KOH, ethanol
Conditions:	Heat
Options:	NaOH as a base

This reaction constitutes an elimination mechanism as follows:

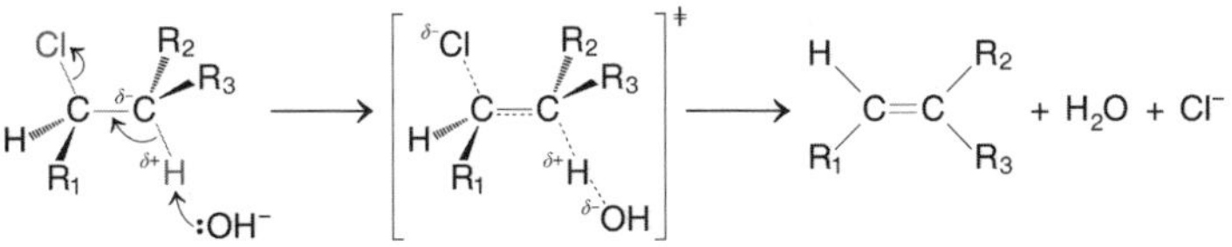

R$_1$, R$_2$, and R$_3$ = alkyl / aryl group

The highly electronegative Cl atom withdraws electron density inductively from the neighboring atoms. Being electron-deficient and thus acidic here, the α-H atom, on the C atom adjacent to the C–X bond, can be easily extracted by a strong base.

In a concerted move, the following can be thought to occur:

- a strong base forms a dative covalent bond to the acidic H atom;
- both the C–H and C–Cl bonds undergo heterolytic cleavage, producing the H^+ and Cl^- ions as the leaving groups;
- a pi bond is formed between these two C atoms.

The simultaneous bond breaking and bond forming is depicted in the activated complex that has the participating atoms lying on the same plane to facilitate formation of the C=C bond.

From kinetics study of the reaction rate, the derived rate equation is: rate $= k[RX][base]$. Hence the acronym E2, which stands for elimination mechanism with a bimolecular rate-determining step.

Q: Since a nucleophile is electron-rich, does it mean that it is also a base?

A: Yes, a nucleophile can function as a base and participate in the elimination reaction. But whether the elimination reaction would occur or not depends very much on the relative reactivity of the two reactants. Similarly, a base can also acts as a nucleophile, but the nucleophilic reaction may not occur if the reactivities of the reactants are low.

Q: So substitution and elimination are two competing reactions involving a nucleophile?

A: Yes, both reactions will occur, but to a different extent. Generally, to have elimination as the predominant reaction, a strong base is used. But if the nucleophile is a weak base or non-basic, then the main reaction will be nucleophilic substitution instead.

Q: Why is the OH^- ion acting as a base in an alcoholic medium causing elimination to occur whereas in an aqueous medium, it acts as a nucleophile?

A: The OH^- ion can be both a nucleophile and a base. In an aqueous medium, the OH^- ion is a very bulky species because it is surrounded by many layers of water molecules. The first layer of water molecules is attracted to the OH^- ion via ion-dipole interaction and subsequent layers are attracted to each other via hydrogen bonding. But for the OH^- ion in ethanol, it is a much smaller solvated species. The first layer of ethanol molecules is attracted to the OH^- ion via ion-dipole interaction, as a result, the hydrophobic alkyl group would be pointing outward. Subsequent layers of the ethanol molecules are attracted via the weak id–id interaction.

Now, in a bulky hydrated OH^- ion, the negative charge is further away from the periphery than in the weakly solvated OH^- ion in ethanol. This would mean that the charge density of the hydrated OH^- ion would be much smaller than the solvated OH^- ion in ethanol. As a result, the ability of the bulky hydrated OH^- ion to "extract" the acidic hydrogen is lower than that of the solvated OH^- ion in an ethanol medium, thus rendering the bulky hydrated OH^- ion a weaker base.

Q: Now, based on your explanation, wouldn't the bulky hydrated OH^- ion face greater steric hindrance when trying to make a nucleophilic attack than the weakly solvated OH^- ion?

A: You are right here. The bulky OH^- ion would face greater steric hindrance when making a nucleophilic attack. If you have looked closely at the mechanism for elimination and compared it against nucleophilic substitution (see Chapter 3), you would notice that the approach of the nucleophile or base is at the opposite direction to the the halogen atom for both mechanisms, but the "attacking species" are approaching the alkyl halide in the same direction relative to the position of the acidic hydrogen. So before the "attacking species" reach the electron-deficient carbon center, be it the highly hydrated OH^- ion or the weakly solvated OH^- ion in ethanol, the "first" atom to be encountered would be the acidic hydrogen. Thus, the weakly solvated OH^- ion would make elimination more likely to happen than nucleophilic substitution. Whereas for the highly hydrated OH^- ion, because the "first" contact with the acidic hydrogen would not result in elimination due to its lower charge density, further approach towards the alkyl halide molecule would result in nucleophilic substitution instead. Thus, take note that in a nucleophilic substitution reaction, there is bound to be a minor amount of elimination products formed. Likewise, for the elimination reaction of alkyl halide, there is bound to be minor nucleophilic substitution reaction.

Just as in the dehydration of alcohols to obtain alkenes (see Chapter 5), when an unsymmetrical halogenoalkane undergoes elimination, the products obtained are a mixture of alkenes in unequal proportions as shown in the case for 2-chlorobutane:

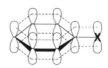

The relative proportions of the products are in accordance to Saytzeff's rule: A more stable alkene is formed in a larger proportion and greater stability is found in the alkene with more alkyl (R) groups attached to the doubly bonded carbons.

7.6 Halogenoalkanes versus Halogenoarenes

Halogenoarenes do not undergo nucleophilic substitution as readily as halogenoalkanes. But under very harsh conditions, such as high temperature and the presence of a strong nucleophile, nucleophilic substitution can be forced to happen.

The lack of reactivity towards nucleophilic attack, where upon the cleavage of the C–X bond has to occur, is due to the following reasons:

1. The C–X bond is strengthened and not as easily cleaved.

The *p* orbital of the halogen can make a side-on overlap with the pi electron cloud of the benzene ring. As a result, the lone pair of electrons on the halogen atom can delocalize into the benzene ring. This results in the C–X bond bearing partial double bond character, and it is thus stronger than the C–X bond in the halogenoalkane.

2. The decrease in the polarity of the C–X bond. Due to the extended network of delocalized electrons over the halogen and the six sp^2 hybridized carbon atoms, a nucleophile is not as strongly drawn to the C atom of the C–X bond as you would expect in the case of a halogenoalkane. This is because the electron deficiency on the carbon atom has been diminished as shown in the following resonance structures:

3. A nucleophile is an electron-rich species, thus the inter-electronic repulsion between the electron-rich benzene ring and nucleophile hinders the approach of the nucleophile.

Q: Do the above reasons also explain why vinylic halides (containing a halogen bonded to a doubly bonded carbon atom) also do not undergo nucleophilic substitution easily?

A: Yes, precisely. The following resonance structures would explain it:

The difference in the strength of the C–X bond of a halogenoarene and that of a halogenoalkane can be highlighted using the following test:

Example 7.2: Describe a chemical test to distinguish between chlorobenzene and 1-chlorohexane.

Solution:

Test: Add NaOH(aq) to each of these compounds and heat. Next, cool the mixture followed by adding HNO_3(aq). Lastly, add $AgNO_3$(aq).

Rationale: When NaOH(aq) is added, only 1-chlorohexane can undergo hydrolysis with cleavage of the C–Cl bond. This reaction releases chloride ions into the solution.

$$CH_3(CH_2)_4CH_2Cl + OH^-(aq) \longrightarrow CH_3(CH_2)_4CH_2OH$$
$$+ Cl^-(aq)$$
$$Ag^+(aq) + Cl^-(aq) \longrightarrow AgCl(s)$$

Observations: For 1-chlorohexane, a white precipitate of AgCl is observed.

For chlorobenzene, no white precipitate is observed.

7.7 Preparation Methods for Halogenoarenes

7.7.1 *Formation of Halogenoarene — Electrophilic Substitution*

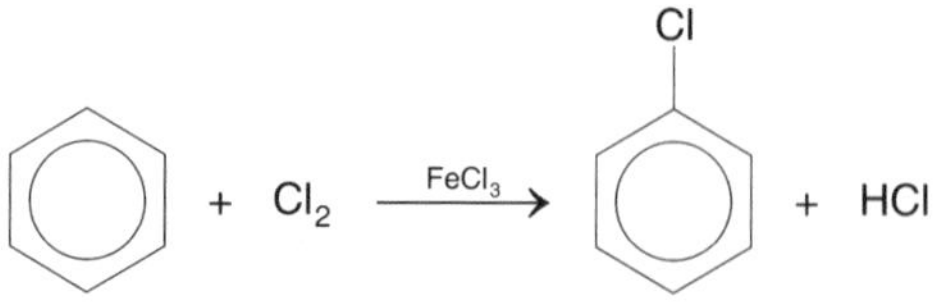

Reagents:	X_2 with FeX_3 catalyst (where X = Br or Cl)
Conditions:	Room temperature; Anhydrous
Options:	AlX_3 or Fe as catalyst

7.8   Chemical Properties of Halogenoarenes

7.8.1   *Formation of Benzene Derivatives —*
Electrophilic Substitution

When compared to the electrophilic substitution of a non-substituted benzene, those of halogenoarenes occur at a slower rate. This is because, overall, the halogen substituents are ring deactivating groups (see Chapter 6). However, the orientation of the attack is still targeted at positions 2 and 4 as this leads to the formation of a more stable carbocation (see Chapter 6).

7.9   Summary

Figure 7.1 depicts the different directing properties of various types of substituents on the benzene ring. Take note that the order of placing two different substituents, each with different directing effect, would produce totally different types and amount of products. For example, with benzene as the common starting reactant, the synthesis route in obtaining 2-methylnitrobenzene (or 4-methylnitrobenzene) is different from that used in producing 3-methylnitrobenzene as the desired product.

Figure 7.2 depicts both the preparation methods and the reactions of halogenoalkanes as exemplified by chloroethane:

My Tutorial

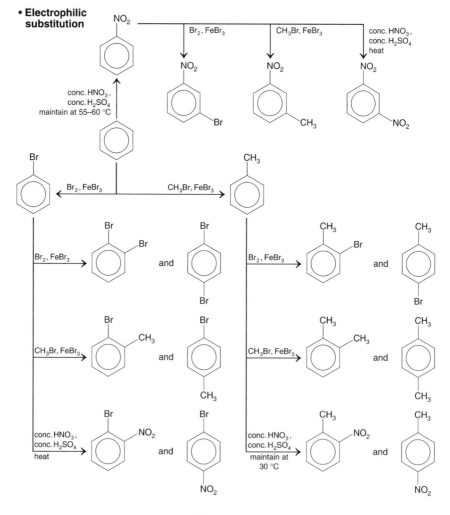

Fig. 7.1.

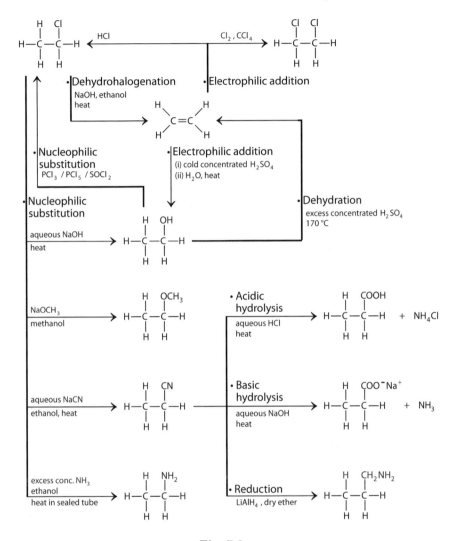

Fig. 7.2.

1. (a) Outline, by giving reagents and conditions for each step, the sequence of reactions by which you would convert a three-carbon-atom organic molecule such as 1-bromopropane into the next higher member of the same homologous series. Give a mechanistic account for the first step in the conversion process.

(b) With appropriate reagents and conditions, describe a simple chemical test to differentiate the following pair of compounds:

  (i) hexane and hex-1-ene
  (ii) 3-chloromethylbenzene and (chloromethyl)benzene
  (iii) (chloromethyl)benzene and (bromomethyl)benzene

(c) (i) 3-chloromethylbenzene can be synthesized from methyl-benzene. Give an account of the mechanism for the conversion, stating the reagents and conditions.

  (ii) (chloromethyl)benzene can also be synthesized from methylbenzene under a different set of conditions. Give an account of the mechanism for the conversion, stating the reagents and conditions.

  (iii) (chloromethyl)benzene yields a white precipitate when warmed with aqueous silver nitrate solution, whereas 3-chloromethylbenzene does not. Explain the observations.

(d) Gaseous compound **P** has a density of $46 \, \text{kg} \, \text{m}^{-3}$ at s.t.p. **P** contains only the elements carbon, hydrogen, oxygen and chlorine. On combustion, $0.158 \, \text{g}$ of **P** gave $0.225 \, \text{g}$ carbon dioxide and $0.089 \, \text{g}$ of water. On treatment with aqueous silver nitrate solution, $0.158 \, \text{g}$ of silver chloride evolved.

  (i) Determine the empirical and molecular formulae of **P**.
  (ii) Give two possible structural formulae for **P**.

(e) Explain why when chloroethane is treated with aqueous potassium hydroxide, it undergoes substitution of Cl by OH, whereas CH_3CH_2CN does not undergo replacement of CN by OH.

2. The three-dimensional structure of cyclobutane is shown below:

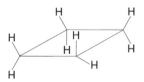

(a) Explain why the C–C bond energy for butane is greater than that for cyclobutane.

(b) Give the three-dimensional structure of the compound 1,2-dibromobutane, like for cyclobutane above.

(c) (i) Use the diagram in (a) to explain how 1, 2-dibromocyclobutane exhibits *cis-trans* (geometrical) isomerism.

   (ii) State, with reasons, whether the *cis* or *trans* isomer would have the higher boiling point.

   (iii) Both *cis-trans* (geometrical) isomers react with alcoholic KOH to give one product of the same molecular formula C_4H_4. Write balanced equations to show their formation.

(d) One of the *cis-trans* (geometrical) isomers in (b) also exhibits enantiomerism. Draw the pair of enantiomers and explain why the enantiomers have different optical activity.

CHAPTER 8

Alcohols and Phenol

8.1 Introduction

Water, H—O—H, is a molecule that is capable of forming hydrogen bonds, and the hydrogen atom is only acidic enough to react with reactive metals, such as sodium, to liberate hydrogen gas. This acid–metal reaction is a characteristic property of an acid, provided the acid is indeed acidic enough to react with the metal. What does it mean? Water is not acidic enough to react with zinc, but hydrochloric acid is. Thus, if one uses zinc metal to determine whether water is an acid or not, then water would fail the acidic test. Other characteristic properties, if they can be observed, include reaction with a carbonate/hydrogencarbonate to give off carbon dioxide gas and reaction with a base to give salt and water. Again, note that an acid may not demonstrate all three characteristics at the same time.

If one of the hydrogen atoms of the H—O—H molecule is replaced by a sp^3 hybridized carbon atom, one would get a homologous group of organic molecules known as alcohol. These compounds have physical and chemical properties similar to that of water. Alcohols, usually abbreviated as ROH, are organic compounds that contain the hydroxyl functional group, which is basically an $-OH$ group bonded to a C atom. They have the general formula $C_nH_{2n+1}OH$. The following structure shows the dipoles that are being created in an ethanol molecule due to the more electronegative oxygen atom.

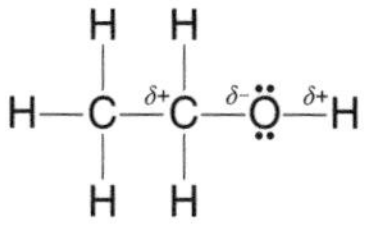

$$H-\underset{\underset{H}{|}}{\overset{\overset{H}{|}}{C}}-\underset{\underset{H}{|}}{\overset{\overset{H}{|}}{C}}\overset{\delta+}{-}\overset{\delta-}{\ddot{O}}\overset{\delta+}{-}H$$

Q: Can we replace both the hydrogen atoms of the H—O—H molecule with carbon atoms?

A: Certainly! We would get a class of compound known as ether, which has the C—O—C functional group. Ether is a functional group isomer of alcohol, and it shares the same molecular formula, $C_nH_{2n+2}O$, as alcohol. Notice that the C:H ratio of both the alcohol and ether are same as alkane, C_nH_{2n+2}? This is because the O atom in both alcohol and ether serves a bridging purpose between two atoms.

Q: Does the replacing carbon atom need to be always a sp^3 hybridized carbon atom?

A: No, it is not necessary. If the replacing carbon atom is a sp^2 hybridized carbon atom, we would get classes of compound known as phenol and carboxylic acid. Subsequently, we will be looking at these two compounds in the book.

8.2 Nomenclature

Alcohols are named with the suffix–*ol*. Other substituents are named as prefixes accompanied by the appropriate positional numbers. If the hydroxyl group is considered a substituent, then it is named as the — *hydroxy* prefix (notice that the prefix is not "hydroxyl").

Alcohols can also be classified as primary (1°), secondary (2°) or tertiary (3°) alcohols, as shown below:

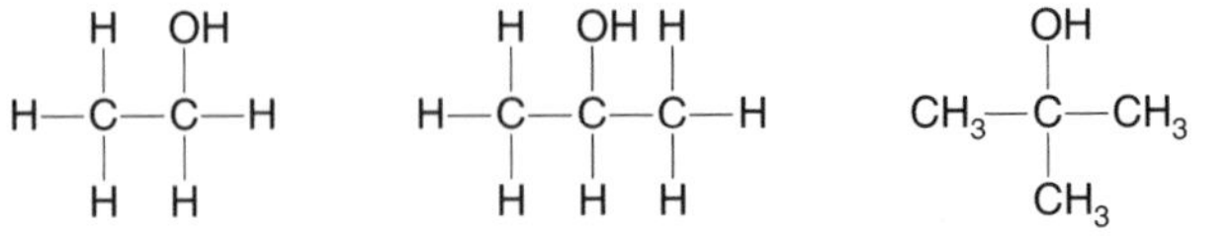

Ethanol	Propan-2-ol	2-Methylpropan-2-ol
(1° alcohol)	(2° alcohol)	(3° alcohol)

Phenol is a special name given to the benzene derivative that contains the hydroxyl group directly attached to the benzene ring. Note that all the carbon atoms in the ring are sp^2 hybridized. For substituted phenols, the substituents are named as prefixes accompanied by positional numbers. The carbon atom with the −OH group is numbered as carbon 1. For example:

Phenol   2,4,6-Tribromophenol

8.3 Physical Properties

8.3.1 *Melting and Boiling Points*

Melting and boiling points increase with increasing carbon chain length and decrease with an increase in the degree of branching. The trend can be accounted for by the differential strength of the instantaneous dipole–induced dipole (id–id) attractive forces arising due to the increase in the number of electrons that accompanies the increasing carbon chain length, rather than due to hydrogen bonding.

Table 8.1

Alcohol	Methanol	Ethanol	Propan-1-ol	Butan-1-ol
Structural formula				
Boiling point (°C)	64.7	78.4	97.1	117.2

Q: Why is the boiling point of methanol so much different from that of water?

A: A water molecule, on average, is capable of forming two hydrogen bonds per water molecule. Replacing one of the hydrogen atoms of a water molecule with a carbon atom, as in alcohol, results in only one hydrogen bond possible per alcohol molecule. In addition, due to the presence of an electron–releasing alkyl group, the oxygen atom of an alcohol molecule is less electron-withdrawing on the remaining hydrogen atom. Therefore, the electron deficiency of the hydrogen atom is less than that of a water molecule. This results in the formation of a weaker hydrogen bond.

Q: The boiling point of propan-2-ol is about 82.5°C. Why is it lower than that of propan-1-ol, which is about 97.1°C?

A: In propan-2-ol, the −OH group is more "hidden" than in propan-1-ol, or more sterically hindered towards the formation of hydrogen bond. Thus, the intermolecular forces between the propan-2-ol molecules are weaker, resulting in a lower boiling point. Note that such reasoning actually accounts for how an increase in the degree of branching affects the physical properties. But in the cases of the alkane and alkene, the effect is different. An increase in the degree of branching leads to a more spherical structure. This in turn decreases the surface area of contact between the molecules, which results in a decrease in the extensivity of intermolecular forces between the molecules. Hence, take note of the actual reasoning behind the degree of branching factor for different types of compounds.

Nonetheless, it is the stronger hydrogen bond that accounts for the higher boiling point of alcohols as compared to their polar isomeric ether (containing the C—O—C functional group) counterpart.

Example:

H—C—C—OH   H—C—O—C—H

Ethanol   Dimethyl ether
(b.p. 79 °C)   (b.p. −23 °C)

The higher boiling point of ethanol is due to the greater amount of energy needed to overcome the stronger hydrogen bonding between its molecules as compared to the permanent dipole–permanent dipole interactions between the molecules of dimethyl ether.

8.3.2   *Solubility*

The ability to form hydrogen bonds accounts for the high solubility of short-chain alcohols in water. However, as the length of the carbon chain increases, solubility of the compound in water decreases due to the increasing size of the hydrophobic alkyl chain. Yet, it is the presence of the alkyl chain that renders alcohols to be also miscible with most organic compounds.

Q: So, the low solubility of the higher molecular weight alcohol in water is due to the increasing hydrophobicity of the alkyl side-chain and not because of the inability of the higher molecular weight alcohol to form hydrogen bonds with the water molecule?

A: Now, when a higher molecular weight alcohol is amongst the water molecules, the -OH group of the alcohol molecule can still form hydrogen bonds with water molecules. So, you can assume that the amount of hydrogen bonds that are formed between a higher molecular weight alcohol molecule is the same as that of a lower molecular weight alcohol. But as a higher molecular weight alcohol is bigger, it needs more space in-between the water molecules. So, you actually need to break more hydrogen bonds in order to dissolve a higher molecular weight alcohol molecule than a lower one. Where would the energy that is required come from? In addition, the id-id interaction between the higher molecular weight alcohol molecules also needs substantial amount of energy to overcome. Thus, the low solubility of higher molecular weight alcohol is because of the insufficient amount of energy released when the alcohol molecule interacts with the water molecules. This energy is needed to overcome the hydrogen bonds between the water molecules AND the predominantly id-id interaction between the alcohol molecules.

We have seen how alcohols are used as solvents in nucleophilic reactions of halogenoalkanes (RX). For instance, in the reaction between NaCN and RX, ethanol helps to improve the miscibility and hence the mixing of the two reactants — water helps to dissolve NaCN while ethanol helps to dissolve RX, and since both ethanol and water mix well together, the solvated species are brought into close proximity to each other.

Q: How could we visualize ethanol helping to improve the miscibility between NaCN and RX?

A: Imagine the ethyl group of the ethanol molecules interacting with the less polar RX molecule. This would result in exposing the more polar $-OH$ groups outward. The exposed $-OH$ groups would now able to interact with the hydrated ions via hydrogen bonding. This would thus bring the different species closer to each other.

Q: Since short-chain alcohols such as methanol and ethanol are miscible with water, coupled with the fact that they can form hydrogen bonds, can they dissolve ionic compounds too?

A: Yes, in fact, short-chain alcohol is capable of dissolving ionic compounds. When an ionic compound is mixed with water, two possible processes can occur, depending on the nature of the ions: (1) it dissolves in water (hydration) and/or (2) it reacts with water (hydrolysis).

For instance, when NaCl(s) is added to water, the Na^+ and Cl^- ions become surrounded by water molecules. We say that hydration occurs when these ions are attracted to the surrounding water molecules through ion–dipole interactions. Similarly, if methanol or ethanol is used as the solvent, the ions would also be attracted to the solvent molecules via ion–dipole interactions. Such process is termed — solvation.

$$NaCl(s) + aq \longrightarrow Na^+(aq) + Cl^-(aq).$$

The process of an ionic solid dissolving is analogous to the peeling of an onion. The positive end of the dipole of a water molecule would be attracted to the anion on the surface of the ionic solid, and *vice-versa* for the negative end of the dipole towards the cations. The formation of the various ion–dipole interactions releases energy. The energy released is transferred to the cations and anions, which thus increases the vibrational energy of these ions. As more ion–dipole interactions form, more energy is released, and the greater amount of vibrational energy would enable the ions to be freed from the lattice.

Q: Why would energy be evolved when a bond forms? Where does this energy come from?

A: Imagine — before the formation of a bond, the water molecules would mainly have kinetic energy only. When ion–dipole bonds are formed between water molecules and the ions on the surface of the ionic solid, the speed of the water molecules slows down. So, where does the decrease in kinetic energy of the water molecules go? The decrease in kinetic energy goes to the vibrational kinetic energy of the solid particles.

Subsequently, after the ions are freed from the lattice, hydrogen bonds would be formed between subsequent layers of water molecules. The greater the charge density of the ions, the more layers of water molecules would be attracted to each other. Hence, the hydration process would make the hydrated ions very bulky in nature. But in the case of methanol or ethanol as a solvent, the solvation sphere is not as big as in the case of water due to weaker van der Waals forces acting between the alkyl groups. This limits the amount of energy released during the solvation process, therefore affecting the solubility. Thus, the solubility of an ionic compound in ethanol is expected to be lower than if water is the solvent used.

In some cases, if we mix certain ionic salts such as Na_2O with water, not only does hydration occur, we have hydrolysis too, i.e. the water molecule is "broken up," as shown in the following reaction:

$$Na_2O(s) + H_2O(l) \longrightarrow 2NaOH(aq).$$

8.4 Hydrolysis and Acidity

When Na_2O, is added to water, the highly electron-rich O^{2-} ion has such a strong affinity for H^+ that it actually extracts a proton from a H_2O molecule:

$$O^{2-} + H_2O \longrightarrow 2OH^-$$

As for the cation, Na^+, its charge density is too low to break up water molecules. Hence, it does not undergo hydrolysis. In order to achieve this, a cation must have high charge density, as it will then be able to distort the electron cloud of the surrounding H_2O molecules and weaken the O–H bonds to the extent that they cleave. Cations considered as having high charge density are typically small and highly charged, such as Fe^{3+}, Al^{3+} and Cr^{3+}. Most doubly charged transition metal cations are also mildly acidic in water due to their small cationic sizes.

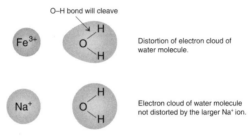

From another perspective, the hydrolysis reaction between the oxide ion and water can also be considered as an acid–base reaction wherein O^{2-} acts as the base and H_2O the acid.

Q: But isn't water neutral? How can we call it an acid?

A: Water is neutral when the pH is measured with respect to itself as the autoionization of water ($H_2O(l) + H_2O(l) \rightleftharpoons OH^-(aq) + H_3O^+(aq)$) produces the same amount of H_3O^+ as OH^-. There are three important characteristics that an acid may display:

- Acid reacts with metal to give off hydrogen gas.
- Acid reacts with carbonate/hydrogencarbonate to give off carbon dioxide gas.
- Acid reacts with base to give salt and water.

Note that each of the above characteristics is a particular reaction of an acid with respect to substances such as a metal, a carbonate or a base. And it is not necessary for all acids to display all the above three properties. Water is considered an acid because it can react with sodium metal to give off hydrogen gas even though it does not react with carbonates or bases. The other important thing to take note of is that what characterizes an acid is not whether it has a pH that is less than 7. Rather, does it demonstrate one of the above characteristics. The pH scale is a measurement of the level of acidity, not a definition of an acid.

So, based on the above characteristics, there are not one but three common definitions for acids and bases:

- the Arrhenius theory of acids and bases,
- the Brønsted-Lowry theory of acids and bases and
- the Lewis theory of acids and bases

The narrowest of the three definitions is the Arrhenius theory of acids and bases, which defines an acid to be a substance that releases H^+ ions when dissolved in water (e.g. HCl) and a base as a substance that releases OH^- ions in the presence of water (e.g. NaOH). There are a few limitations to the Arrhenius theory. For one, there are some acid–base reactions that do not occur in water but in other mediums. For instance, the reaction between gaseous hydrogen chloride and ammonia can be considered an acid–base reaction as it produces a salt. However, NH_3 is not considered a base if we apply the Arrhenius theory since it does not contain the OH^- ion, but it does behave as a base.

An alternative definition for acids and bases is offered under the Brønsted–Lowry theory. Based on the theory, **an acid is a proton donor and a base is a proton acceptor**. Under this theory, the basic properties of substances such as NH_3 can be accounted for; NH_3 is a base which accepts a proton from HCl to form NH_4^+ in the acid–base reaction.

To be considered a proton donor, a substance must have a hydrogen atom that can be lost.

To be considered a proton acceptor, a substance must have a lone pair of electrons to form a dative covalent bond with the proton.

A substance can be either a base or an acid depending on what it reacts with. For instance, H_2O functions as the base in the presence of HCl but behaves as an acid when in the presence of O^{2-}. Such a substance is known as an amphiprotic compound — referring to a substance that can be both a proton donor and proton acceptor.

$$HCl(aq) \quad\quad + H_2O(l) \longrightarrow Cl^-(aq) + H_3O^+(aq),$$
acid   *base*
proton donor   proton acceptor

$$H_2O(l) \quad\quad + O^{2-}(aq) \longrightarrow 2OH^-(aq).$$
acid   *base*
proton donor   proton acceptor

Note that in reality, hydrogen ions do not exist in solution. The H^+ ion has such a high charge density that it is actually bonded to at least one water molecule when in aqueous solution, i.e. H^+ binds with H_2O to form H_3O^+ (hydronium ion).

It is common to find the symbol "$H^+(aq)$" used for simplicity's sake in many texts, including this one. It is alright to use it, but we must bear in mind that when we write "$H^+(aq)$", we are actually referring to "$H_3O^+(aq)$."

A further generalisation of the acid–base definition involves the perspective of electron flow to account for how bases actually accept protons. According to the Lewis theory of acids and bases, NH_3 accepts a proton by actually donating a pair of electrons to it, and as a result, a dative covalent bond is formed.

A Lewis base is an electron-pair donor, and a Lewis acid is a substance that accepts a pair of electrons from a base, hence, it is an electron-pair acceptor. A dative covalent bond results from the sharing of the pair of electrons.

In general, the Lewis theory of acids and bases is the broadest of the three theories since it includes all the possible acids and bases ascribed by the other two theories. In addition, the Lewis definition allows us to understand the mechanism of an acid–base reaction, all from the perspective of electron moving from an electron-rich center to an electron–deficient one. Of course, this movement of electrons is only possible if there is a significant difference in the electron density between the two centers.

Alcohols are also considered acids, albeit weaker acids than water. This conclusion is supported by the fact that they react with sodium, $Na(s)$, to give off hydrogen gas. However, being a weak acid, they do not react with carbonates or bases.

Do you know?

- A **strong** acid **completely dissociates** in aqueous solution.
  1 mol of HCl will provide 1 mol of H_3O^+ and 1 mol of Cl^- upon dissociation in water, i.e. $[HCl] = [H_3O^+]$,

$$HCl(aq) + H_2O(l) \longrightarrow H_3O^+(aq) + Cl^-(aq).$$

  Strong acids include inorganic acids such as HCl, HBr, HI and HNO_3.

- A **weak** acid undergoes **partial dissociation** in aqueous solution (represented by the reversible arrow '$\rightleftharpoons$' in the equation below). The reversible arrow in the chemical equation indicates two important concepts: (i) the reaction is incomplete, and (ii) the system is in a dynamic equilibrium.

  1 mol of CH_3COOH will provide less than 1 mole of H_3O^+ and CH_3COO^- each upon partial dissociation in water, i.e. $[CH_3COOH]_{before\ dissociation} > [H_3O^+]_{at\ equilibrium}$ but $[H_3O^+]_{at\ equilibrium} = [CH_3COO^-]_{at\ equilibrium}$. The latter relationship is true for a pure weak acid in water alone.

$$CH_3COOH(aq) + H_2O(l) \rightleftharpoons CH_3COO^-(aq) + H_3O^+(aq).$$

  We did not use alcohol as an example here to discuss the dissociation simply because alcohol is a weaker acid than water. The dissociation reaction equation would be a misrepresentation of a dissociation that is insignificant in water.

Q: What determines the strength of an acid?

A: The strength of an acid depends on the extent of its dissociation in aqueous solution and this extent of dissociation is indicated by the acid dissociation constant, K_a. Consider the partial dissociation of a weak acid, CH_3COOH:

$$CH_3COOH(aq) + H_2O(l) \rightleftharpoons CH_3COO^-(aq) + H_3O^+(aq),$$

Acid dissociation constant, $K_a = \dfrac{[CH_3COO^-(aq)][H_3O^+(aq)]}{[CH_3COOH(aq)]}$.

The magnitude of K_a is fixed at a particular temperature and is independent of the concentration of the weak acid. The greater the magnitude of the K_a, the higher would be the concentration of the H_3O^+ ions at equilibrium. Although the magnitude of K_a serves to measure the strength of acids, it is more convenient to use the corresponding pK_a value for comparison. The relationship between K_a and pK_a is given as follows:

$$\mathbf{p}K_a = -\mathbf{log}\ K_a$$

The smaller the pK_a value, the stronger the acid.

Q: Can we associate a pK_b for a weak acid such as CH_3COOH?

A: Yes, certainly. The $-COOH$ group of CH_3COOH does have a pK_b value too, as there is lone pair of electrons ready to accept protons, just like H_2O is amphiprotic and has both pK_a and pK_b values. But take note that if the pK_a value of a weak acid is small (the smaller the pK_a, the stronger the weak acid), its pK_b value would be greater. This is logical as if the pK_b value of the weak acid is smaller than its pK_a value, then it wouldn't be an acid, right? It should be acting as a base instead! In addition, the $pK_a + pK_b \neq 14$ for an amphiprotic species because the conjugate base (CH_3COO^-) of the weak acid (CH_3COOH) is not the same as the conjugate acid $(CH_3COOH_2^+)$ of the weak acid when it is acting as a base (CH_3COOH).

Q: Why would K_a be independent of the concentration of the acid? Shouldn't a higher concentration of the acid lead to a greater amount of dissociation?

A: Higher concentration of the acid does lead to a greater amount of dissociation, thus more $CH_3COO^-(aq)$ and $H_3O^+(aq)$ at equilibrium. But at the same time, the higher concentrations of $CH_3COO^-(aq)$ and $H_3O^+(aq)$ at equilibrium would also lead to a higher rate for the backward reaction. Therefore, the increase in concentration terms for the numerator and denominator of the K_a expression would cancel each other out, resulting in a constant K_a.

Q: So, is there a way to predict the initial pH of a weak acid given a specific initial concentration of c mol dm^{-3}?

A: Yes, the following derivation would help:

$$CH_3COOH(aq) + H_2O(l) \rightleftharpoons CH_3COO^-(aq) + H_3O^+(aq)$$

Initial conc./mol dm^{-3} c 0   0

At equilibrium $c - x$ x x

$$K_a = \frac{[CH_3COO^-(aq)][H_3O^+(aq)]}{[CH_3COOH(aq)]} = \frac{x.x}{(c-x)}.$$

Rearrangement gives $[H_3O^+] = \sqrt{K_a(c-x)} \Rightarrow$ pH $= \frac{pK_a}{2} - \frac{1}{2}\log(c-x)$.

Thus, if the initial concentration of the weak acid is 1 mol dm^{-3} and $x \approx 0$ such that $c - x = 1$, then

$$pH \approx \frac{pK_a}{2}.$$

This is a very important relationship because if you are given the K_a value of a weak acid, you can roughly estimate the initial pH of the acid solution and hence predict the position of the equilibrium of dissociation at a specific pH. For example, if the pK_a of a weak acid is 5, then its initial pH is roughly 2.5 for an initial concentration of 1 mol dm^{-3}. Which means that at a pH of 4 (less acidic than the initial pH or more alkaline than the initial pH), the position of dissociation is more to the right, forming more of the conjugate base. In addition, the smaller the pK_a of a weak acid, the lower its initial pH as compared to another weak acid of the same concentration but is of higher pK_a value.

Although a weak acid does not fully dissociate in water, the extent of dissociation increases with dilution. Nevertheless, the amount of base required to completely neutralize the weak acid is the same as that needed to neutralize a strong acid of the same concentration. This is because as the base that is added reacts with the H_3O^+, the decrease in the concentration of the H_3O^+ ion would result in an increase of the dissociation of the weak acid molecules, shifting the position of the equilibrium towards the right in accordance to the prediction by Le Chatelier's principle. The dissociation would only stop when all the weak acid molecules have been reacted.

Q: Since the dissociation of the weak acid molecules increases as dilution is being carried out, this would lead to a greater amount of H_3O^+ being produced. Does it mean that the pH of the diluted weak acid would decrease (i.e. the solution becomes more acidic) with dilution?

A: The amount of H_3O^+ does increase with dilution because of greater degree of dissociation. This is because once the acid molecule dissociates, the ions would be separated, and this separation is made more pronounced when the solution is more diluted. But at the same time, the volume of the solution also increases. Hence, the concentration of the H_3O^+, which is the ratio of the amount of H_3O^+ (in moles) over the volume of the solution, actually decreases. This causes an increase in pH (i.e. the solution becomes less acidic), since $pH = -\log[H_3O^+]$.

Q: Does this mean that we can't use pH values to deduce the strength of two weak acids?

A: Actually, you can, provided the initial concentrations of the two weak acids are the same. With the same initial concentration, the weak acid that gives a lower pH, i.e. a higher concentration of H_3O^+, would be the stronger of the two weak acids.

Q: The pH of the weak acid increases with dilution. Is this phenomenon similar to the increase in pH when diluting a strong acid, such as HCl?

A: There is a difference here! For a strong acid such as HCl, it fully dissociates, which means that the amount of H_3O^+ (in moles) is a constant with respect to dilution. With dilution, more water is added. The increase in the volume of water causes $[H_3O^+]$ to decrease for the strong acid, hence an increase in pH. But the rate of increase of the pH for the dilution of a strong acid is greater than that for diluting a weak acid. This is because for the dilution of weak acid, both the amount of H_3O^+ and volume increase at the same time, but at different rates. Hence, overall, the increase in pH for a weak acid with dilution is not as drastic as that for a strong acid.

In addition to alcohols, weak organic acids also include the carboxylic acids (RCOOH) and phenols. The relative acid strength for these organic compounds when compared to water is shown below:

Increasing acid strength: $ROH < H_2O <$ Phenol $< RCOOH$.

The relative acidity of these compounds can be observed from their reactions or non–reactions with the following bases and reactive metal:

✔

	Na	NaOH	Na_2CO_3 or $NaHCO_3$
RCOOH	☑ $H_2(g)$ + salt	☑ H_2O + salt	☑ $CO_2(g)$ + H_2O + salt
Phenol	☑ $H_2(g)$ + salt	☑ H_2O + salt	—
water	☑ $H_2(g)$ + salt	—	—
ROH	☑ $H_2(g)$ + salt	—	—

Comparing NaOH and Na_2CO_3, NaOH is a stronger base while Na_2CO_3 is a weaker one. Thus, the fact that only carboxylic acid can react with Na_2CO_3 indicates that it is the strongest acid among the above. Hence, we can use Na_2CO_3 to differentiate the above acid.

Q: How does one know that NaOH is a stronger base than Na_2CO_3?

A: When NaOH is dissolved in water, full dissociation takes place. But when Na_2CO_3 is dissolved in water, the following hydrolysis takes place:
$$CO_3^{2-}(aq) + H_2O(l) \rightleftharpoons HCO_3^-(aq) + OH^-(aq).$$
The concentration of OH^- generated is lower than when dissolving NaOH. This makes Na_2CO_3 a weaker base than NaOH.

Q: But the CO_3^{2-} is doubly charged, shouldn't it be more likely to extract a H^+ from a H_2O molecule?

A: Due to resonance, the two negative charges are actually delocalized over three oxygen atoms. As a result, the average negative charge per oxygen atom is about two–third, which is less than a full negative charge, hence accounting for the low ability of the CO_3^{2-} ion to extract a H^+ from a H_2O molecule.

Q: Can I therefore say that the conjugate acid, HCO_3^-, is thus a weaker acid than the conjugate acid of OH^-, which is H_2O?

A: Certainly you can, and it is easy to understand why. This is because it is more difficult to remove a H^+ ion from the already negatively charged HCO_3^- than from a neutral H_2O species. Note that the removal of the H^+ is facilitated by a lone pair of electrons from a base (from Lewis theory of acid–base). The inter–electronic repulsion between the lone pair of electrons from the base and the negatively charged HCO_3^- would make the removal of H^+ more difficult.

Q: Can I say that $Cu(OH)_2$ is a stronger base than Na_2CO_3?

A: No, you can't. This is because $Cu(OH)_2$ is insoluble in water, thus the comparison is an invalid one. In order to determine the strength of the acid or base, solubility is a prerequisite.

On the other hand, both water and alcohols are so weakly acidic that they can only react with the very reactive sodium metal. Note that since alcohol is a weaker acid than water, the alcohol molecule is not acidic enough to protonate the water molecules to cause an increase in the concentration of H_3O^+ ions. Thus, an alcohol-water mixture would still have a neutral pH. The relative acidity of these organic compounds can be accounted for by looking at the relative stability of their conjugate bases. A more stable conjugate base will remain in its present state, with lower tendency to react with H_3O^+ to form back its conjugate acid (the weak organic acid).

Q: As the alcohol is not acidic enough to protonate the water molecules, then the water molecule should be acidic enough to protonate the alcohol molecule. Shouldn't this cause an increase in the concentration of the OH^-, and thus a change in pH?

A: Yes, indeed. Theoretically, the water molecule is acidic enough to protonate the alcohol molecule. In addition, the lone pair of electrons on the oxygen atom of the alcohol molecule would be more available for protonation than the water molecule itself. This is due to the intensification of electron density by the electron-releasing alkyl group. Still, all this would not cause a change in the pH because the protonation of alcohol by water molecules is rather insignificant due to the similarity in the basic strength of these two molecules. You simply can't detect the difference, if there is any.

Given the reversible reaction:

$$CH_3COOH(aq) + H_2O(l) \rightleftharpoons CH_3COO^-(aq) + H_3O^+(aq)$$
$$\text{acid} \qquad\qquad \text{base} \qquad\quad \text{conjugate base} \quad \text{conjugate acid}$$

- The acid, CH_3COOH, donates a proton, leaving behind CH_3COO^-.
  CH_3COO^- is known as the conjugate base of the acid, CH_3COOH.
  In the backward reaction, CH_3COO^- acts as a base, accepting a proton from H_3O^+ to form CH_3COOH. The CH_3COOH is known as the conjugate acid of the base, CH_3COO^-.
- The base, H_2O, accepts the proton, forming H_3O^+.
  H_3O^+ is known as the conjugate acid of the base, H_2O.
  In the backward reaction, H_3O^+ acts as an acid, donating a proton to CH_3COO^- and forming H_2O. The H_2O is known as the conjugate base of the acid, H_3O^+.

Hence, a conjugate acid–base pair differs by only a H^+ ion. All weak acids have a conjugate base, as shown in the respective equations below:

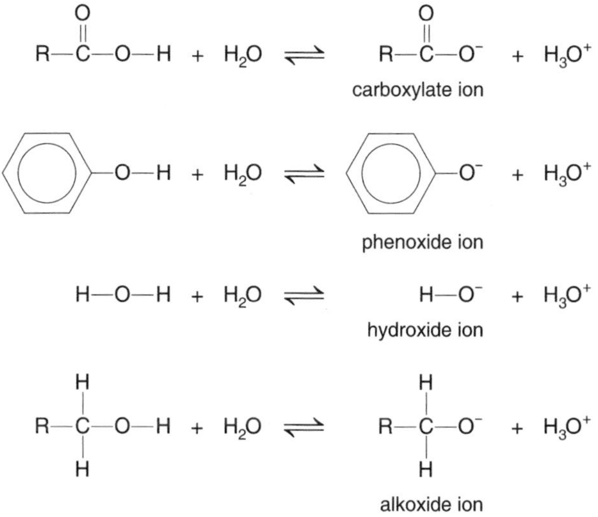

carboxylate ion

phenoxide ion

hydroxide ion

alkoxide ion

If we acknowledge that a strong acid is one that completely dissociates in water, and that $[H_3O^+]_{\text{after dissociation}} = [\text{strong acid}]_{\text{initial}}$, then when comparing the strength of the two weak acids with the same concentration value, the one that can dissociate to a greater extent will be the stronger of the weak acids.

Q: How can we link the degree of dissociation to the strength of weak acids?

A: The degree of dissociation, denoted by α, is the ratio of the concentration of the conjugate base formed over the initial concentration of the acid molecule. For example, $\alpha =$ [CH$_3$COO$^-$]$_{\text{at equilibrium}}$/[CH$_3$COOH]$_{\text{initial}}$. As a weak acid only dissociates partially, the reversibility of the reaction means that the backward reaction does occur, i.e. the conjugate base "extracts" a proton from another acid and forms back the original weak acid. If there is lower tendency for the conjugate base to form back the original weak acid, then it goes to say that the backward reaction occurs to a smaller extent than the forward reaction. This would mean that the dissociation of the weak acid would have proceeded to a greater extent — which is the deciding factor of acid strength. But note that using α to compare acid strength is only valid when the initial concentrations of two weak acids are the same!

Q: Can we use $\alpha =$ [H$_3$O$^+$]$_{\text{at equilibrium}}$/[CH$_3$COOH]$_{\text{initial}}$ to denote the degree of dissociation of the weak acid instead?

A: No, it may not be appropriate. It is better to use the concentration of the conjugate base, especially if the weak acid is being dissolved in the presence of a strong acid. This is because if we dissolve a weak acid in the presence of a strong acid, the high concentration of the H$_3$O$^+$ from the strong acid would suppress the dissociation of the weak acid. This is known as the common ion effect. The overall [H$_3$O$^+$]$_{\text{at equilibrium}}$ is the sum of the H$_3$O$^+$ from both the strong and weak acid. Similarly, $\alpha =$ [CH$_3$COO$^-$]$_{\text{at equilibrium}}$/[CH$_3$COOH]$_{\text{initial}}$ is inappropriate for denoting the degree of dissociation of the weak acid in the presence of high concentration of the conjugate base from another source.

Thus, if the acid strength increases in the order: ROH $<$ H$_2$O $<$ Phenol $<$ RCOOH, then the strength of the conjugate base would decrease in the order: RO$^-$ $>$ OH$^-$ $>$ C$_6$H$_5$O$^-$ $>$ RCOO$^-$. As a result, the stability of the conjugate base should increase in the order: RO$^-$ $<$ OH$^-$ $<$ C$_6$H$_5$O$^-$ $<$ RCOO$^-$.

Q: What affects the stability of the conjugate bases of the above weak acids?

A: Remember, for a charged species, its stability depends on the extent of charge dispersal. For these anionic species, if the negative charge can be made less concentrated on the oxygen atom bearing it, there will be less tendency for it to be attacked by H_3O^+, rendering it more stable. We would then expect that electron-withdrawing groups could help to disperse this negative charge on the O atom and thus make the anion less basic and hence more stable. On the other hand, electron-donating groups would tend to destabilize the anion as these groups could increase the electron density on the O atom, making it more susceptible to attack by H_3O^+.

Looking at the structures of the conjugate bases of the weak acids below, we will find that the alkoxide ion is the most unstable due to the electron-donating alkyl group being directly attached to the O atom.

alkoxide ion   hydroxide ion   phenoxide ion   carboxylate ion

But for both the phenoxide and carboxylate ions, these species are resonance stabilized, as the p orbital of the O atom, which bears the negative charge, is able to interact with the π electron cloud of the benzene ring and that of the carbonyl group respectively. This would lead to a decrease in charge density on this O atom (see Fig. 8.1). As for the hydroxide ion, neither stabilizing nor destabilizing effect is present, thus accounting for its position in the relative stability trend.

Q: If there is delocalization of electrons over a greater network of atoms, shouldn't this make the phenoxide ion more stable than

Fig. 8.1. (left) The extensive network of delocalized π-electrons over seven atoms in the phenoxide ion. (right) The delocalization of π-electrons over the carbonyl group and the O atom bearing the negative charge.

the carboxylate ion and hence make phenol a stronger acid than RCOOH?

A: The carboxylate ion is more resonance-stabilized than the phenoxide because the negative charge is dispersed over two highly electronegative O atoms, which exert a stronger electron-withdrawing effect. In the case of the phenoxide, the negative charge is dispersed over six carbon atoms which have lower electronegativity than the O atom.

The above five resonance structures are non-equivalent. The resultant resonance hybrid has the greatest contribution from the first structure. This would mean that the negative charge would be concentrating more on the highly electronegative oxygen atom most of the time than on the six carbon atoms.

For the carboxylate ion, on the other hand, the above two resonance structures are equivalent. The negative charge would be evenly distributed among the two highly electronegative oxygen atoms.

Q: How does the acid strength compare for the following substituted phenols?

A: The CH$_3$– group is ring activating (i.e. making the benzene ring more electron-rich as compared to an unsubstituted benzene molecule), while the Br– group is ring deactivating; as a result 4-methylphenol is less acidic than 4-bromophenol. The electron-donating methyl group actually hinders the delocalization of the negative charge on the O atom into the benzene ring of the 4-methylphenoxide ion. This makes it less stable than the

4-bromophenoxide ion. Thus, 4-methylphenol dissociates to a lesser extent, rendering it a weaker acid.

Q: The more stable the conjugate base, the more acidic would be the acid. How can the stability of the product (conjugate base) affect the dissociation? Isn't this like telling people that an outcome which has yet to take place can affect the process of it occurring?

A: You are right that an outcome that is yet to take place cannot affect the process of it occuring. Let us use 4-methylphenol and 4-bromophenol to try to rationalize the fallacy here. Due to the electron-donating effect of the CH_3- group, the benzene ring is more electron-rich. As a result of the electron-rich benzene ring, the highly electronegative O atom has an "alternative" source of electrons to distort from rather than to just distort it from the H atom. Thus, the electron-deficiency of the H atom is much lower than that of phenol. This would make the H atom of 4-methylphenol less susceptible to extraction by a base, hence less acidic. As for 4-bromophenol, the reverse is true. The less electron-rich benzene ring would make the highly electronegative O atom "bully" the H atom more than that in phenol. The higher electron deficiency of the H atom of 4-bromophenol would then be more acidic in nature. Therefore, note that although we have been constantly using the stability of the conjugate base to account for the acidity, the reality is that the strength of the acidity actually lies on the level of electron deficiency in the H atom.

Q: How then do you explain the acidic trend using the electron deficiency of the H atom?

A: It is not difficult. Alcohol is less acidic than water because the electron-donating alkyl group intensifies the electron density on the O atom, making it less polarizing, thus resulting in a smaller degree of electron deficiency on the H atom. Phenol is more acidic than water because the lone pair of electrons on the O atom can delocalize into the benzene ring; this makes the O atom more polarizing than the O atom of water, which means greater electron deficiency on the H atom of phenol. As for carboxylic acid, the lone pair of electrons on the O atom of the –OH group can delocalize onto another highly electronegative O atom. This causes the O atom of the –OH group to be more polarizing as compared to phenol, thus making carboxylic acid a stronger acid than phenol.

We have seen how the nature of a substituent affects the stability of carbocations (see Chapter 5 on Alkenes and Chapter 6 on Arenes). Likewise, the nature of a substituent can also affects the stability of the anions but in the reverse order. Take, for instance, the same substituents, the ethyl group and the Br atom, found on a carbocation and an alkoxide ion:

- The electron-donating ethyl group *stabilizes* the carbocation by *dispersing* the positive charge on it. On the other hand, the ethyl group *destabilizes* the alkoxide ion by *intensifying* the negative charge on the O atom.

-

$$
\begin{array}{cc}
CH_2CH_3 & CH_2CH_3 \\
\downarrow & \downarrow \\
H-C^+ & O^- \\
| & \\
H &
\end{array}
$$

As a result of this, we would expect a tertiary alcohol to be less acidic than a secondary alcohol, which would in turn be less acidic than a primary alcohol — the more alkyl groups that are present, the greater would be the intensification of electron density on the O atom, thus the greater the destabilization of the conjugate base.

- The electron-withdrawing Br atom *destabilizes* the carbocation by *intensifying* the positive charge on it. On the other hand, the Br atom *stabilizes* the anion by *dispersing* the negative charge on the O atom.

-

$$
\begin{array}{cc}
Br & Br \\
\uparrow & \uparrow \\
H-C^+ & H-C-O^- \\
| & | \\
H & H
\end{array}
$$

Q: How does the acid strength compare for the following carboxylic acids that have the same substituents?

$$
\begin{array}{cccc}
H & H & Br & O \\
| & | & | & \| \\
H-C-C-C-C-OH \\
| & | & | \\
H & H & H
\end{array}
\qquad
\begin{array}{cccc}
Br & H & H & O \\
| & | & | & \| \\
H-C-C-C-C-OH \\
| & | & | \\
H & H & H
\end{array}
$$

2-Bromobutanoic acid   4-Bromobutanoic acid

A: The point to note here is that the electron-withdrawing effect of the Br atom is influenced by its proximity to the electron-rich $-COO^-$ group. Since the Br group is nearer to the $-COO^-$ group for 2-bromobutanoic acid than for 4-bromobutanoic acid, the electron density on the $-COO^-$ group of 2-bromobutanoic acid would be more dispersed. Thus, the conjugate base of 2-bromobutanoic acid would be more stabilized.

Q: Phenol, with the hydroxyl group directly attached to the benzene ring, is more acidic than ethanol. Is benzoic acid more acidic than ethanoic acid?

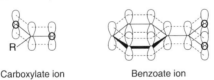

Ethanoic acid   Benzoic acid

A: The benzoate ion is more electronically stabilized than a carboxylate ion due to a greater surface area for charge dispersal (Fig. 8.2). Hence, we should expect benzoic acid to be a stronger acid than ethanoic acid. Yes! The pK_a values indicate so — ethanoic acid (4.75) and benzoic acid (4.20). The higher the pK_a value, the weaker the acid.

Carboxylate ion   Benzoate ion

Fig. 8.2. (left) The delocalization of π-electrons over three atoms in the carboxylate ion. (right) The extensive network of delocalized π-electrons over nine atoms in the benzoate ion.

Q: But wouldn't the benzoic acid be less soluble than the ethanoic acid, hence making benzoic acid less acidic?

A: Yes, the more hydrophobic benzene ring would limit the solubility of benzoic acid in water. But it seems that this does not lower its acidity.

Moving back to our discussion on hydrolysis, organic compounds such as esters and amides undergo hydrolysis whereby their molecules are split into two parts by the addition of a water molecule. One part of the molecule gains a proton (H^+), and the other fragment gains a hydroxide ion (OH^-). But in the hydrolysis of a nitrile, the nitrile is not split into two different parts. In fact, the nitrile is first converted to an amide, and then the amide is further converted to a carboxylic acid.

Based on the hydrolysis of the amide, the following can be thought to occur in a concerted move:

- the C–N bond of the amide functional group undergoes heterolytic cleavage with the N atom acquiring both of the bonding electrons.
- the fragment containing the carbonyl $(-C=O)$ group gains a $-OH$ group, forming a carboxylic acid.
- the other fragment, containing the amino $(-NH_2)$ group, gains a hydrogen atom, forming NH_3.

Acids or bases are added as catalysts to speed up the hydrolysis process of the ester, amide and nitrile.

- If it is an acid-catalyzed hydrolysis, the presence of acid will cause products that are basic (e.g. NH_3) to be protonated:

$$\underset{\text{acid}}{H^+} + \underset{\text{base}}{NH_3} \longrightarrow NH_4^+ .$$

- If it is a base-catalyzed hydrolysis, the presence of the base will cause products that are acidic (e.g. $RCOOH$) to be deprotonated:

$$\underset{\text{acid base}}{RCOOH + OH^-} \longrightarrow \underset{\text{salt}}{RCOO^-} + \underset{\text{water}}{H_2O} .$$

8.5 Preparation Methods for Alcohols

8.5.1 *Electrophilic Addition of Alkenes*

	For Step (i),		For Step (ii),
Reagents:	Concentrated H_2SO_4	Reagents:	Water
Conditions:	Cold temperature	Conditions:	Heat
Options:	—	Options:	—

8.5.2 *Nucleophilic Substitution of Halogenoalkanes*

$$H-\overset{\overset{\displaystyle H}{|}}{\underset{\underset{\displaystyle H}{|}}{C}}-\overset{\overset{\displaystyle Cl}{|}}{\underset{\underset{\displaystyle H}{|}}{C}}-H \;+\; OH^- \xrightarrow{\text{heat}} H-\overset{\overset{\displaystyle H}{|}}{\underset{\underset{\displaystyle H}{|}}{C}}-\overset{\overset{\displaystyle OH}{|}}{\underset{\underset{\displaystyle H}{|}}{C}}-H \;+\; Cl^-$$

Reagents:	NaOH(aq)
Conditions:	Heat
Options:	KOH(aq)

8.5.3 *Reduction of Carboxylic Acids and Carbonyl Compounds*

Carboxylic acids and aldehydes are reduced to primary alcohols.

$$R-\overset{\overset{\displaystyle O}{\|}}{C}-OH \;+\; 4[H] \xrightarrow[\text{dry ether}]{LiAlH_4} R-\overset{\overset{\displaystyle OH}{|}}{\underset{\underset{\displaystyle H}{|}}{C}}-H \;+\; H_2O$$

$$R-\overset{\overset{\displaystyle O}{\|}}{C}-H \;+\; 2[H] \xrightarrow[\text{dry ether}]{LiAlH_4} R-\overset{\overset{\displaystyle OH}{|}}{\underset{\underset{\displaystyle H}{|}}{C}}-H$$

Ketones are reduced to secondary alcohols.

$$R-\overset{\overset{\displaystyle O}{\|}}{C}-R \;+\; 2[H] \xrightarrow[\text{dry ether}]{LiAlH_4} R-\overset{\overset{\displaystyle OH}{|}}{\underset{\underset{\displaystyle H}{|}}{C}}-R$$

	For Step (i),		For Step (ii),
Reagents:	LiAlH$_4$ in dry ether	Reagents:	Water
Conditions:	Room temperature	Conditions:	Heat
Options:	NaBH$_4$(aq)* or H$_2$/Pt/Heat$^\#$	Options:	—

*Being a milder reducing agent than LiAlH$_4$, NaBH$_4$(aq) is not able to reduce RCOOH and its derivatives.

$^\#$H$_2$/Pt/Heat cannot be used to reduce carboxylic acid, thus catalytic hydrogenation can be used to reduce carbon–carbon unsaturated bonds with the acid functional group present.

Refer to Chapter 9 on Carbonyl Compounds for further information on the reduction mechanism.

Q: Can a tertiary alcohol be obtained through similar reduction reactions?

A: No. The reason is that aldehyde, ketone and carboxylic acid can be obtained by oxidizing the respective alcohol — aldehyde and carboxylic acid can be obtained by oxidation of primary alcohol while ketone is from secondary alcohol. Tertiary alcohol cannot be oxidized. As reduction is simply the reversal of oxidation, you cannot obtain a tertiary alcohol from reduction.

8.6   Chemical Properties of Alcohols

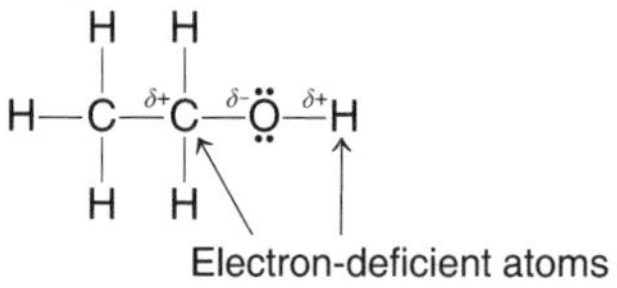

Electron-deficient atoms

Alcohols are polar molecules with not one but two polar bonds, the C–O bond and the O–H bond. With the O atom being the most electronegative, both the C atom and H atom acquire a partial positive charge. Consequently, nucleophiles are attracted to the electron-deficient carbon atom while a base would be attracted towards the electron-deficient hydrogen atom. Hence, the reactions of alcohols involve the cleaving of either the C–O bond or the O–H bond.

The following reactions involve the cleaving of the O–H bond:

• Acid-base reaction with strong bases and reactive metals

Reactions involving the cleaving of the C–O bond are:

• Esterification
• Acylation
• Dehydration
• Halogenation

8.6.1   *Acid-Base Reaction with Strong Bases and Reactive Metals*

Being weak acids, alcohols can only react with strong bases or reactive metals. The reaction of an alcohol with the strong base NaH or reactive metals such as Na and K will yield the alkoxide ion along with evolution of $H_2(g)$. One can thus determine for the presence of alcohol by testing the H_2 gas that is evolved with a lighted splint — a pop sound will be heard.

$$H\text{—}\underset{\underset{H}{|}}{\overset{\overset{H}{|}}{C}}\text{—}\underset{\underset{H}{|}}{\overset{\overset{H}{|}}{C}}\text{—}O\text{—}H \;+\; Na \longrightarrow H\text{—}\underset{\underset{H}{|}}{\overset{\overset{H}{|}}{C}}\text{—}\underset{\underset{H}{|}}{\overset{\overset{H}{|}}{C}}\text{—}O^- Na^+ \;+\; \tfrac{1}{2}H_2$$

$$H\text{—}\underset{\underset{H}{|}}{\overset{\overset{H}{|}}{C}}\text{—}\underset{\underset{H}{|}}{\overset{\overset{H}{|}}{C}}\text{—}O\text{—}H \;+\; NaH \longrightarrow H\text{—}\underset{\underset{H}{|}}{\overset{\overset{H}{|}}{C}}\text{—}\underset{\underset{H}{|}}{\overset{\overset{H}{|}}{C}}\text{—}O^- Na^+ \;+\; H_2$$

Q: Can the H^- an ion from NaH act a nucleophile and attack the electron-deficient carbon atom?

A: Yes, certainly there is this possibility. But before the H^- ion reaches the electron-deficient carbon atom, which is more "hidden," it would first have encountered the acidic hydrogen. Thus, an acid-base reaction would be more likely to take place than nucleophilic substitution.

The reaction with the strong base sodium amide ($NaNH_2$) will also cause the cleavage of the O–H bond, producing an alkoxide ion and $NH_3(g)$.

$$H\text{—}\underset{\underset{H}{|}}{\overset{\overset{H}{|}}{C}}\text{—}\underset{\underset{H}{|}}{\overset{\overset{H}{|}}{C}}\text{—}O\text{—}H \;+\; NaNH_2 \longrightarrow H\text{—}\underset{\underset{H}{|}}{\overset{\overset{H}{|}}{C}}\text{—}\underset{\underset{H}{|}}{\overset{\overset{H}{|}}{C}}\text{—}O^- Na^+ \;+\; NH_3$$

The acidity of alcohols can also be observed in the isotopic exchange reaction that occurs with heavy water, i.e. deuterium oxide (D_2O), at room temperature. With this possibility of proton exchange, the O–H bond is constantly being broken and formed, even in the absence of D_2O (refer to Chapter 17 on NMR). This shows that covalent bond can be non-permanent in nature even at room temperature.

$$H\text{—}\underset{\underset{H}{|}}{\overset{\overset{H}{|}}{C}}\text{—}\underset{\underset{H}{|}}{\overset{\overset{H}{|}}{C}}\text{—}O\text{—}H \;+\; D\text{—}O\text{—}D \rightleftharpoons H\text{—}\underset{\underset{H}{|}}{\overset{\overset{H}{|}}{C}}\text{—}\underset{\underset{H}{|}}{\overset{\overset{H}{|}}{C}}\text{—}O\text{—}D \;+\; H\text{—}O\text{—}D$$

Though alcohols are acidic, they are not strongly acidic enough to react with hydroxides and carbonates.

Q: How does one explain that alcohol cannot react with hydroxide?

A: Alcohol is a weaker acid than H_2O. This would mean that the alkoxide ion, RO^-, is a stronger base than the OH^-. Hence, a weaker acid (alcohol as compared to water) cannot react with a weaker base (OH^-) as this base is the conjugate base of an acid (H_2O) that is a stronger acid than alcohol. Vice versa, RO^- would be able to react with H_2O. This is because RO^-, being the conjugate base of an acid (alcohol) that is weaker than H_2O, is itself a strong base and hence would be able to react with the stronger acid (H_2O). The same also goes to explain why alcohol is not able to react with carbonates. This is because alcohol is a weaker acid as compared to the conjugate acid of carbonates, the hydrogen carbonates (HCO_3^-). Hence, the weaker acid cannot react with the weaker base as carbonates (CO_3^{2-}) are a weaker base than OH^-. The following trends help one to decide which acids can react with which bases:

Acid strength increases in the order: $ROH < H_2O < Phenol < RCOOH$

Basic strength decreases in the order: $RO^- > OH^- > C_6H_5O^- > RCOO^-$.

For example, ROH, being the weakest acid among the above, cannot protonate these bases: OH^-, $C_6H_5O^-$ and $RCOO^-$, whereas phenol would be able to protonate both RO^- and OH^- but not $RCOO^-$. The phenoxide, $C_6H_5O^-$, would be able to react completely with RCOOH, but with regards to the weaker acids such as H_2O, an equilibrium would be established. No reaction would occur between $C_6H_5O^-$ and ROH as the ROH is too weak an acid to react with $C_6H_5O^-$.

8.6.2 *Formation of Esters with Carboxylic Acids — Esterification via Nucleophilic Acyl Substitution/Condensation*

Reagents:	RCOOH
Conditions:	Conc. H_2SO_4 as catalyst/Heat
Options:	Can use other acid catalyst

Ethyl methanoate

from ethanol   from methanoic acid

Alcohols react with carboxylic acids to form esters. The esterification process is a reversible reaction and it occurs at such a slow rate that equilibrium is attained only after a few hours. As such, concentrated H_2SO_4 or other acids are used as a catalyst to speed up the reaction. The reverse reaction is the hydrolysis of esters, which will be covered in Chapter 10.

Q: How does H_2SO_4 speed up the rate of esterification?

A: The H^+ would first protonate the carbonyl oxygen (C=O) of the –COOH group. As a result of the positive charge residing on the O atom, the carbonyl O atom becomes even more electron-withdrawing. This causes the carbonyl carbon to be even more electron-deficient, hence more susceptible to attack by the alcohol molecule.

Q: If H_2SO_4 is a catalyst, does that mean that increasing its concentration would not affect the rate of the reaction?

A: With more H_2SO_4, there would be more H^+ to protonate the carboxylic acid, hence a higher rate of reaction. Note that the concentration of catalyst does affect the rate of reaction. But if the amount of H^+ provided is more than the amount of carboxylic acid to be protonated, further increase in the concentration of H^+ would not increase the rate of the reaction. This is known as saturation kinetics in which the concentration of catalyst (or enzymes) is greater than the concentration of the reactant (or substrates).

Q: Since the carboxylic acid is acidic, can the acid itself catalyze the esterification reaction?

A: Theoretically it can. But the carboxylic acid is also a reactant. If it is to dissociate and provide the H^+ needed for catalysis, the concentration of the reactant would decrease, hence affecting the rate. In addition, the carboxylic acid is a weak acid, which does not provide much H^+ in the first place. If the product that is formed can act as a catalyst, such reaction is known as autocatalysis. For such a reaction, the rate of reaction would increase with the formation of the product (because more catalyst is formed) but would later on decrease because the concentration of reactant has decreased to a level that is much lower than the concentration of the catalyst. Basically, the rate of reaction decreased because the concentration of reactant has decreased.

Q: Why can't the O atom of the hydroxyl group (–OH) of the –COOH make a nucleophilic attack on the electron-deficient carbon atom of the alcohol molecule instead?

A: This is because the lone pair of electrons on the O atom of the hydroxyl group of the –COOH delocalizes into the carbonyl group (C=O). As a result, it is less available to make a nucleophilic attack than the lone pair of electrons on the –OH group of the alcohol. In addition, the electron-donating effect of the alkyl group on the alcohol molecule intensifies the electron density on the –OH group of the alcohol, making it more nucleophilic in nature.

Esterification is an example of a condensation reaction, otherwise known as an addition-elimination reaction, wherein two molecular reactants react to form a single product with the elimination of a small molecule (e.g. H_2O in esterification). This result is in fact a nucleophilic acyl substitution in nature. The term addition-elimination depicts the mechanism for the formation of the ester, i.e. an addition step followed by an elimination stage.

To obtain a much better yield of an ester product, the reaction of alcohols with acid chlorides is recommended.

Q: Is there no other way to improve the yield of esters through esterification?

A: Actually you can, by constantly distilling out the water as it is formed. This would drive the position of the equilibrium towards the right in accordance to Le Chatelier's principle. In addition, by checking the amount of water distilled out, you can check for the extent of the esterification process, i.e. how much of ester has been formed. Such a technique is normally used in tracking the amount of polyester formed in a large reactor during the industrial manufacturing of polyester.

8.6.3 *Formation of Esters with Acid Chlorides — Acylation via Nucleophilic Substitution/ Condensation*

$$H-\underset{\underset{H}{|}}{\overset{\overset{H}{|}}{C}}-\underset{\underset{H}{|}}{\overset{\overset{H}{|}}{C}}-O-H \;+\; H-\overset{\overset{O}{\|}}{C}-Cl \;\longrightarrow\; H-\underset{\underset{H}{|}}{\overset{\overset{H}{|}}{C}}-\underset{\underset{H}{|}}{\overset{\overset{H}{|}}{C}}-O-\overset{\overset{O}{\|}}{C}-H \;+\; H-Cl$$

Reagents:	RCOCl
Conditions:	Room temperature/Anhydrous
Options:	—

Treating an alcohol with an acid chloride will give a better yield of an ester. This reaction is a condensation reaction, similar to that of esterification, but the small HCl molecule is eliminated instead of H_2O. This particular reaction with an acid chloride is termed acylation since the reaction involves the introduction of an acyl group (R–C=O) onto the alcohol molecule.

The reaction is rapid due to the high reactivity of the acid chloride. As such, the reaction proceeds to completion and gives a high yield of the ester without the need of catalyst or heating. The possible shortcomings for this reaction as compared to esterification are the corrosive nature of the HCl gas and the requirement of an anhydrous state to carry out the reaction.

Q: Why is the acid chloride more reactive than a carboxylic acid molecule towards a nucleophilic substition reaction?

A: We will discuss this in more detail in Chapter 10 under Carboxylic Acids.

8.6.4 *Formation of Alkenes — Dehydration/Elimination*

$$H-\underset{\underset{H}{|}}{\overset{\overset{H}{|}}{C}}-\underset{\underset{H}{|}}{\overset{\overset{OH}{|}}{C}}-H \;\xrightarrow[\text{170 °C}]{\text{excess concentrated } H_2SO_4}\; \overset{H}{_H}\!\!>\!C=C\!<\!\overset{H}{_H} \;+\; H_2O$$

Reagents: Excess concentrated H_2SO_4
Conditions: 170°C
Options: Al_2O_3, 400°C or H_3PO_4, 200–250°C

This reaction involves the cleavage of the C–O bond of the alcohol. For the dehydration of an unsymmetrical alcohol, remember to consider **Saytzeff's rule** to identify the major product, i.e. the *trans*-isomer of the most substituted alkene (see Chapter 5).

Q: Is the dehydration of alcohol an elimination reaction similar to the dehydrohalogenation of halogenoalkanes?

A: Both reactions are indeed elimination reactions. The elimination of water from an alcohol is termed dehydration, catalyzed by the presence of an acid which would seek out a basic site (note that the basic site is a Brønsted–Lowry base — a proton acceptor). The term dehydration appropriately tells us what has happened in the reaction — loss of a water molecule — hence it is more appropriate to be used to describe the elimination of a water molecule from alcohol. The elimination of a water molecule is an intramolecular phenomenon as compared with esterification, whereby the water molecule is formed from two different molecules — an intermolecular reaction. As for the elimination reaction of a halogenoalkane, it is mediated by the presence of a base (a Lewis base) which would attack an acidic site (a Brønsted–Lowry acid). Thus, the nature of the two reaction mechanisms would be different. In addition, take note that the term "condensation" refers to two molecules coming together to form a single molecule with the expulsion of a small molecule.

Q: Why must the H_2SO_4 be in excess?

A: Excess H_2SO_4 ensures that all the alcohol molecules are being protonated at one go. This would encourage dehydration to proceed first and diminish other side reactions such as ether formation — an intermolecular dehydration. If the alcohol molecules are not all protonated at one go, the –OH group of some of the unprotonated alcohol molecules can make a nucleophilic attack at the electron-deficient site of the protonated alcohol molecules, resulting in the formation of ether (C–O–C) molecules.

Q: Can dilute H_2SO_4 do the job of dehydrating the alcohol?

A: No, it can't. No doubt there are numerous $H^+(aq)$ ions in dilute H_2SO_4, but $H^+(aq)$ would be too hydrated with multiple layers of water molecules. This would decrease the ability of the $H^+(aq)$ to protonate the –OH group of the alcohol molecule due to the smaller charge density after hydration.

If an alcohol is to proceed via the same elimination mechanism as for a halogenoalkane, the C–O bond would have to be cleaved, which would result in the loss of an OH^- ion. But as we know it, the C–O bond in alcohol is stronger and more difficult to be cleaved than the C–Cl bond in a halogenoalkane. Hence, the –OH group is a poorer leaving group as compared to the Cl group. To make the –OH group a better leaving group, in the dehydration of alcohols, an acid is used to protonate the hydroxyl group in the alcohol as shown in the mechanism below:

Q: Why is the C–O bond in alcohol stronger than the C–Cl bond in halogenoalkanes?

A: The valence orbital that the Cl atom (from $n = 3$ Principal Quantum Shell) used to form a bond with the C atom is bigger and more diffuse than the one used by the O atom (from $n = 2$ Principal Quantum Shell). Hence, the overlap of the orbitals during covalent formation is less effective for the C–Cl bond than the C–O bond, accounting for its weaker strength and making it a better leaving group. In addition, the C–O bond is made stronger because of the additional attractive force between the dipoles formed on the C and O atoms due to the more electronegative O atom.

Q: The electron-deficiency of the carbon atom of the C–OH group is greater than that of the one in the C–Cl group. Shouldn't the higher level of electron-deficiency cause the carbon atom of the C–OH group to be more susceptible to nucleophilic attack, hence more likely to undergo an elimination reaction?

A: Yes, indeed the carbon atom of the C–OH group is more susceptible to nucleophilic attack. But unfortunately, this cannot be used to explain the observed phenomenon that the –OH group is a poorer leaving group as compared to the Cl group.

The highly electronegative oxygen atom which possesses a positive charge after protonation is highly electron withdrawing. This would weaken the C–O bond due to a "thinner" electron cloud and make it more susceptible to breakage. The acid-catalyzed dehydration of alcohols is actually the reverse of the acid-catalyzed hydration of alkenes, as shown in the simplified mechanism below:

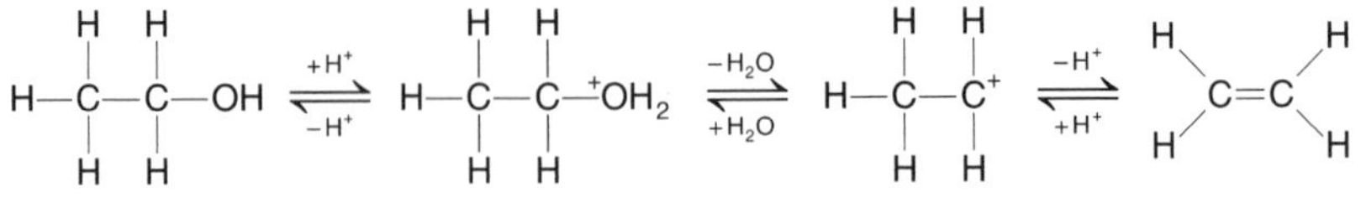

Suitable acids include conc. H_2SO_4 or H_3PO_4, but not hydrogen halides. The use of the latter, such as HCl, would result in the formation of conjugate bases (eg. Cl^-) that are strongly nucleophilic in nature and hence may instead initiate nucleophilic substitution, which would be the side reactions.

Then again, care must also be observed even in the use of conc. H_2SO_4. Different reaction conditions may result in a different reaction altogether. For example, if excess alcohol is treated with conc.

H_2SO_4 at a lower temperature of 140°C, the alcohol will be converted to an ether via nucleophilic substitution, rather than to an alkene through elimination.

Q: Why would excess alcohol lead to ether formation?

A: More alcohol molecules increase the statistical chances of an alcohol molecule attacking the carbocation that has been formed in the above mechanism. Isn't excess alcohol a contrast to the condition of excess conc. H_2SO_4?

8.6.5 *Formation of Halogenoalkanes — Nucleophilic Substitution*

Reagents:	$PCl_5(s)$
Conditions:	Room temperature
Options:	$PCl_3(l)$; $PBr_3(l)$; (red P + I_2) or $SOCl_2(l)$

The use of different reagents produces different side products as shown in the following general equations:

- $ROH + PCl_5 \rightarrow RCl + POCl_3 + HCl$
- $3ROH + PCl_3 \rightarrow 3RCl + H_3PO_3$
- $ROH + SOCl_2 \rightarrow RCl + SO_2 + HCl$

The formation of white fumes of HCl can be useful for determining the presence of the alcohol functional group. But it cannot be used to distinguish an alcohol from a carboxylic acid as the latter also reacts with PCl_5 and $SOCl_2$ to form HCl (see Chapter 10).

Alcohols can also react with *dry* HX(g) to yield halogenoalkanes.

- $ROH + HCl \rightarrow RCl + H_2O$

Q: If alcohols can undergo nucleophilic substitution reaction just like halogenoalkanes, why can't they simply react with the $X^-(aq)$ nucleophile from, let's say, NaX(aq)? Why must it be dry HX(g)?

A: In water, HX dissociates to give $H^+(aq)$ and $X^-(aq)$. The $X^-(aq)$ is surrounded by many layers of water molecules, which makes it a weaker nucleophile. In addition, note that the –OH group is a relatively bad leaving group, and the heavily hydrated $H^+(aq)$ would be less likely to protonate the –OH group and make it a good leaving group. Moreover, even if the $X^-(aq)$ nucleophile can successfully displace the –OH group, the resultant $OH^-(aq)$ ion is also a strong nucleophile, which can "re-attack" the halogenoalkane that is formed. The latter is the reaction to convert halogenoalkane into alcohol by heating it with NaOH(aq). Overall, a solution containing $X^-(aq)$ nucleophiles can't just undergo nucleophilic substitution on the alcohol.

In all, these reactions that convert alcohol to halogenolkanes also involve the cleavage of the C–O bond of the alcohol. Do note, reactions involving the cleavage of the C–O bond will first involve the conversion of the hydroxyl group into a better leaving group which is more stable. For instance, the reaction with $SOCl_2$ yields a chlorosulfite ester which contains the good leaving group — the chlorosulfite ion (SO_2Cl^-). The mechanism consists of the formation of the chlorosulfite ester, which will then undergo nucleophilic substitution with the Cl^- nucleophile produced in the esterification reaction to finally yield the halogenoalkane.

chlorosulfite ester

Q: Why is the chlorosulfite ion a better leaving group than the hydroxyl group?

A: By forming the chlorosulfite ester, the O atom of the –OH group of alcohol is now bonded to a few highly electronegative atoms, including S, O and Cl. Such highly electron-withdrawing groups further weaken the C–O bond of the alcohol by withdrawing electron density away from the C–O bond.

8.6.6 Formation of Carbonyl Compounds and Carboxylic Acids — Oxidation

Oxidation of alcohols involves the removal of at least one α-H atom that is bonded to the α-C atom bearing the hydroxyl group.

$$
\begin{array}{ccc}
\text{OH} & \text{OH} & \text{OH} \\
| & | & | \\
R-\overset{\alpha}{C}-H & R-\overset{\alpha}{C}-H & R-\overset{\alpha}{C}-R \\
| & | & | \\
H & R & R \\
\text{1}^\circ \text{ alcohol} & \text{2}^\circ \text{ alcohol} & \text{3}^\circ \text{ alcohol}
\end{array}
$$

Since tertiary alcohols do not contain any α-H atom to be removed, they cannot be oxidized.

Q: What is an α-carbon atom?

A: An α-carbon atom refers to the carbon atom that is bonded to the functional group. Since the functional group for alcohol is an –OH group, then the α-carbon atom is the carbon atom directly bonded to the –OH group.

Q: Why can't the tertiary alcohol be oxidized?

A: It has to do with the mechanism of how an alcohol is oxidized. In the oxidation process of an alcohol, a chromate ester intermediate $(C-O-CrO_3H)$ is formed if acidified $K_2Cr_2O_7$ is used. Subsequent reaction involves the extraction of an acidic α-H by a H_2O molecule which is acting as a base. There is no α-H present in a tertiary alcohol, hence it can't be oxidized through such a mechanism.

$$
-\underset{|}{\overset{\overset{\displaystyle CrO_3H}{\displaystyle O}}{C}}-\overset{\delta+}{H} + H_2\ddot{O} \longrightarrow \overset{O}{\underset{}{\overset{||}{C}}} + H_3O^+ + HCrO_3^-
$$

The type of oxidation products obtainable depends on the type of alcohol and the strength of the oxidizing agent. Common oxidizing agents are potassium manganate(VII) and potassium dichromate(VI); the latter is a weaker oxidizing agent as compared to potassium manganate(VII). The following standard reduction potential values explain why:

$$E^{\ominus}/V$$

$$MnO_4^- + 8H^+ + 5e^- \rightleftharpoons Mn^{2+} + 4H_2O \quad +1.52$$
$$Cr_2O_7^{2-} + 14H^+ + 6e^- \rightleftharpoons 2Cr^{3+} + 7H_2O \quad +1.33$$

Primary alcohols are readily oxidized to aldehydes which, in turn, are easily oxidized to carboxylic acids.

$$R-\underset{\underset{H}{|}}{\overset{\overset{OH}{|}}{C}}-H \ + \ [O] \ \xrightarrow[\substack{\text{heat with} \\ \text{immediate distillation}}]{K_2Cr_2O_7,\ H_2SO_4} \ R-\overset{\overset{O}{||}}{C}-H \ + \ H_2O$$

$$R-\underset{\underset{H}{|}}{\overset{\overset{OH}{|}}{C}}-H \ + \ 2[O] \ \xrightarrow[\text{heat}]{K_2Cr_2O_7,\ H_2SO_4} \ R-\overset{\overset{O}{||}}{C}-OH \ + \ H_2O$$

Reagents:	$K_2Cr_2O_7(aq)$, $H_2SO_4(aq)$
Conditions:	Heat with immediate distillation
Options:	$MnO_2(s)$

Hence, if the desired product is the aldehyde, the milder oxidizing agent $K_2Cr_2O_7$ should be used, and the aldehyde would be immediately distilled off from the reaction vessel to prevent it from being further oxidized. It is also important to note that the oxidizing agent is being added dropwise into the alcohol. This way of addition is to ensure that the added oxidizing agent is limiting with respect to the alcohol and it is immediately being consumed once it comes in contact with the alcohol. There would not be any excess oxidizing agent present to further oxidize the aldehyde that is formed into carboxylic acid. The set-up for the oxidation is shown below:

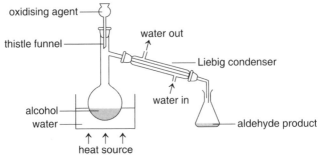

Q: When we apply immediate distillation during the oxidation process, wouldn't this cause the alcohol to be distilled out too?

A: The aldehyde that is formed from the alcohol would have a lower boiling point than the alcohol due to the stronger hydrogen bonding in alcohol. As for the aldehyde, it is polar and has permanent dipole-permanent dipole interaction. Thus, heating at the boiling point of aldehyde would not distill the alcohol out.

Q: Can we use acidified $KMnO_4$ with immediate distillation instead?

A: No, you can't. Acidified $KMnO_4$ is a much more powerful oxidizing agent than acidified $K_2Cr_2O_7$. The primary alcohol would be immediately oxidized to the carboxylic acid straight away without having the chance to isolate the aldehyde. Thus, the immediate distillation would not help at all.

Secondary alcohols are oxidized to the corresponding ketones. As ketones do not contain a H atom bonded to the carbonyl carbon atom, they are resistant to oxidation.

Reagents:	$K_2Cr_2O_7(aq)$, $H_2SO_4(aq)$
Conditions:	Heat
Options:	$KMnO_4(aq)$, $H_2SO_4(aq)$

These reactions are essentially redox reactions wherein one reactant (alcohol) is oxidized while the other reactant (the oxidizing agent) is reduced:

- When a 1° or 2° alcohol is oxidized by acidified $K_2Cr_2O_7(aq)$, the orange $K_2Cr_2O_7(aq)$ is reduced to green $Cr^{3+}(aq)$ ions. For instance, the oxidation of ethanol can be viewed to proceed in stages depicted by the following redox equations:

$$3CH_3CH_2OH + Cr_2O_7^{2-} + 8H^+ \rightarrow 3CH_3CHO + 2Cr^{3+} + 7H_2O,$$
$$3CH_3CHO + Cr_2O_7^{2-} + 8H^+ \rightarrow 3CH_3COOH + 2Cr^{3+} + 4H_2O.$$

- When a 1° or 2° alcohol is oxidized by acidified $KMnO_4(aq)$, the purple $KMnO_4(aq)$ is reduced to pink $Mn^{2+}(aq)$ ions, which essentially seems colorless. Redox equation for the oxidation of ethanol:

$$5CH_3CH_2OH + 4MnO_4^- + 12H^+ \rightarrow 5CH_3COOH + 4Mn^{2+} + 11H_2O.$$

As the oxidized and reduced forms of the oxidizing agents are of different colors, they can be used as characteristics tests to differentiate between the three different classes of alcohols. In addition, the redox reaction between ethanol and acidified $K_2Cr_2O_7(aq)$ to generate the green $Cr^{3+}(aq)$ ions, has become a characteristics reaction in the breathalyzer. The higher the concentration of ethanol in the breath, the greener the coloration shown by the indicator.

Example: To distinguish between ethanol and 2-methylpropan-2-ol.

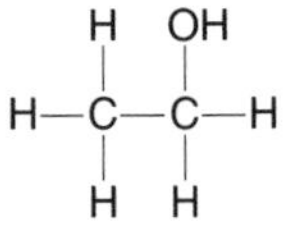

1° alcohol 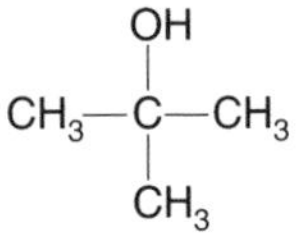   3° alcohol

Approach:

Test: Add $KMnO_4(aq)$ acidified with $H_2SO_4(aq)$ to each of these compounds and heat.

Observations: For ethanol, decolorization of purple $KMnO_4(aq)$ is observed.
For 2-methylpropan-2-ol, no decolorization of purple $KMnO_4(aq)$ is observed.

OR

Test: Add $K_2Cr_2O_7(aq)$ acidified with $H_2SO_4(aq)$ to each of these compounds and heat.

Observations: For ethanol, green coloration is observed.
For 2-methylpropan-2-ol, no green coloration is observed.

8.6.7 *Formation of Tri-Iodomethane (Iodoform) — Oxidation*

Only alcohols containing the following bonding pattern (i.e. 1-hydroxyethyl functional group) can undergo this oxidation reaction.

The carbon atom here must be bonded to either a H atom or another C atom	$\begin{array}{c} OH \\ | \\ -C-CH_3 \\ | \\ H \end{array}$

When such alcohols are warmed with alkaline $I_2(aq)$, a carboxylate salt will be obtained along with yellow crystals of tri-iodomethane:

$$H-\underset{\underset{H}{|}}{\overset{\overset{OH}{|}}{C}}-CH_3 \ + \ 4I_2 \ + \ 6OH^- \ \xrightarrow{\text{warm}} \ H-\overset{\overset{O}{\|}}{C}-O^- \ + \ CHI_3 \ + \ 5I^- \ + \ 5H_2O$$

Reagents:	$I_2(aq)$ with NaOH(aq)
Conditions:	Warm
Options:	KOH(aq)

Since an alkaline medium is used in this reaction, the oxidized product would be the carboxylate ion. Any carboxylic acid formed will be deprotonated as an acid-base reaction is inevitable in such an alkaline medium. In addition, unlike previous oxidation reactions, the formation of the tri-iodomethane oxidation reaction is a step-down reaction, meaning one carbon less in the product as compared to the reactant.

This reaction is useful in distinguishing certain 2° alcohols from 1° alcohols. Carbonyl compounds with the following bonding pattern (i.e. containing a methyl ketone functional group) also give a positive result for the tri-iodomethane test:

The carbonyl carbon atom here must be bonded to either a H atom or another C atom	$\begin{array}{c} O \\ \| \\ -C-CH_3 \end{array}$

The redox reaction occurs in the following steps:

- Step 1: Alcohol is oxidized to the corresponding carbonyl compound.

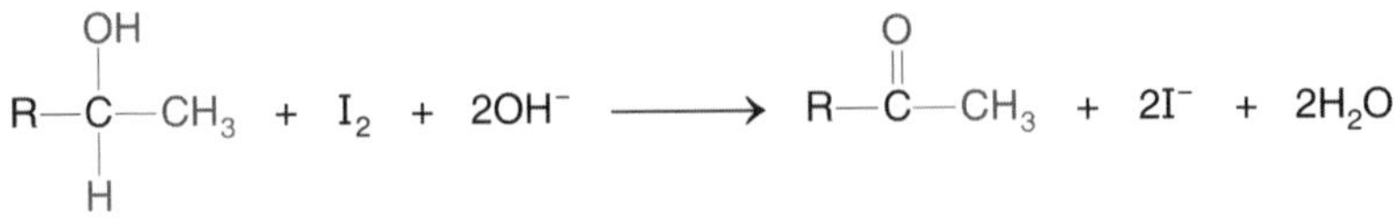

$$\underset{\underset{H}{|}}{\overset{\overset{OH}{|}}{R-C}}-CH_3 + I_2 + 2OH^- \longrightarrow R-\overset{\overset{O}{\parallel}}{C}-CH_3 + 2I^- + 2H_2O$$

- Step 2: Carbonyl compound is further oxidized.

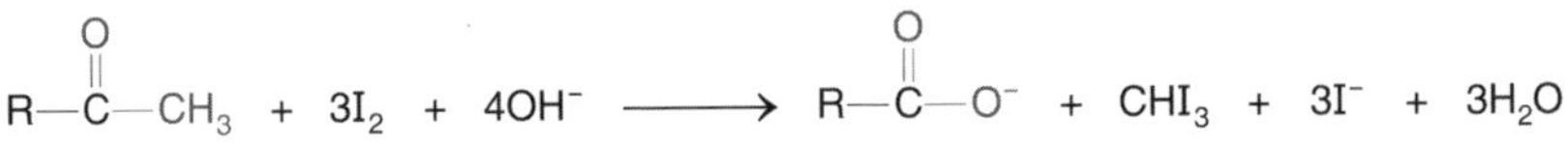

$$R-\overset{\overset{O}{\parallel}}{C}-CH_3 + 3I_2 + 4OH^- \longrightarrow R-\overset{\overset{O}{\parallel}}{C}-O^- + CHI_3 + 3I^- + 3H_2O$$

The overall redox reaction is a summation of both steps as shown by the following equation:

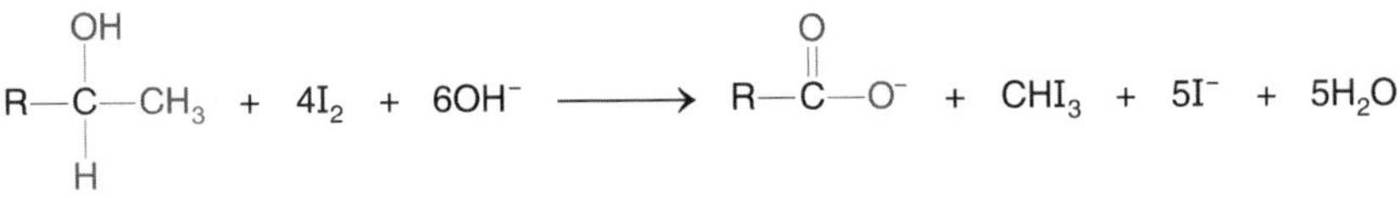

$$\underset{\underset{H}{|}}{\overset{\overset{OH}{|}}{R-C}}-CH_3 + 4I_2 + 6OH^- \longrightarrow R-\overset{\overset{O}{\parallel}}{C}-O^- + CHI_3 + 5I^- + 5H_2O$$

8.7 Chemical Properties of Phenol

Alcohols are mostly neutral in water. Phenol, on the other hand, forms a slightly acidic solution in water. Although phenol can form hydrogen bonds with water molecules, it is only partially soluble in water because of its large hydrophobic benzene ring. Solubility in water can be enhanced by adding an alkali which converts phenol to a phenoxide ion. The phenoxide ion is more soluble in water as it can form stronger ion-dipole interactions with the water molecules.

Phenol exhibits a lack of reactivity towards reactions that involve the cleavage of the C–O bond. The explanation for this is similar to that used to account for the lack of reactivity of halogenoarenes towards nucleophilic substitution, as given below:

Reason: The *p* orbital of the O atom can make a side-on overlap with the π-electron cloud of the benzene ring. As a result, the lone pair of electrons on the O atom can delocalize into the benzene ring. This results in the C–O bond bearing a partial double bond character and thus being stronger than the C–O bond in an aliphatic alcohol. In addition, due to this delocalization, the carbon atom that bears the

–OH group is less electron-deficient, hence less susceptible to nucleophilic attack.

The reactions that phenol undergoes can be classified into one of the following:

* Reactions involving cleavage of O–H bond
* Electrophilic substitution of the parent benzene ring

8.7.1 *Acid-Base Reaction with Strong Bases and Reactive Metals*

Phenol reacts with strong bases such as NaOH and reactive metals. However, it is not strongly acidic enough to react with carbonates.

8.7.2 *Formation of Esters with Acid Chlorides — Acylation vis Nucleophilic Substitution/ Condensation*

Reagents:	RCOCl
Conditions:	Room temperature/Anhydrous
Options:	Convert phenol to phenoxide first

Phenol reacts with acid chloride to form a phenolic-ester, but it does not react with carboxylic acids to generate the phenolic-ester.

Q: Why can alcohol react with carboxylic acid to form an ester but not phenol?

A: The reaction of an alcohol with carboxylic acid to form an ester is a nucleophilic substitution as shown:

$$
\begin{array}{c}
\overset{\displaystyle O}{\underset{}{\|}} \\
R\overset{\delta+}{-}C-OH \\
\diagup \\
R-\overset{..}{O} \\
\diagdown \\
H
\end{array}
$$

An alchohol's lone pair of electrons is especially available due to the electron donating effect of the alkyl group. As for phenol, the lone pair of electrons on the O atom is not readily available since they can delocalize into the benzene ring. Hence, phenol is not a good nucleophile as compared to alcohol. But by converting the carboxylic acid to a more reactive acid chloride, we can make phenol react with acid chloride to form an ester. In addition, phenol can be made a better nucleophile by converting it to the phenoxide ion upon treatment with NaOH(aq). So, take note that to obtain phenolic ester, it is not necessary to convert the phenol to phenoxide first!

Q: Since the phenoxide is a stronger nucleophile than phenol itself, can we form the ester by just reacting the phenoxide with carboxylic acid instead?

A: No, you can't. If phenoxide is mixed with carboxylic acid, an acid-base reaction would take place as the phenoxide ion is a stronger conjugate base than a carboxylate ion (Remember: Carboxylic acid is a stronger acid than phenol, thus the carboxylate conjugate base must be a weaker base as compared to phenoxide). Hence, it would be protonated by the carboxylic acid instead of making a nucleophilic attack on the carboxylic acid molecule.

8.7.3 *Formation of Halogenophenol — Electrophilic Substitution*

When compared to the rate of electrophilic substitution of benzene, that for phenol occurs at a much faster rate. This is because the hydroxyl substituent is a strongly activating group that enhances the electron density in the benzene ring, making it more susceptible to electrophilic attack. As such, milder reaction conditions are required and, even in such instances, polysubstitution is still observed.

Orientation of electrophilic attack is targeted at positions 2- and 4- as this leads to the formation of a more stable carbocation (refer to Chapter 6 for more details).

Reagents:	$Br_2(aq)$
Conditions:	Room temperature
Options:	—

Q: Why are 3 moles of Br_2 needed? Why not just 1.5 moles?

A: The requirement of 3 moles of Br_2 has to do with the mechanism of electrophilic substitution. The first step involves a Br_2 molecule attacking the π-electron cloud of the benzene ring, generating a Br^-. The second step involves the regeneration of the resonance stabilized benzene ring. Thus, the Br^- would be left as HBr, and this accounts for why there are 3 moles of HBr formed too.

Q: If one of the 2-, 4- and 6-positions is occupied by other substituents, can the Br_2 still bring about the electrophilic substitution?

A: Obviously not. The Br_2 would only attack the remaining sites of the 2-, 4- and 6-positions that are still available.

The reaction with $Br_2(aq)$ occurs readily at room temperature without the need for a Lewis acid catalyst to create a stronger electrophile. Brown $Br_2(aq)$ is decolorized and a white precipitate of 2,4,6-tribromophenol is obtained due to the low solubility of the organic product that is formed. These observations serve as characteristics test for the presence of the phenolic –OH group.

Halogenation can also be carried out using Br_2 in a non-polar solvent such as CCl_4. However, only monosubstituted products are obtained.

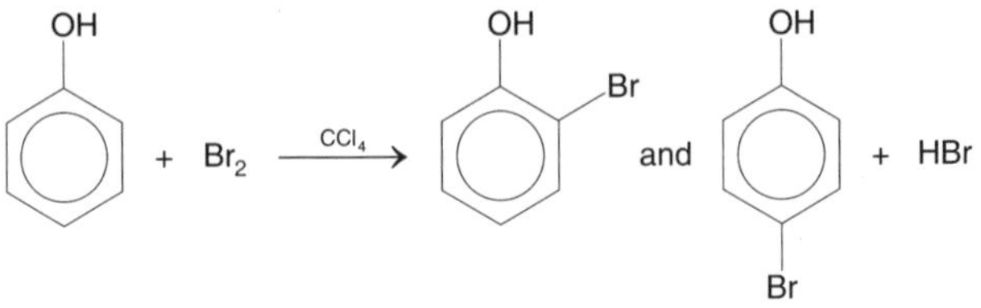

Reagents:	Br_2 in CCl_4
Conditions:	Room temperature
Options:	—

Q: Why are the bromination products different when carried out in different mediums?

A: In an aqueous medium, the phenol is partially ionized, generating the phenoxide ion, which is absence in the CCl_4 medium. The benzene ring of the phenoxide ion is much more electron rich than a phenol molecule as shown below:

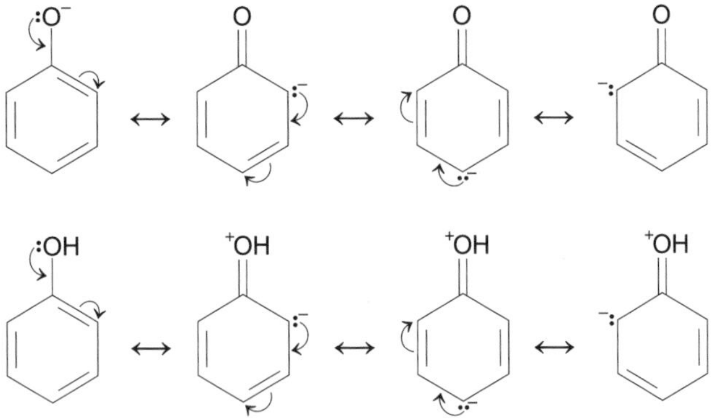

The more electron-rich the benzene ring, the more likely it is to polarize the electron cloud of the Br–Br molecule, making the Br_2 molecule a better electrophile. The separation of charges during the delocalization of electrons in the phenol hinders the effectiveness of the delocalization. As a result, the benzene ring of phenol is not as electron rich as that of the phenoxide ion, but note that it is still more electron-rich than an unsubstituted benzene. Hence, phenol becomes less susceptible to further electrophilic attack after the first halogenation process. Remember that a halogen group is a ring-deactivating group; it would thus oppose the ring-activating effect of the –OH group.

8.7.4 *Formation of Nitrophenol — Electrophilic Substitution*

Reagents:	conc. $HNO_3(aq)$
Conditions:	Room temperature
Options:	—

Nitration of phenol using conc. HNO_3 occurs readily at room temperature as compared to benzene. There is no need for the use of conc. H_2SO_4 to generate a stronger electrophile. This can be accounted for by the highly electron-rich benzene ring due to the delocalization of the lone pair of electrons from the –OH group into the ring.

If mono-nitrated products are preferred, nitration of phenol can be carried out just by using dilute $HNO_3(aq)$. This is because the concentration of the nitronium electrophile, NO_2^+, is low in dilute $HNO_3(aq)$.

Reagents:	Dilute $HNO_3(aq)$
Conditions:	Room temperature
Options:	—

8.7.5   *Test for Phenol*

Phenol reacts with neutral aqueous iron (III) chloride to form a violet complex as shown:

violet complex

This reaction is used to test for the presence of phenol. The color of the complex may vary if other substituents are bonded to the ring.

Q: Why is neutral aqueous iron (III) chloride being used? Can it be non-neutral?

A: If the aqueous iron (III) chloride is an acidified one, then the phenoxide ions would not be able to be generated, thus the violet complex would not form. If alkaline aqueous iron (III) chloride is used, then the insoluble $Fe(OH)_3$ would be precipitated out instead.

Q: Can't the lone-pair of electrons on the phenol molecule cause the phenol molecules to form a complex with the aqueous iron (III) chloride instead?

A: Do not forget that the lone-pair of electrons on the phenol is delocalized into the benzene ring, hence it would be less available to be used for complex formation with the aqueous iron (III) chloride.

8.8   Summary

The mind maps below depict the reactions of a 3-carbon alcohol and phenol.

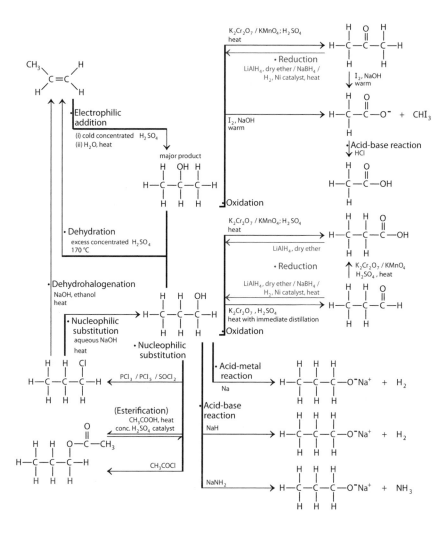

My Tutorial

1. (a) The organic compound **P** is known to contain bromine and no other halogen. Describe how you would determine its percentage by mass of bromine.

   (b) The plan for a sequence of reactions is as follows:

$$RCH_2Br \xrightarrow{\ I\ } RCH_2CN \xrightarrow{\ II\ } RCH_2CH_2NH_2$$

$$\downarrow {\scriptstyle III}$$

$$RCH=CH_2 \xleftarrow{\ IV\ } RCH_2CH_2OH$$

   (i) Describe briefly the reagents and conditions you would carry out for each of the conversions: I, II, III, and IV.

   (ii) When $RCH=CH_2$ was analyzed at r.t.p., $60\,cm^3$ of $RCH=CH_2$, weighing $0.14\,g$, were found to react with an equal volume of hydrogen gas. Determine the relative molecular mass and hence the structural formula of $RCH=CH_2$.

(iii) With appropriate reagents and conditions, describe a simple chemical test to differentiate the following pair of compounds:

(I) RCH_2Br and RCH_2CN

(II) RCH_2Br and RCH_2CH_2OH

(III) RCH_2Br and $RCH=CH_2$

(IV) RCH_2CH_2OH and $RCH=CH_2$

(c) An alcohol, C_4H_9OH, has two stereoisomers, **R** and **R'**, which both have the same melting and boiling points. If **R** and **R'** are heated with concentrated sulfuric acid at 170°C, a hydrocarbon, C_4H_8, is produced, which can also exist as two stereoisomers, **S** and **S'**. Both **S** and **S'** have different melting and boiling points. Deduce the structures of the two stereoisomers and explain the type of isomerism illustrated by each pair.

(d) The two –OH groups in morphine are chemically different, but both can be esterified in the same way to make heroin.

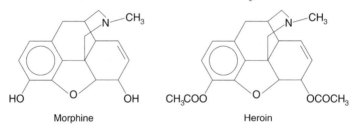

Morphine   Heroin

(i) Other than the –OH groups, name the other functional groups that are present in each of the above molecules.

(ii) Explain, in terms of the structure of the –OH group and the surrounding molecule, why the two OH groups in morphine show different chemical behaviors.

(iii) Give reagents and conditions as to how the two –OH groups in morphine could be esterified using the same method.

(iv) Suggest a simple chemical test to differentiate between morphine and heroin.

(e) Hydroquinone can be catalyzed to form benzoquinone by the enzyme peroxidase. The reaction is exothermic.

Hydroquinone   Benzoquinone

(i) Explain why the reaction is exothermic.

(ii) Name the functional groups that are present in both hydroquinone and benzoquinone.

(iii) Suggest a simple chemical test to differentiate between hydroquinone and benzoquinone.

(iv) Hydroquinone has K_a value that is smaller than 2-methylphenol. Explain.

CHAPTER 9

Carbonyl Compounds

9.1 Introduction

Carbonyl compounds are organic compounds that contain the following carbonyl functional group.

$$\underset{\diagup C\diagdown}{\overset{\overset{\textstyle O}{\|}}{}}$$

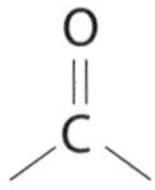

Aldehydes have the carbonyl functional group on the terminal carbon and hence are commonly abbreviated as RCHO and not as RCOH. Note that if the "OH" appears in the abbreviation, it may indicate that the molecule possesses the hydroxyl functional group (–OH group). Thus, to avoid confusion, the aldehyde functional group is abbreviated as –CHO. Ketones have the carbonyl group flanked by adjacent alkyl or aryl groups and are thus abbreviated as RCOR.

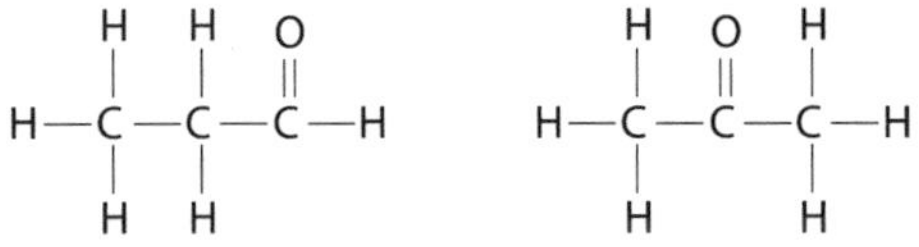

Propanal
(aldehyde)

Propanone
(ketone)

Collectively, aldehydes and ketones have the general formula $C_nH_{2n}O$, which makes them isomeric pairs. The presence of a C=O double bond is equivalent to the loss of two hydrogen atoms, which is also equivalent to an alkene double bond. Carbonyl compounds thus belong to a class of unsaturated compound.

9.2 Nomenclature

Aldehydes and ketones are named with the suffixes *–al* and *–one*, respectively. Other substituents are named as prefixes accompanied by the appropriate positional numbers. If the carbonyl group is considered a substituent, then it is named as the *–oxo* prefix for the ketone functional group and the *–formyl* prefix for aldehyde.

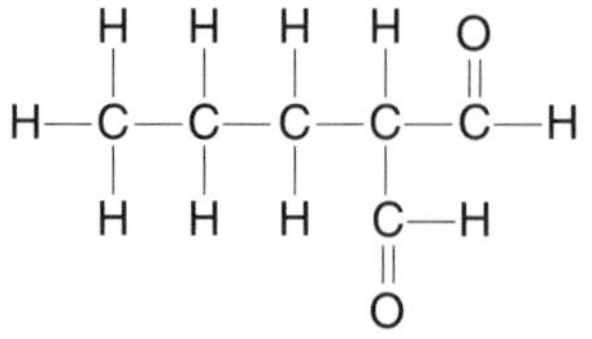

2-formylpentanal

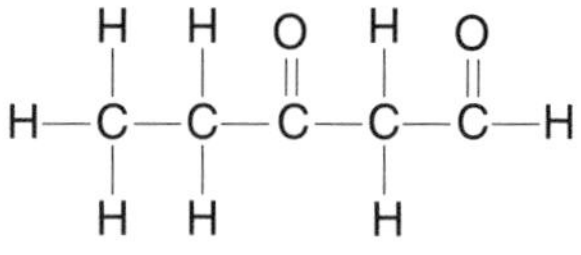

3-oxopentanal

9.3 Physical Properties

9.3.1 *Melting and Boiling Points*

As polar compounds, aldehydes and ketones have higher boiling points than alkanes of similar relative molecular mass (M_r).

Example:

Table 9.1

	Pentane	Butanal	Butanone
Structural formula	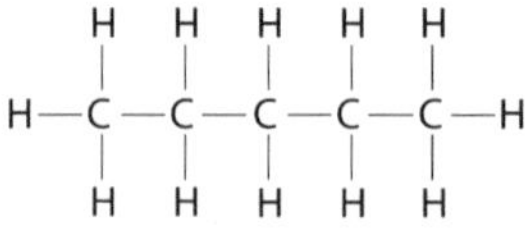	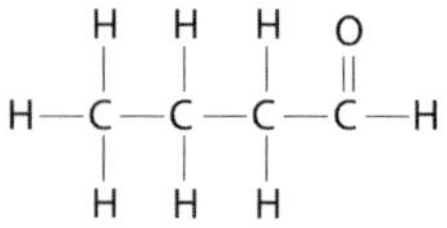	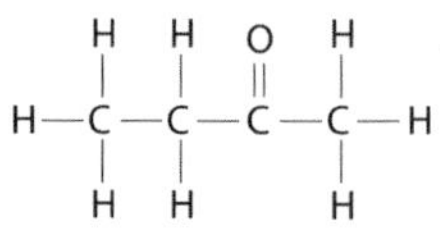
Boiling point ($^\circ$C)	36.0	76.0	80.0

The higher boiling points of both butanal and butanone are due to the greater amount of energy needed to overcome the stronger permanent dipole–permanent dipole (pd–pd) interactions between the polar carbonyl molecules compared to the instantaneous dipole–induced dipole interactions between the non-polar pentane molecules.

Q: The boiling point of diethylether ($CH_3CH_2OCH_2CH_3$, M_r = 74.0) is about 34°C. Why is it so much lower than that of butanal and butanone? Why is the pd–pd interaction of diethylether so much weaker than those of butanal and butanone?

A: The carbonyl functional group of aldehyde and ketone consists of a C=O double bond. In this double bond, the pi electron cloud is accumulated out of the inter-nuclei region. Therefore, it is more polarizable than the sigma electron cloud, which is accumulated within the inter-nuclei region. As a result, the dipole moment of the C=O double bond is much greater, accounting for its stronger pd–pd interaction.

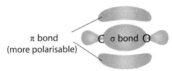

π bond
(more polarisable)

C σ bond O

9.3.2 *Solubility*

The lone pair of electrons on the carbonyl oxygen atom allows carbonyl compounds to function as hydrogen bond acceptors. This ability to form hydrogen bonds accounts for the appreciable solubility of short-chain carbonyl compounds in water. However, as the length of the carbon chain increases, solubility in water decreases due to the increasing size of the hydrophobic alkyl chain. The longer the carbon chain length, the more non-polar the molecule essentially becomes, bearing the instantaneous dipole–induced dipole interaction as the predominant intermolecular force.

Hydrogen bond

Carbonyl compounds have good solubility in non-polar solvents. In fact, propanone or acetone itself is widely used as an organic solvent for cleaning purposes in laboratories and as the active ingredient in nail polish remover.

9.4　Preparation Methods for Carbonyl Compounds

9.4.1　*Oxidation of Alcohols*

The type of oxidation product obtainable depends on the type of alcohol and the strength of the oxidizing agent (refer to Chapter 8 for more details).

For example:

Oxidation of primary alcohol to form aldehyde

Reagents:	$K_2Cr_2O_7(aq)$, $H_2SO_4(aq)$
Conditions:	Heat with immediate distillation
Options:	$MnO_2(s)$

Oxidation of secondary alcohol to form ketone

Reagents:	$K_2Cr_2O_7(aq)$, $H_2SO_4(aq)$
Conditions:	Heat
Options:	$KMnO_4(aq)$, $H_2SO_4(aq)$

9.4.2　*Oxidation of Alkenes*

Reagents:	$KMnO_4(aq)$, $H_2SO_4(aq)$
Conditions:	Heat
Options:	—

Ketones can be obtained from the oxidation of alkenes but not aldehydes as the aldehydes formed are easily further oxidized to carboxylic acids (refer to Chapter 5 for more details). If the intention is to obtain an aldehyde from the oxidation of an alkene, ozonolysis is employed.

9.4.3 *Ozonolysis — Oxidation of Alkenes*

For Step (i),		For Step (ii),	
Reagents:	$O_3(g)$	Reagents:	$Zn(s)$, $H_2O(l)$
Conditions:	—	Conditions:	Heat
Options:	—	Options:	—

Under the experimental conditions used, aldehydes formed from the oxidative cleavage of the C=C bond are not further oxidized.

9.5 Chemical Properties of Carbonyl Compounds

Carbonyl compounds are polar molecules with the C atom, of the –C=O group, bearing the partial positive charge. As a result, nucleophiles are attracted to this electron-deficient site. But unlike halogenoalkanes and alcohols, carbonyl compounds do not undergo substitution reactions. Instead, they undergo nucleophilic addition reactions. A typical nucleophilic addition reaction is the addition of a HCN molecule to the carbonyl functional group (refer to Chapter 3).

Q: Why don't carbonyl compounds undergo nucleophilic substitution?

A: The carbonyl functional group is either bonded to two other carbon atoms (as in a ketone) or a carbon atom and a hydrogen atom (as in an aldehyde). Thus, if the carbonyl compound is to undergo nucleophilic substitution, then either the C—C bond or the C—H bonds has to be broken. The breaking of these two types of bonds is energetically demanding. In addition, if the C—C bond or the C—H bond is broken, then we would have a leaving group such as a hydride ion (H$^-$ for aldehyde) or a carbanion ion for ketone. These two leaving groups are themselves very strong nucleophiles and highly unstable.

9.5.1 *Formation of Cyanohydrins — Nucleophilic Addition*

$$
\begin{array}{c}
\overset{\displaystyle O}{\overset{\displaystyle \|}{R-C-H}} + HCN \xrightarrow{\text{small amount of KOH}} R-\overset{\displaystyle CN}{\underset{\displaystyle H}{\overset{\displaystyle |}{\underset{\displaystyle |}{C}}}}-OH
\end{array}
$$

$$
\begin{array}{c}
\overset{\displaystyle O}{\overset{\displaystyle \|}{R-C-R}} + HCN \xrightarrow{\text{small amount of KOH}} R-\overset{\displaystyle CN}{\underset{\displaystyle R}{\overset{\displaystyle |}{\underset{\displaystyle |}{C}}}}-OH
\end{array}
$$

Reagents:	HCN, a small amount of KOH
Conditions:	10–20°C
Options:	HCN, NaCN or KCN, H$_2$SO$_4$

Q: Why is a low temperature required for the addition reaction with HCN?

A: The boiling point of HCN is about 26°C. Thus, to prevent the poisonous HCN from escaping into the environment, the reaction has to be performed below the boiling point of HCN.

Q: Are aldehyde and ketone equally reactive to nucleophilic addition?

A: No. Generally, aldehyde is more reactive than a ketone that has similar number of carbon atoms. As aldehyde has a H atom and an alkyl group bonded to the carbonyl functional group, a nucleophile would face less steric hindrance when approaching the carbonyl carbon of an aldehyde as compared to a ketone. There are two bulky electron-releasing alkyl groups bonded to the carbonyl group of ketone. As a result of this electronic effect, the carbonyl carbon atom of ketone is less electron-deficient than an aldehyde, thus less susceptible to nucleophilic attack.

Cyanohydrins are known also as 2-hydroxynitriles. Just like nitriles, cyanohydrins are useful reagents in organic synthesis as their formation leads to an extension in carbon chain length by one C atom. In addition, they serve as an intermediate in the synthesis of other organic compounds such as carboxylic acids and amines:

- Cyanohydrins can undergo acidic hydrolysis to form 2-hydroxycarboxylic acids, an important component in cosmetics:

$$R-\underset{\underset{H}{|}}{\overset{\overset{CN}{|}}{C}}-OH \ + \ HCl \ + \ 2H_2O \ \xrightarrow{\text{heat}} \ R-\underset{\underset{H}{|}}{\overset{\overset{COOH}{|}}{C}}-OH \ + \ NH_4Cl$$

If basic hydrolysis is used instead, the 2-hydroxycarboxylic acid will react with the base to form a carboxylate salt:

$$R-\underset{\underset{H}{|}}{\overset{\overset{CN}{|}}{C}}-OH \ + \ NaOH \ + \ H_2O \ \xrightarrow{\text{heat}} \ R-\underset{\underset{H}{|}}{\overset{\overset{COO^-Na^+}{|}}{C}}-OH \ + \ NH_3$$

- Cyanohydrins can undergo reduction, using different reagents, to form amines:

$$R-\underset{\underset{H}{|}}{\overset{\overset{CN}{|}}{C}}-OH \ + \ 4[H] \ \xrightarrow[\text{dry ether}]{\text{LiAlH}_4} \ R-\underset{\underset{H}{|}}{\overset{\overset{CH_2NH_2}{|}}{C}}-OH$$

9.5.2 *Condensation Reaction (Addition–Elimination Reaction)*

$$\underset{H}{\overset{R}{>}}C=O \ + \ H-\underset{\underset{}{\overset{H}{|}}}{N}-X \ \longrightarrow \ \underset{H}{\overset{R}{>}}C=N-X \ + \ H_2O$$

$$\underset{R}{\overset{R}{>}}C=O \ + \ H-\underset{\underset{}{\overset{H}{|}}}{N}-X \ \longrightarrow \ \underset{R}{\overset{R}{>}}C=N-X \ + \ H_2O$$

Carbonyl compounds react with compounds containing the $-NH_2$ group in what is known as a condensation reaction, since it entails the elimination of water.

When "X" represents an alkyl or aryl group, the compound is known as a primary amine. Its reaction with a carbonyl results in the formation of an imine, which is characterized by the presence of a carbon–nitrogen double bond ($-C=N-$).

The condensation reaction is also known as an addition–elimination reaction. A simplified view of the mechanism consists first of a nucleophilic addition reaction, whereby the amine acts as a nucleophile attacking the electron-deficient carbonyl C atom. Subsequently, the elimination of water gives the iminium ion, which then undergoes deprotonation to form the neutral imine product.

When "X" represents an $-OH$ group, the compound is known as hydroxylamine. Its reaction with a carbonyl results in the formation of an oxime.

When "X" represents another $-NH_2$ group, the compound is known as hydrazine. Its reaction with a carbonyl results in the formation of a hydrazone.

If there is excess ethanal present, it can further react with the hydrazone:

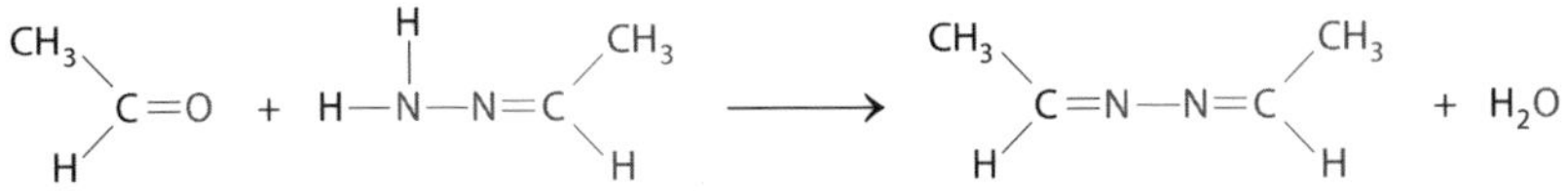

$$CH_3\!\!\diagdown\!\!\underset{H}{\overset{}{C}}=O \ + \ H\!-\!\underset{}{\overset{H}{N}}\!-\!N\!=\!C\!\!\diagup^{CH_3}_{\diagdown H} \ \longrightarrow \ CH_3\!\!\diagdown\!\!\underset{H}{\overset{}{C}}=N\!-\!N\!=\!C\!\!\diagup^{CH_3}_{\diagdown H} \ + \ H_2O$$

When a carbonyl compound reacts with 2,4-dinitrophenylhydrazine (2,4-DNPH), a 2,4-dinitrophenylhydrazone is formed.

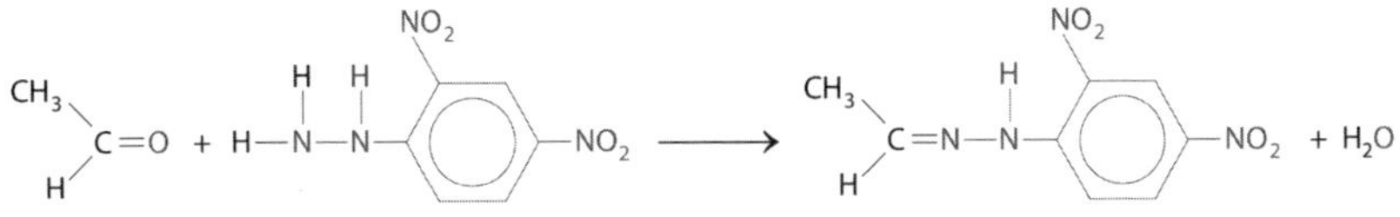

2,4-dinitrophenylhydrazine   a 2,4-dinitrophenylhydrazone
(orange precipitate)

The distinctive appearance of the red (for aromatic carbonyl compound) or yellow (for aliphatic carbonyl compound) precipitate is a qualitative determinant for the presence of aldehydes and ketones. Dissolving solid 2,4-dinitrophenylhydrazine in methanol and some concentrated sulfuric acid gives Brady's reagent, which is used in the detection of these carbonyl groups.

The condensation product that is formed can have a unique stereoisomeric property at the — C=N–, as shown below:

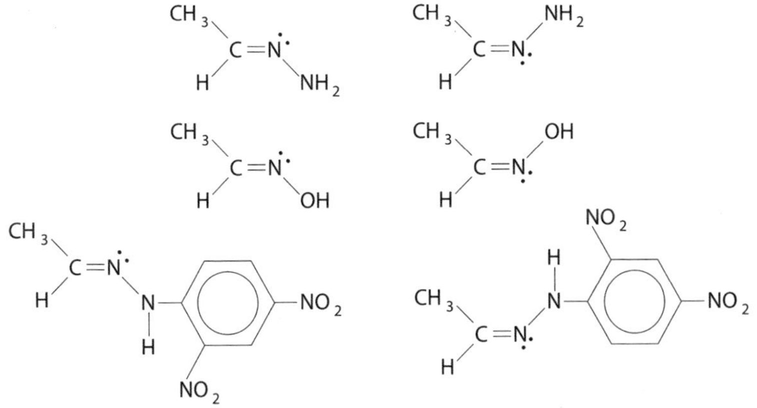

Q: There is also a carbonyl functional group in the –COOH group. Thus, can the carboxylic acid also undergo an addition–elimination reaction with 2,4–DNPH?

A: No, it can't. This is because we would expect an acid–base reaction between carboxylic acid and 2,4–DNPH instead, as the –NH$_2$ group of 2,4–DNPH is basic in nature. In addition, the lone pair of electrons of the –OH group of the –COOH functional group is delocalized into the C=O group, making the molecule relatively more stable. If an addition–elimination reaction were to take place, the resonance stabilization effect would be lost.

Q: How about for functional groups such as esters, amides and acyl halides?

A: No. There is a better leaving group, such as –OR, –NH$_2$ and –Cl, bonded to the carbonyl carbon atom. Thus, the condensation mechanism won't work here; nucleophilic substitution will take place instead. Similarly, if an addition–elimination reaction were to take place, the resonance stabilization effect mentioned above would be lost.

Q: What happens when we reduce an imine, an oxime or hydrazone? What do we get?

A: Imines, oximes and hydrazones can undergo reduction via LiAlH$_4$ in dry ether to produce an amine or substituted amine. For example:

$$\underset{R}{\overset{R}{>}}C=N-X \ + \ 2[H] \ \xrightarrow[\text{dry ether}]{\text{LiAlH}_4} \ R-\overset{\overset{\displaystyle R}{|}}{\underset{\underset{\displaystyle H}{|}}{C}}-\overset{\overset{\displaystyle H}{|}}{N}-X$$

9.5.3 *Formation of Carboxylic Acid — Oxidation*

$$R-\overset{\overset{\displaystyle O}{\|}}{C}-H \ + \ [O] \ \xrightarrow[\text{heat}]{K_2Cr_2O_7,\, H_2SO_4} \ R-\overset{\overset{\displaystyle O}{\|}}{C}-OH$$

Reagents:	$K_2Cr_2O_7(aq)$, $H_2SO_4(aq)$
Conditions:	Heat
Options:	$KMnO_4(aq)$, $H_2SO_4(aq)$

In Chapter 8, it was mentioned that ketones are more resistant to oxidation compared to aldehydes. This oxidation reaction is one method that can be used to distinguish between aldehydes and ketones. However, one must note that some other functional groups may also give positive results. Therefore, the presence of these may mask either the absence or presence of a specific carbonyl group.

For instance, if we subject an unknown compound solely to this oxidation reaction hoping to determine whether an aldehyde or ketone functional group is present, can we ascertain that the decolorization of purple $KMnO_4(aq)$ is due to the presence of an aldehyde group?

If the unknown compound happens to have the following structure, which contains both the ketone and primary alcohol functional groups, you will expect the alcohol to be oxidized by $KMnO_4$.

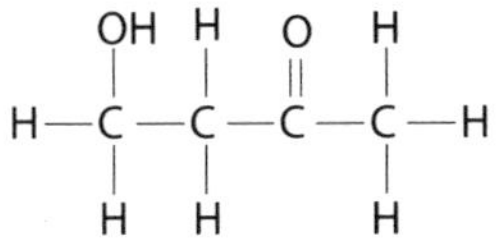

In this case, we cannot attribute the decolorization of the oxidizing agent to the presence of an aldehyde, can we?

Q: Is it really true that a ketone can't be oxidized at all?

A: Not really. If a very strong oxidizing agent is used, the ketone can still be oxidized, but in a destructive way, with the cleavage of C–C bonds. This is possible because of a phenomenon known as keto-enol tautomerism.

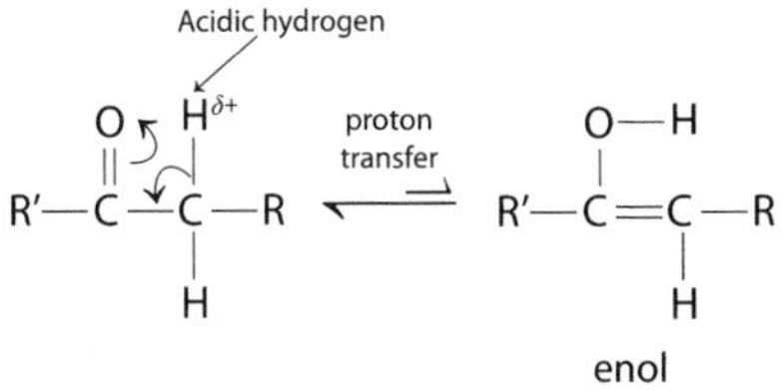

The two tautomers, ketone and enol, are isomers, and tautomerism is an equilibrium process. Tautomerism arises because of the proximity of an acidic α-hydrogen to a lone pair of electrons on the carbonyl oxygen atom. The enol is known as such because there is an alkene functional group together with a hydroxyl group. The presence of this alkene functional group makes the enol susceptible to oxidative cleavage like what would happen to an alkene molecule (refer to Chapter 5). As the concentration of the enol is very minimal, indicated by a shorter forward arrow, oxidation of the ketone is not very prominent. But in the presence of stronger oxidizing agent, the position of the equilibrium can be driven to the right.

9.5.4 *Formation of Carboxylate Salts Using Tollens' Reagent — Oxidation*

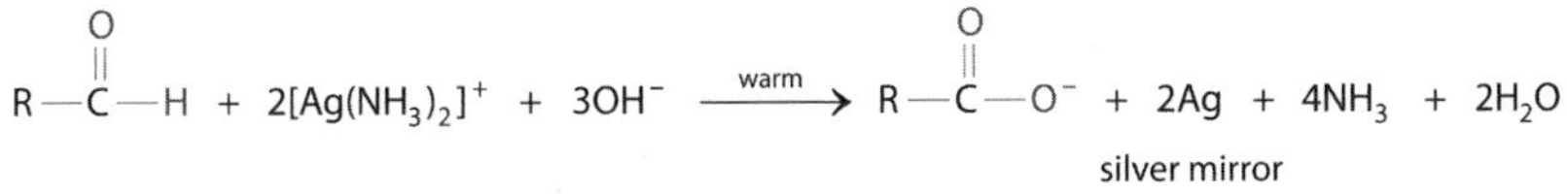

Reagents:	Tollens' reagent
Conditions:	Heat
Options:	—

Tollens' reagent contains aqueous diamminesilver(I) ions, [Ag $(NH_3)_2]^+$, which function as an oxidizing agent in this redox reaction. The diamminesilver(I) ion oxidizes aldehyde to carboxylic acid and is itself reduced to silver metal. It should be noted that the actual organic product is the carboxylate ion since the carboxylic acid that is formed deprotonates in the alkaline medium.

This reaction is also known as the "silver mirror" test, as a positive result is the formation of a silver mirror achieved under carefully controlled conditions.

This distinctive feature is used to test for the presence of aldehydes and to distinguish them from ketones, since the latter give a negative result with Tollens' reagent.

Q: Is there a way to help remember the balanced equation for the reaction of an aldehyde with Tollens' reagent?

A: Well, you can try formulating both the oxidation and reduction half-equations as follows:

Oxidation half-equation:

$RCHO \rightarrow RCOO^-$

$RCHO + H_2O \rightarrow RCOO^-$ (Add H_2O to balance O)

$RCHO + H_2O \rightarrow RCOO^- + 3H^+$ (Add H^+ to balance H)

$RCHO + H_2O + 3OH^- \rightarrow RCOO^- + 3H^+ + 3OH^-$

(Add OH^- to remove the H^+ as solution is alkaline)

$RCHO + 3OH^- \rightarrow RCOO^- + 2H_2O$

$RCHO + 3OH^- \rightarrow RCOO^- + 2H_2O + 2e^-$

(Add electron to balance charge).

Reduction half-equation:

$[Ag(NH_3)_2]^+ \rightarrow Ag$

$[Ag(NH_3)_2]^+ \rightarrow Ag + 2NH_3$

$[Ag(NH_3)_2]^+ + e^- \rightarrow Ag + 2NH_3$

(Add electron to balance charge).

Overall balanced equation:

$$RCHO + 3OH^- + 2[Ag(NH_3)_2]^+$$
$$\rightarrow RCOO^- + 2H_2O + 2Ag + 4NH_3.$$

9.5.5 *Formation of Carboxylate Salts using Fehling's Reagent — Oxidation of Aliphatic Aldehydes*

$$R-\overset{\overset{\text{O}}{\|}}{C}-H + 2Cu^{2+} + 5OH^- \xrightarrow{\text{warm}} R-\overset{\overset{\text{O}}{\|}}{C}-O^- + Cu_2O + 3H_2O$$

reddish-brown
precipitate

Reagents:	Fehling's solution
Conditions:	Heat
Options:	—

The oxidizing agent in the alkaline Fehling's solution is a copper(II) complex. In this redox reaction, the aldehye is oxidized to the carboxylic acid. However, as the reaction occurs in an alkaline medium, the actual oxidation product obtained is the carboxylate ion. The copper(II) ion in the Fehling's solution is reduced to the reddish brown Cu_2O, which contains copper in the $+1$ oxidation state. This oxidative reaction is similar to the action of Benedict's solution on reducing sugar, which also contains a copper(II) complex but one dissimilar to the one in Fehling's solution. The reducing sugar possesses an aldehydic functional group too.

Q: How does one formulate the balanced equation between Fehling's solution and an aldehyde?

A: The oxidation equation would be the same as the one for Tollens' reagent. As for deriving the reduction half-equation, it would be as follows:

$$2Cu^{2+} \rightarrow Cu_2O$$
$$2Cu^{2+} + H_2O \rightarrow Cu_2O$$
$$2Cu^{2+} + H_2O \rightarrow Cu_2O + 2H^+$$
$$2Cu^{2+} + H_2O + 2OH^- \rightarrow Cu_2O + 2H^+ + 2OH^-$$
$$2Cu^{2+} + 2OH^- \rightarrow Cu_2O + H_2O$$

$$2Cu^{2+} + 2OH^- + 2e^- \rightarrow Cu_2O + H_2O$$
(Balanced reduction half-equation).

Only aliphatic aldehydes give positive results. Aromatic aldehydes and ketones do not react with Fehling's solution. As such, this reaction can be used to tell an aliphatic aldehyde apart from the other two classes of carbonyl compounds.

Q: If Fehling's solution can only oxidize an aliphatic aldehyde, while Tollens' reagent can oxidize both, does it mean that Tollen's reagent is a stronger oxidizing agent?

A: Yes, Tollen's reagent is a stronger oxidizing agent than Fehling's solution. This would also mean that the aromatic aldehyde is more resistant to oxidation than the aliphatic aldehyde.

Q: Why is the aromatic aldehyde more resistant to oxidation than the aliphatic aldehyde?

A: Due to the delocalization of electrons (see below), the oxidation number of the carbonyl carbon atom can be perceived to be zero. For the aliphatic aldehyde, the oxidation number of the carbonyl carbon is $+1$. The carbonyl carbon of the carboxylic acid has an oxidation number of $+3$. It is thus more difficult to go from an oxidation number of 0 to $+3$ as compared to from $+1$ to $+3$ (for aliphatic aldehyde). Hence, a stronger oxidizing agent is required to "move" it from 0 to $+3$.

Example: To determine the identity of liquids **A**, **B** and **C** given that these are propanal, propanone and benzaldehyde, but not necessarily in that order.

Propanal   Propanone   Benzaldehyde

Approach:

Since there are three colorless unknown compounds, at least two different tests need to be carried out to correctly identify them, as shown below:

	Observations recorded for:		
	A	**B**	**C**
Test 1: AddTollens' reagent and warm	Silver mirror	No silver mirror	Silver mirror
Test 2: Add Fehling's solution and warm	Reddish-brown ppt	No reddish-brown ppt	No reddish-brown ppt
Deductions made:	propanal	propanone	benzaldehyde

9.5.6 Formation of Tri-Iodomethane (Iodoform) — Oxidation

Only carbonyl compounds containing the following bonding pattern can undergo this oxidation reaction. This group consists of ethanal and all methyl ketones.

$$\boxed{\text{The carbonyl carbon atom here must be bonded to either a } H \text{ atom or another } C \text{ atom}} \quad \overset{\overset{O}{\|}}{-C-CH_3}$$

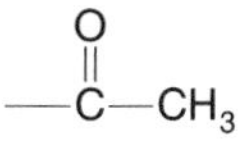

When such carbonyl compounds are warmed with alkaline $I_2(aq)$, a carboxylate salt will be obtained along with yellow crystals of tri-iodomethane:

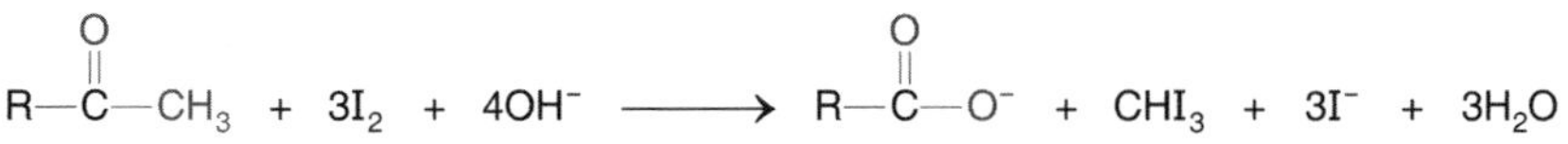

$$R-\overset{\overset{O}{\|}}{C}-CH_3 + 3I_2 + 4OH^- \longrightarrow R-\overset{\overset{O}{\|}}{C}-O^- + CHI_3 + 3I^- + 3H_2O$$

Reagents:	$I_2(aq)$ with $NaOH(aq)$
Conditions:	Warm
Options:	$KOH(aq)$

Since an alkaline medium is used in this reaction, the oxidized product is the carboxylate ion.

Q: Is there a way to derive the balanced equation from the basic one?

A: Yes, if you can't memorize it, just know how to derive it as follows:

$$RCOCH_3 \rightarrow RCOO^- + CHI_3$$

$$RCOCH_3 + H_2O \rightarrow RCOO^- + CHI_3$$

$$RCOCH_3 + H_2O + 3I_2 \rightarrow RCOO^- + CHI_3$$

$$RCOCH_3 + H_2O + 3I_2 \rightarrow RCOO^- + CHI_3 + 3I^-$$

$$RCOCH_3 + H_2O + 3I_2 \rightarrow RCOO^- + CHI_3 + 3I^- + 4H^+$$

$$RCOCH_3 + H_2O + 3I_2 + 4OH^- \rightarrow RCOO^- + CHI_3 + 3I^-$$
$$+ 4H^+ + 4OH^- \text{ (Alkaline medium used)}$$

$$\mathbf{RCOCH_3 + \underline{3}I_2 + \underline{4}OH^- \rightarrow RCOO^- + CHI_3 + \underline{3}I^- + \underline{3}H_2O.}$$

Another way to help remember is that the I_2: OH^-: I^-: H_2O ratio is $3:4:3:3$.

Alcohols with the following bonding pattern also give a positive result for the tri-iodomethane test:

The carbon atom here must be bonded to either a H atom or another C atom

$$\begin{array}{c} OH \\ | \\ -C-CH_3 \\ | \\ H \end{array}$$

Q: Is there a way to derive the balanced equation from the basic one?

A: Yes, take note that the alcohol is being oxidized to the methylketone first, as follows:

Oxidation half-equation:

$$RCH(OH)CH_3 \rightarrow RCOCH_3$$

$$RCH(OH)CH_3 \rightarrow RCOCH_3 + 2H^+$$

$$RCH(OH)CH_3 + 2OH^- \rightarrow RCOCH_3 + 2H^+ + 2OH^-$$
$$\text{(Alkaline medium used)}$$

$$RCH(OH)CH_3 + 2OH^- \rightarrow RCOCH_3 + 2H_2O + 2e^-.$$

Reduction half-equation:

$$I_2 + 2e^- \rightarrow 2I^-.$$

Overall:

$$\mathbf{RCH(OH)CH_3 + 2OH^- + I_2 \rightarrow RCOCH_3}$$
$$\mathbf{+2H_2O + 2I^-} \qquad \text{(Equation 1).}$$

Then, adding Equation 1 with the following equation:

$$RCOCH_3 + 3I_2 + 4OH^- \rightarrow RCOO^- + CHI_3 + 3I^- + 3H_2O,$$

giving: $RCH(OH)CH_3 + \underline{4}I_2 + \underline{6}OH^- \rightarrow RCOO^- + CHI_3 + \underline{5}I^- + \underline{5}H_2O$.

Another way to help remember is that the I_2: OH^-: I^-: H_2O ratio is $4 : 6 : 5 : 5$.

Apart from I_2, both Cl_2 and Br_2 can also be used in similar reactions to yield chloroform ($CHCl_3$) and bromoform ($CHBr_3$). However, their non-distinctive appearances render them not useful as distinguishing tests. In addition, both $CHCl_3$ and $CHBr_3$ are liquids at room temperature, which makes them inconspicuous as detected products.

Q: Can you show us how CHI_3 is formed?
A: The mechanism is as follows:

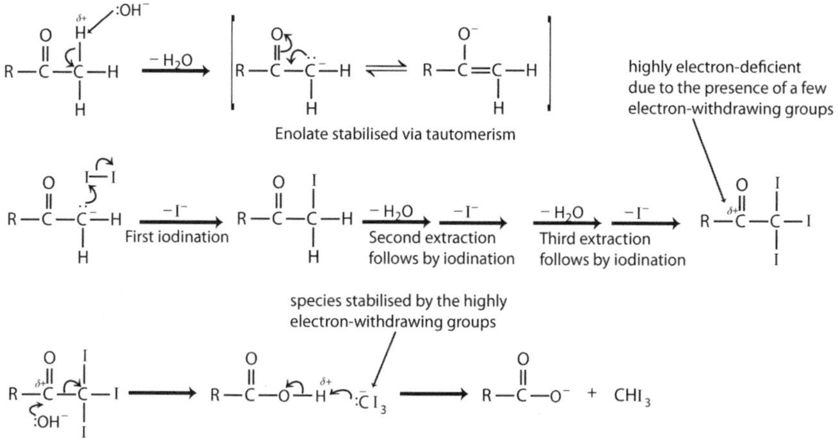

9.5.7 *Formation of Alcohols — Reduction*

Aldehydes are reduced to primary alcohols whereas ketones are reduced to secondary alcohols.

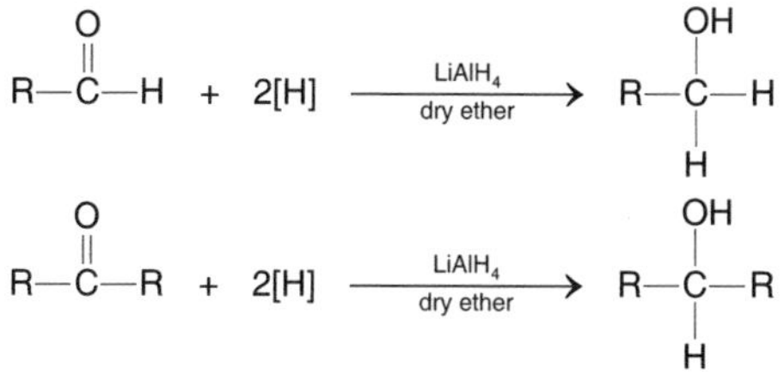

For Step (i),		For Step (ii),	
Reagents:	LiAlH$_4$ in dry ether	Reagents:	Water
Conditions:	Room temperature	Conditions:	Heat
Options:	NaBH$_4$(aq) or H$_2$/Pt/Heat	Options:	—

Any of the three reagents can be used to reduce the carbonyl compounds. The selection of a particular method over the others should factor in the reactivity, if any, of other functional groups present in the intended compound. For instance, if an organic compound contains both the aldehyde group and alkene functional groups, the use of H$_2$ with Ni is not feasible unless the objective is to reduce both groups.

Among the three reactions, LiAlH$_4$ (lithium aluminium hydride) can reduce a wider range of organic compounds that include the following:

- Carboxylic acids, esters, and aldehydes — these are reduced to primary alcohols;
- Ketones — these are reduced to secondary alcohols;
- Amides — these are reduced to amines;
- Nitriles — these are reduced to primary amines.

LiAlH$_4$ does not affect both double and triple carbon–carbon bonds (i.e. C=C and C≡C) as it only attacks polar bonds. It is to be noted that the reduction cannot be carried out in the presence of water as LiAlH$_4$ reacts vigorously with it:

$$LiAlH_4 + 4H_2O \longrightarrow LiOH + Al(OH)_3 + 4H_2.$$

Reaction with NaBH$_4$ (sodium borohydride) can be carried out in aqueous solution. Since it is a milder reducing agent than LiAlH$_4$ and less sensitive to water, NaBH$_4$ is commonly used as a laboratory reducing agent.

The reduction mechanism, involving either LiAlH$_4$ or NaBH$_4$, is rather complex, but it can be thought to proceed via the following mechanism:

aldehyde   alkoxide   primary alcohol

In the first step, the reducing agent acts as a hydride ion donor (H^-), which attacks the electron-deficient carbonyl C atom in a nucleophilic addition reaction. Water or aqueous acid is then added in the subsequent step to protonate the alkoxide intermediate to yield the alcohol.

Exercise:
What is the organic product formed from the reduction of propanone with $LiAlD_4$, in dry ether, followed by D_2O?

Solution:

$$CH_3 - \overset{\overset{\displaystyle OD}{|}}{\underset{\underset{\displaystyle CH_3}{|}}{C}} - D$$

9.6 Aromatic Carbonyl Compounds

Benzaldehyde and phenylethanone are the simplest examples of aromatic carbonyl compounds.

Benzaldehyde   Phenylethanone

They both undergo similar reactions to their aliphatic counterparts. One exception is the difference in behavior of benzaldehyde and aliphatic aldehydes towards Fehling's solution.

These aromatic carbonyl compounds are less reactive towards nucleophilic attack as the carbonyl C atom is less electron deficient due to the interaction of the π electron cloud of the carbonyl group and those of the adjacent benzene ring. In addition, the bulky benzene ring poses steric hindrance to the approaching nucleophile.

Being derivatives of benzene, these compounds can also undergo electrophilic substitution reactions whereby the carbonyl group is considered to be ring deactivating and 3-directing.

9.7 Summary

The mind map below focuses on the reactions of propanal and propanone. Propene is used as a starting material so as to integrate the chemistry of aldehyde and ketone with other functional groups.

My Tutorial

1. A hydrocarbon with a relative molecular mass of 56 and composition by mass: 85.7% carbon and 14.3% hydrogen undergoes the following reaction:

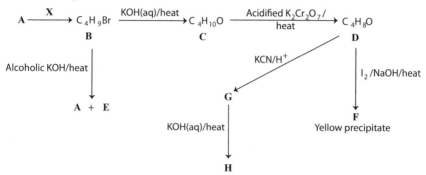

(a) Deduce the molecular formula of **A**. Draw all the possible structural formulae for this molecular formula.

(b) Deduce the structures of compounds **A** to **H**. Hence, give the reagents and condition for step **X**.

(c) Using the structural formulae, write balanced equations for the:

    (i) conversion of **A** to **B**.
    (ii) conversion of **B** to **C**.
    (iii) conversion of **C** to **D**.
    (iv) conversion of **D** to **F**.
    (v) conversion of **D** to **G**.
    (vi) conversion of **G** to **H**.
    (vii) conversion of **B** to **E**.

(d) (i) Give the mechanism for the conversion of **A** to **B**.

    (ii) Give possible reasons why two products **A** and **E** are possibly formed when **B** is heated with alcoholic KOH solution.

    (iii) Both compounds **B** and **E** exhibit stereoisomerism. State the type of stereoisomerism in each of the compounds and draw the stereoisomers.

(e) (i) If compound **B** is optically active but compound **C** is not, give the mechanism for the conversion of **B** to **C**.

    (ii) Both compounds **B** and **C** are optically active. Give the mechanism for the conversion of **B** to **C**.

(iii) Describe how the rate of each of the reactions in (i) and (ii) are affected when the concentration of KOH is doubled. Explain your reasoning.

(iv) By applying your understanding of the organic reaction mechanisms described in (e)(i) and (e)(ii), predict the organic products that you would expect when **B** reacts with potassium hydrogen sulfide, KHS.

(v) Explain briefly the different roles played by KOH in the conversion of **B** into **C** as compared to the conversion of **B** into **A** and **E**.

(f) (i) What structural features of **D** enables it to form **F**.

(ii) State the role played by iodine and sodium hydroxide.

(iii) Give the name and structural formula of the other organic product of this reaction.

(iv) Compound **C** would also be able to form **F**. Give the structural feature of **C** that gives the possible result.

(v) With appropriate reagents and conditions, describe a simple chemical test to differentiate **C** from **D**.

(g) (i) Give the mechanism for the conversion of **D** to **G**. Hence, draw the stereoisomers formed in **G**.

(ii) Compound **D** reacts with 2,4–DNPH to give its 2,4–dinitrophenylhydrazone derivatives.

(A) Write a balanced equation for the reaction between **D** and 2,4–DNPH.

(B) The 2,4–dinitrophenylhydrazone derivative exhibits stereoisomerism. Draw the possible stereoisomers and name the type of stereoisomerism.

(C) Suggest why the two stereoisomers have different melting points.

(h) Heating of the organic content would usually be carried out in a water-bath rather than over a Bunsen burner. Give a possible reason for such a practice.

2. (a) $5\,cm^3$ of an alkene **P** was completely burnt with $50\,cm^3$ of pure oxygen in a sealed vessel, and the resulting mixture showed a reduction in volume of $20\,cm^3$ when shaken with barium hydroxide solution, leaving a resultant volume of $20\,cm^3$.

P reacts with bromine water to give isomeric products **Q** and **R**. On treatment with concentrated sulfuric acid, **Q** does not react, but **R** is dehydrated to two isomeric bromoalkenes **S** and **T**. On oxidation at the double bond, **S** gives propanone among the products, while **T** gives methanal. Give the structural formulae of the compounds **P** to **T**. Write balanced equations for each of the reactions.

(b) Explain why propanone is not hydrolyzed to sodium ethanoate by aqueous barium hydroxide, but 1,1,1-trichloropropanone is.

(c) When $50\,\text{cm}^3$ of the gaseous CH_3COCH_3 was exploded with excess oxygen, there was an overall contraction of $x\,\text{cm}^3$. A further contraction of $y\,\text{cm}^3$ took place when excess calcium hydroxide was added. Deduce the values of x and y. (All measurements were done under r.t.p.)

(d) Account for why the reactivity of aldehydes and ketones towards nucleophiles falls along each of the series.

$$H_2CO > RCHO > R_2CO$$

$> (CH_3CH_2)_2CO > ((CH_3)_2CH)_2CO$

3. Propanone can be converted into MIBK, a solvent that is commonly used in adhesive, via the following synthetic pathway:

(a) Draw the displayed structural formula of compound **U**.

(b) Suggest a name for the reaction in Step **I** and draw the species that reacts with propanone.

(c) What is the role played by Al_2O_3?

(d) Suggest reagents and conditions for Step **III**.

(e) Suggest a simple chemical test to differentiate between propanone and compound **U**.

4. (a) Aldehydes and ketones both contain the carbonyl group, C=O, but differ in their positions in the hydrocarbon skeleton. The main reaction that these carbonyl compounds undergo is nucleophilic addition.

   (i) It is observed that aldehdyes are usually more reactive toward nucleophilic addition than ketones. State two reasons to account for this observation.

   (ii) When heated with an alcohol (which serves as the nucleophile) in the presence of aqueous hydrochloric acid, an aldehyde produces a hemiacetal, a functional group consisting of one –OH and one –OR group bonded to the same carbon. For example, the reaction of methanol and ethanal produces the following hemiacetal:

$$CH_3CHO + CH_3OH \longrightarrow CH_3-\underset{\underset{OCH_3}{|}}{\overset{\overset{OH}{|}}{C}}-H$$

hemiacetal

Propose the mechanism for the reaction described above.

(b) In addition to nucleophilic addition, aldehydes and ketones can also undergo a condensation reaction due to the unusual acidity of the hydrogen atom attached to the carbon adjacent to the carbonyl group. Such a hydrogen atom is referred to as α-hydrogen. With the aid of suitable diagrams, account for the unusual acidity of a α-hydrogen.

(c) Aldehydes that have α-hydrogen can react with themselves or another carbonyl compound when mixed with an aqueous acid or base. Such a reaction is known as aldol condensation and the resulting compounds, β-hydroxyl aldehydes, are known as aldol compounds as they possess both an aldehyde and an alcohol functional group. For example:

3-hydroxybutanal

(i) Suggest a suitable aldehyde as the starting material for the synthesis of the compound **X**, $CH_3CH_2CH(CHO)$ $CH(CH_2CH_2CH_3)OH$.

(ii) Hence, outline the reaction scheme for the conversion of the aldehyde in c(i) to compound **X**, stating clearly the reagents and conditions required and the structure of the intermediates formed.

(d) When a dicarbonyl compound is treated with a base, intramolecular aldol condensation can occur, leading to the formation of cyclic products as exemplified by 2,5-hexanedione below:

Another possible product of the aldol condensation of 2,5-hexanedione is

Suggest briefly why this compound is not formed.

CHAPTER 10

Carboxylic Acids and Their Derivatives

10.1 Introduction

Carboxylic acids are organic compounds that contain the following carboxyl functional group, denoted as –COOH.

$$\underset{\underset{\displaystyle OH}{\diagup}}{\overset{\overset{\displaystyle O}{\parallel}}{C}}$$

From the constitutional/structural formula, the carboxyl functional group is a composite of a carbonyl functional group and a hydroxyl group. Similar to that in carbonyl compounds, the carbon atom of the carboxyl functional group is sp^2 hybridized and trigonal planar in shape. Thus, it is not surprising that carboxylic acids possess physical and chemical properties similar to both the _carbo_nyl and hydro_xyl_ compounds.

Derivatives of carboxylic acids have the –OH group substituted. For instance, esters contain the functional group –COOR, and acid chlorides the –COCl group.

10.2 Nomenclature

Carboxylic acids are named with the suffix _–oic acid_. Other substituents are named as prefixes accompanied by the appropriate positional numbers. If the carboxylic acid group is considered a substituent, then it is named as the _–carboxy_ prefix (notice that the prefix is not "carboxyl").

Ethanoic acid   Benzoic acid

10.3   Physical Properties

10.3.1   *Melting and Boiling Points*

Melting and boiling points increase with increasing carbon chain length (Table 10.1) and decrease with increasing degree of branching. As with alcohols, the trend is accounted for by the increasing strength of instantaneous dipole–induced dipole (id–id) attractive forces due to the greater number of electrons accompanying the increasing carbon chain length, rather than hydrogen-bonding.

Table 10.1

Acid	Methanoic Acid	Ethanoic Acid	Propanoic Acid	Butanoic Acid
Structural formula				
Boiling point ($^\circ$C)	101.0	118.1	141.0	163.5

The boiling points of carboxylic acids are higher than those of their isomeric ester counterparts (Table 10.2). This is attributed to the stronger hydrogen bond between the acid molecules as compared to the van der Waals attractive forces (id–id and pd–pd) between the ester molecules. The same conclusion can be made on the higher boiling point of an alcohol when compared to that of the ester of similar relative molecular mass.

Table 10.2

	Ethanoic Acid	Methyl Methanoate	Propan-2-ol
Structural formula	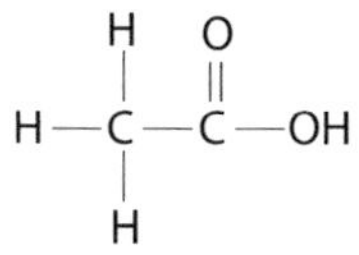	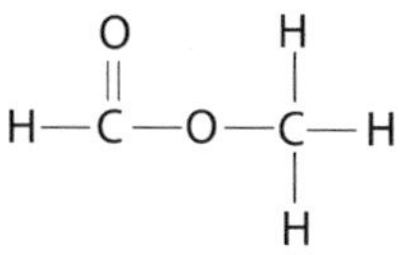	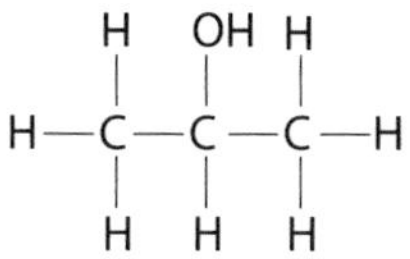
M_r	60.0	60.0	60.0
Boiling point (°C)	118.1	32.0	82.5

Q: Why is the boiling point of propan-2-ol lower than that of ethanoic acid?

A: The hydrogen bonds formed between carboxylic acids are much stronger as the –OH group of the carboxyl functional group is more polarized due to the electron-withdrawing effect of the carbonyl group. The electron deficiency of the hydrogen atom of the –OH group of carboxylic acid is thus greater than that on the –OH group of an alcohol. This results in a stronger attractive force between the electron-deficient hydrogen atom of the –OH group of the carboxyl functional group and the lone pair of electrons on another acid molecule.

In fact, the hydrogen bond is so strong that carboxylic acids exist as dimers (the joining of two molecules) in the liquid phase and in non-polar solvents.

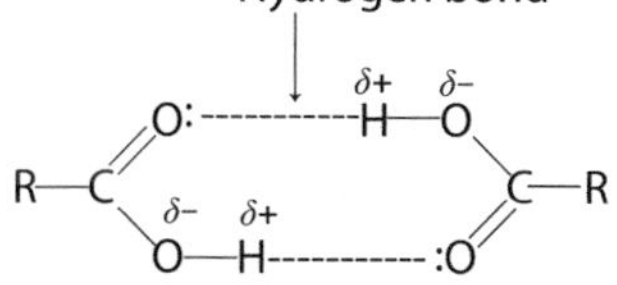

Hydrogen bond

A dimer

But in polar solvents such as water, carboxylic acids exist as singular entities (monomers), and some of these even dissociate in water to form the carboxylate ion. Carboxylic acids are stronger organic acids

as compared to phenol and alcohols but weaker acids as compared
to the mineral acids such as HCl and H_2SO_4. Refer to Chapter 8 for
the discussion of the strength of different organic acids.

$$R-\overset{\overset{O}{\|}}{C}-OH \; + \; H_2O \; \rightleftharpoons \; R-\overset{\overset{O}{\|}}{C}-O^- \; + \; H_3O^+$$

carboxylate ion

Q: Why is dimerization preferred when the carboxylic acid
molecules are in liquid phase or in a non-polar solvent?

A: In the liquid phase, the carboxylic acid molecules are closer
together, therefore dimerization is more feasible than in the
gaseous state, where the molecules are further apart. In a non-
polar solvent, dimerization is again feasible as the formation of
hydrogen bonds allows the polar carboxylic acid molecules to
aggregate among their "likes." This would lower the energy state
of the solution as compared to if the two entities do not mix at all.

Q: Why can't the alcohol molecules dimerize?

A: In order to form a stable dimer molecule, the intermolecular force
must be strong enough to "hold" the two molecules together.
For the carboxylic acid, it is possible for two reasons. First,
the hydrogen bond between two carboxylic acid molecules is
stronger than that between the alcohol ones due to the more
electron-deficient hydrogen atom of the –COOH group. Second,
two hydrogen bonds are being formed between two carboxylic
acid molecules which are absent in alcohol. Thus, the strength
and extensivity of hydrogen bond formed between the carboxylic
acid molecules result in dimerization.

Q: Why are carboxylic acids weaker acids than the mineral acids
such as HCl?

A: It is easier to cleave the H–Cl bond than the O–H bond in the
carboxylic acid, as reflected in their bond energies (BE).

$$BE \; (H–Cl) = 432 \, kJ \, mol^{-1}; \; BE(O–H) = 460 \, kJ \, mol^{-1}.$$

Q: But shouldn't the highly electron-deficient hydrogen atom of
the –COOH group cause it to be more susceptible to attack by
the Lewis base (i.e. H_2O molecule)?

A: Yes, indeed the highly electron-deficient hydrogen atom of the
–COOH group is more susceptible to attack by the Lewis base. But
this factor does not overshadow the factor of the strong O–H bond.

10.3.2 *Solubility*

The ability to form hydrogen bonds accounts for the high solubility of short-chain carboxylic acids in water. However, as the length of the carbon chain increases, solubility in water will decrease due to the increasing size of the hydrophobic alkyl chain. The latter is responsible for carboxylic acid's increasing solubility in non-polar solvents. To enhance the solubility of the higher molecular weight carboxylic acid in water, the molecules aggregate together to form the three-dimensional spherical structures known as micelles. The micelle is much more soluble in water as all the hydrophobic groups aggregate away from the polar water molecules, forming the core of the spherical structure. The hydrophilic –COOH group forms the outer spherical surface of the micelle, thus interacting readily with the polar water molecules.

10.4 Preparation Methods for Carboxylic Acids

10.4.1 *Oxidation of Primary Alcohols*

Primary alcohols are readily oxidized to aldehydes first, which in turn are easily oxidized to carboxylic acids.

$$\text{Reagents:} \quad K_2Cr_2O_7(aq), H_2SO_4(aq)$$

Reagents: $K_2Cr_2O_7(aq)$, $H_2SO_4(aq)$
Conditions: Heat
Options: —

10.4.2 *Oxidation of Aldehydes*

Reagents:	$K_2Cr_2O_7(aq)$, $H_2SO_4(aq)$
Conditions:	Heat
Options:	$KMnO_4(aq)$, $H_2SO_4(aq)$

Oxidation of aldehydes to obtain the corresponding carboxylic acids can be carried out using Tollens' reagent or Fehling's solution (except for benzaldehyde). Refer to Chapter 9 for the chemistry on aldehyde. The carboxylate formed can be acidified to obtain the carboxylic acid product.

10.4.3 *Oxidation of Alkenes*

Reagents:	$KMnO_4(aq)$, $H_2SO_4(aq)$
Conditions:	Heat
Options:	—

Q: Why doesn't the oxidation of buta-1,2-diene yield the following ethanedioic acid instead?

A: Ethanedioic acid is actually being formed initially, but it is further oxidized by the acidified $KMnO_4(aq)$ to form CO_2 and H_2O. Its susceptibility to oxidation is attributed to the ease of cleavage of the C–C bond that has its electron density drawn away by the four highly electronegative O atoms.

10.4.4 *Hydrolysis of Nitriles*

Reagents:	HCl(aq)
Conditions:	Heat
Options:	H_2SO_4(aq)

Nitriles undergo acidic hydrolysis to form carboxylic acids (see Chapter 8 for details). If basic hydrolysis is carried out, the carboxylate product has to be acidified to obtain the carboxylic acid.

$$CH_3-C\equiv N \xrightarrow[\text{heat}]{\text{NaOH(aq)}} CH_3-\overset{\overset{\displaystyle O}{\|}}{C}-O^-Na^+ \xrightarrow{\text{HCl(aq)}} CH_3-\overset{\overset{\displaystyle O}{\|}}{C}-OH$$

Reagents:	NaOH(aq) followed by HCl(aq)
Conditions:	Heat
Options:	KOH(aq) followed by H_2SO_4(aq)

Q: Can the above hydrolysis take place without any heating?

A: Heating is necessary for the above hydrolysis to take place! Thus, if you see base or acid coupled with heating, it must be meant for hydrolysis reaction. Normal acid-base reaction does not need heating!

10.4.5 *Oxidation of Alkylbenzene*

Benzoic acid can be obtained from the oxidation of benzaldehyde or the corresponding aromatic primary alcohol. It can also be obtained from the side-chain oxidation of an alkylbenzene by heating it with acidified $KMnO_4$(aq).

$$\text{(benzene with } CH_3) + 3[O] \xrightarrow[\text{heat}]{KMnO_4, H_2SO_4} \text{(benzene with } COOH) + H_2O$$

Reagents:	$KMnO_4$(aq), H_2SO_4(aq)
Conditions:	Heat
Options:	Alkaline medium using NaOH(aq) followed by acidification

10.5 Chemical Properties of Carboxylic Acids

Ethanoic acid

The presence of the highly electron deficient carbon atom of the carboxyl group makes it susceptible to nucleophilic attack, in a similar fashion to aldehydes and ketones. In fact, the electron deficiency is even greater than in aldehydes and ketones due to the presence of two highly electronegative O atoms. But unlike aldehydes and ketones, carboxylic acids do not undergo addition reaction. In contrast, they prefer to undergo nucleophilic substitution due to the presence of the hydroxyl group, which is a better leaving group relative to the breaking of the C–C or C–H bonds, in ketone or aldehyde respectively, to generate carbanion (R^-) or hydride (H^-) ions as the leaving groups. The hydroxyl group can break off from the carboxyl functional group and exist "comfortably" as a hydroxide ion.

With such close proximity, both the carbonyl and hydroxyl functional groups of the carboxyl actually alter each other's properties. The interaction of the *p*-orbital of the hydroxyl O atom with that of the carbonyl group results in the delocalization of electrons into the carbonyl group. This thus results in a partial double bond characteristic between the O atom of the hydroxyl group and the carbonyl carbon.

Q: If there is a partial double bond characteristic between the O atom of the hydroxyl group and the carbonyl carbon, shouldn't this make the hydroxyl group not likely to break away from the carbonyl group? Like in the case of phenol or chlorobenzene?

A: Yes, indeed it would be more difficult to break the C—OH bond due to the partial double bond characteristic. But unlike in the case of phenol or chlorobenzene, the carbonyl carbon atom is more susceptible to nucleophilic attack as its electron deficiency is increased due to the presence of another highly electron-withdrawing O atom. The latter factor is a more dominant one, thus resulting in an ease of substitution of the –OH group forming other acid derivatives. In addition, one can look at the reaction from the mechanistic perspective. Overall, carboxylic acid undergoes the nucleophilic substitution mechanism. But the mechanism is a nucleophilic addition followed by elimination (refer to Chapter 3 on mechanism). During the addition stage, the intermediate has a tetrahedral configuration. At this stage, the partial double characteristic between the O atom of the hydroxyl group and the carbonyl carbon disappears. Hence, the C–OH bond becomes more susceptible to be broken during the elimination stage. The breakage of the C–OH bond would result in a reestablishment of the resonance stability in the acid derivatives.

In reality, the reactions of the carboxylic acids mostly revolve around the hydroxyl group. Cleavage of the O–H bond gives rise to the acidic nature of carboxylic acids (see Chapter 8 for nature of acidity). Cleavage of the C–O bond allows the carboxylic acid to be converted to acid derivatives, which is only possible because of the carbonyl group present, then nucleophilic acyl substitution mechanism that takes place and the ease of leaving groups (see Chapter 3).

10.5.1 *Acid–Base Reaction with Strong Bases and Reactive Metals*

$$CH_3COOH + Na \longrightarrow CH_3COO^-Na^+ + 1/2H_2$$
$$CH_3COOH + NaOH \longrightarrow CH_3COO^-Na^+ + H_2O$$
$$CH_3COOH + NaHCO_3 \longrightarrow CH_3COO^-Na^+ + H_2O + CO_2$$
$$2CH_3COOH + Na_2CO_3 \longrightarrow 2CH_3COO^-Na^+ + H_2O + CO_2.$$

In qualitative analysis, confirmatory tests are usually carried out to determine the identity of the gas evolved. For instance:

If $H_2(g)$ is present, it will extinguish a lighted splint with a pop sound.

If $CO_2(g)$ is present, it will form a white precipitate with $Ca(OH)_2(aq)$.

Carboxylic acids are stronger acids compared to alcohols, as observed from their behavior towards bases. Refer to Chapter 8 for more details on the acidity of organic compounds.

	Na	NaOH	Na_2CO_3 or $NaHCO_3$
RCOOH	☑ $H_2(g)$ + salt	☑ H_2O + salt	☑ $CO_2(g)$ + H_2O + salt
Phenol	☑ $H_2(g)$ + salt	☑ H_2O + salt	—
water	☑ $H_2(g)$ + salt	—	—
ROH	☑ $H_2(g)$ + salt	—	—

This difference in acidity of carboxylic acids, phenols and alcohols can be used to differentiate them.

Example: To distinguish between ethanoic acid and phenol

Approach:

Test: Add Na_2CO_3.

Observations: For ethanoic acid, a gas is evolved which forms a white precipitate with $Ca(OH)_2(aq)$.
For phenol, no gas is evolved which forms a white precipitate with $Ca(OH)_2(aq)$.

10.5.2 *Formation of Acyl Chlorides — Nucleophilic Acyl Substitution/Condensation*

$$CH_3-\overset{\overset{\displaystyle O}{\|}}{C}-OH \longrightarrow CH_3-\overset{\overset{\displaystyle O}{\|}}{C}-Cl$$

Reagents:	$PCl_5(s)$
Conditions:	Room temperature
Options:	$PCl_3(l)$; $PBr_3(l)$; (red P + I_2) or $SOCl_2(l)$

The use of different reagents produces different side products, as shown in the following general equations:

- $RCOOH + PCl_5 \rightarrow RCOCl + POCl_3 + HCl$
- $3RCOOH + PCl_3 \rightarrow 3RCOCl + H_3PO_3$
- $RCOOH + SOCl_2 \rightarrow RCOCl + SO_2 + HCl.$

The formation of white fumes of HCl can be useful for determining the presence of the carboxyl functional group. But it cannot be used to distinguish an alcohol from a carboxylic acid as the latter also reacts with PCl_5 and $SOCl_2$ to form HCl (see Chapter 8).

To detect a positive result for the reaction with either PCl_5 or $SOCl_2$, it is not sufficient to conclude from just the evolving HCl fumes. A confirmatory test for the gas has to be done. A typical test for an acidic gas such as HCl will be that it turns moist blue litmus red. On a similar note, the litmus test for a basic gas will have it turn moist red litmus blue.

In all, these reactions involve a nucleophilic attack on the carbonyl C atom that leads to the cleavage of the C–O bond. But before this can occur, the hydroxyl group is first converted into a better leaving group that is more stable.

10.5.3 *Formation of Esters with Alcohols — Esterification via Nucleophilic Acyl Substitution/Condensation*

	Ethanol	Ethanoic acid	Ethyl ethanoate

Reagents:	ROH
Conditions:	Conc. H_2SO_4 as catalyst/Heat
Options:	Can use other acid catalyst

As this reaction only reaches equilibrium after a few hours with the need of a catalyst, and yield of ester product is not high, esterification is best done using acyl chlorides (see Section 10.6.4).

10.5.4 *Reduction of Carboxylic Acids to Primary Alcohols*

Carboxylic acids can be reduced to primary alcohols.

For Step (i),		For Step (ii),	
Reagents:	LiAlH$_4$ in dry ether	Reagents:	Water
Conditions:	Room temperature	Conditions:	Heat
Options:	NaBH$_4$(aq)* H$_2$/Pt/Heat$^\#$	Options:	—

*Being a milder reducing agent than LiAlH$_4$, NaBH$_4$(aq) is not able to reduce RCOOH and its derivatives.
$^\#$H$_2$/Pt/Heat cannot be used to reduce carboxylic acid, thus catalytic hydrogenation can be used to reduce carbon−carbon unsaturated bonds with the acid functional group present.

Refer to Chapter 9 on Carbonyl Compounds for further information on the reduction mechanism.

10.6 Derivatives of Carboxylic Acids

Derivatives of carboxylic acids can be obtained by replacing the hydroxyl group with other substituents. This is because the hydroxyl (–OH) group of the carboxyl group is a relatively good leaving group.

For instance, when a chlorine atom takes the place of the hydroxyl group, we have what is called an acid chloride. The formula for acid chlorides is usually written as RCOCl.

$$R-\overset{\overset{O}{\|}}{C}-OH \longrightarrow R-\overset{\overset{O}{\|}}{C}-Cl$$

If the hydroxyl group is replaced by an –OR group, the compound is known as an ester, usually abbreviated as RCOOR.

$$R-\overset{\overset{O}{\|}}{C}-OH \longrightarrow R-\overset{\overset{O}{\|}}{C}-O-R$$

Amides are obtained when the hydroxyl group is replaced by an amino group. Amides consist of the –CONH-functional group, and they can be categorized as follows:

$$R-\overset{\overset{O}{\|}}{C}-\overset{\overset{H}{|}}{N}-H \qquad R-\overset{\overset{O}{\|}}{C}-\overset{\overset{R}{|}}{N}-H \qquad R-\overset{\overset{O}{\|}}{C}-\overset{\overset{R}{|}}{N}-R$$

Primary (1°) amide   Secondary (2°) amide   Tertiary (3°) amide

10.6.1 *Nomenclature*

The names of these derivatives can be formed by substituting the end portion of the name of the corresponding carboxylic acid, as shown in Table 10.3 below.

Table 10.3

Acid & Derivatives	Carboxylic Acid	Acid Chlorides	Esters	Amides
Nomenclature	Name ends with *-ic acid*	Replace *-ic acid* with *-yl chloride*	Replace *-ic acid* with *-ate* preceded by the name of alcohol	Replace *-oic acid* with *-amide*
Examples	$CH_3-\overset{\overset{O}{\|\|}}{C}-OH$ Ethanoic acid	$CH_3-\overset{\overset{O}{\|\|}}{C}-Cl$ Ethanoyl chloride	$CH_3-\overset{\overset{O}{\|\|}}{C}-O-CH_3$ Methyl ethanoate	$CH_3-\overset{\overset{O}{\|\|}}{C}-\overset{\overset{H}{\|}}{N}-H$ Ethanamide
	⬡—$\overset{\overset{O}{\|\|}}{C}$—OH Benzoic acid	⬡—$\overset{\overset{O}{\|\|}}{C}$—Cl Benzoyl chloride	⬡—$\overset{\overset{O}{\|\|}}{C}$—O—$CH_3$ Methyl benzoate	⬡—$\overset{\overset{O}{\|\|}}{C}$—$NH_2$ Benzamide

10.6.2 *Physical Properties*

10.6.2.1 *Melting and boiling points*

Ethanamide is a liquid at r.t.p.; all other 1° amides are solid at r.t.p. The more extensive hydrogen bonding that exists between the 1° amide molecules makes them less volatile. A 1° amide molecule has two electron-deficient H atoms and three lone pairs of electrons, like a water molecule, it is capable of forming two hydrogen bonds per amide molecule. Unlike the 1° amide, the 2° amide has only one electron-deficient hydrogen atom, thus the extensivity of the hydrogen bonding between the 2° amide molecules is less extensive than for the 1° amide.

For the polar 3° amides, acid chlorides and esters, only weak van der Waals attractive forces of the permanent dipole–permanent dipole type exist between their molecules. They are not able to form hydrogen bonds as they lack a H atom bonded to a small and highly electronegative atom such as O, F and N. This also accounts for their higher volatility (relatively lower boiling point) as compared to the parent carboxylic acids which, of course, are capable of hydrogen bonding.

10.6.2.2 *Solubility*

Short-chain 1° and 2° amides have high solubility in water because they can act as both H–bond donor (by using the electron-deficient H atom) and acceptor (by using the lone pair of electrons), just like the parent carboxylic acids. Esters and 3° amides have lower solubility in water since they can only function as hydrogen bond acceptors and hence form less extensive hydrogen bonds with water molecules. Just as for any organic compounds, the solubility of the acid derivatives in polar solvents is compromised by the size of the hydrophobic group. As for acid chlorides, these readily hydrolyze in water (see Section 10.6.4).

10.6.3 *Chemical Properties*

The chemistry of the acid derivatives centers on nucleophilic acyl substitution that involves the replacement of good leaving groups such as –Cl, –OR, and –NH$_2$ by another nucleophile. These reactions mainly involve the conversion of the more reactive acid derivative into a less reactive one and also the conversion of these into the parent carboxylic acid.

Reactivity of these derivatives decreases in the order:

acid chloride   ester   amide

The observed reactivity trend can be accounted for by considering the ease of the nucleophile making a nucleophilic attack on the electron-deficient carbonyl carbon center. Based on the trend of electronegativity, O > N > Cl, we would expect the electron deficiency of the carbonyl carbon to be greatest for the ester, followed by the amide then the acid chloride. This obviously does not account for the observed trend of reactivity. So, what gives rise to the observed trend, especially since the Cl atom is not the most electronegative atom among the three atoms?

There are two different combinational factors, namely the delocalization of the lone pair of electrons of the $-Cl$, $-OR$, and $-NH_2$ groups into the carbonyl functional group and the difference in the electronegativities of the Cl, O, and N atoms.

The delocalization of the lone pair of electrons from the Cl, O, and N atoms into the carbonyl functional group(via the overlap of two p-orbitals) diminishes the electron deficiency on the carbonyl carbon. But the delocalization for acid chloride is the least effective as it involves a $2p$–$3p$ orbitals overlap between the carbonyl carbon and the chlorine atom. Thus, the electron deficiency of the carbonyl carbon for acid chloride is the highest among the three given derivatives. As for the higher reactivity of an ester as compared to an amide, the more electronegative nature of the O atom of the $-OR'$ group as compared to the N atom of $-NH_2$ results in a diminished effect of the delocalization. That is, the more electronegative O atom would "pull back" more electron density that has been delocalized than the N atom. Therefore, a higher level of electron deficiency would make the ester more susceptible to nucleophilic attack than the amide.

In addition, the trend that arises due to susceptibility to nucleophilic attack also mirrors the decreasing ease of the corresponding leaving group:

$$-Cl > -OR > -NH_2$$

After the cleavage of the C—Y bond, where Y = Cl, OR, or NH_2, the stability of the species that is generated would follow the following trend:

$$Cl^- > {}^-OR > {}^-NH_2$$

Q: How is the reactivity of the carboxylic acid towards nucleophilic substitution as compared to the trend above?

A: Carboxylic acid has the $-OH$ group attached to the carbonyl group, whereas an ester has the $-OR$ group bonded. Hence, both functional groups have the same O atom bonded to the carbonyl functional group. From the delocalization of electrons and electronegativity considerations, we would expect both the ester and the carboxylic acid to have the same reactivity. But we can rationalize the ester to be less reactive than the acid due to the electron-releasing effect of the alkyl group from the alcohol group in the ester; the electron-withdrawing effect of the O atom that is bonded to the carbonyl carbon is less than the O atom of the –OH group that is attached to the carbonyl group. As a result, the delocalization of electrons into the carbonyl functional group in the ester functional group is more effective. This would diminish the electron deficiency of the carbonyl carbon atom to a greater extent for the ester functional group as compared to the carboxylic acid. In addition, from the leaving group perspective, –OH is a better leaving group than the –OR group as the stability of the $^-$OH ion is higher than that of $^-$OR.

Q: Can I say that the partial double bond character for acid chloride < carboxylic acid < ester < amide?

A: Yes, you can. Basically this trend can be accounted for by considering the same factors that affect the electron deficiency of the carbonyl carbon atom.

Q: Can I say that the carboxylate ion is less reactive to nucleophilic substitution than the carboxylic acid molecule because the electron deficiency of the carbonyl carbon is lower for the carboxylate ion?

A: Yes. This is because in the carboxylate ion, the delocalization of the negative charge (which signifies an extra electron) into the carbonyl group decreases the electron deficiency of the carbon atom to a much greater extent than for carboxylic acid. In addition, the partial double bond characteristic is greater in the carboxylate ion than for the carboxylic acid molecule.

10.6.4 *Chemistry of Acid Chlorides*

10.6.4.1 *Formation of esters with alcohols — acylation via nucleophilic substitution/condensation*

$$CH_3-\overset{\overset{O}{\|}}{C}-Cl \ + \ H-O-\overset{\overset{H}{|}}{\underset{\underset{H}{|}}{C}}-\overset{\overset{H}{|}}{\underset{\underset{H}{|}}{C}}-H \ \longrightarrow \ CH_3-\overset{\overset{O}{\|}}{C}-O-\overset{\overset{H}{|}}{\underset{\underset{H}{|}}{C}}-\overset{\overset{H}{|}}{\underset{\underset{H}{|}}{C}}-H \ + \ HCl$$

ethanoyl chloride   ethanol   ethyl ethanoate

Reagents:	ROH
Conditions:	Room temperature
Options:	—

The advantages of this method over esterification involving carboxylic acid and alcohol include a good yield of ester, the doing-away of heating and the use of a catalyst. All this is because acid chlorides are highly reactive towards nucleophilic substitution.

This method can also be used for the preparation of phenyl carboxylate, $RCOOC_6H_5$. For example, one can synthesize phenyl benzoate by treating benzoyl chloride with phenol:

benzoyl chloride   phenol   phenyl benzoate

Phenyl benzoate cannot be synthesized by simply reacting benzoic acid and phenol in the presence of conc. H_2SO_4 as a catalyst. In fact, phenyl carboxylate cannot be synthesized using the following method:

carboxylic acid   phenol   phenyl carboxylate

Q: Why we can synthesize an alcoholic ester simply by reacting an alcohol with a carboxylic acid but not a phenyl carboxylate by reacting phenol with a carboxylic acid?

A: The reaction between an alcohol and carboxylic acid to form an ester is based on the nucleophilic acyl substitution mechanism. In this reaction, it is the O atom of the alcohol that makes a nucleophilic attack at the electron-deficient carbonyl carbon. This means that the oxygen atom in the $-COOR$ functional group originated from the alcohol.

This oxygen atom originates from the alcohol, ROH

Now if we compare the oxygen atom of phenol with an alcohol, the lone pair of electrons of the phenolic oxygen atom is less likely to make a nucleophilic attack than that of the alcohol because the lone pair is delocalized into the benzene ring. Thus, one way to solve the problem is to convert the carboxylic acid to the more reactive acid chloride. To further enhance the reactivity, one can convert the phenol into a phenoxide, which is a better nucleophile than the phenol as the oxygen atom of phenoxide is negatively charged.

$$C_6H_5OH + NaOH \longrightarrow C_6H_5O^-Na^+ + H_2O,$$
$$C_6H_5COCl + C_6H_5O^-Na^+ \longrightarrow C_6H_5COOC_6H_5 + NaCl.$$

Q: If the phenoxide ion is a better nucleophile than phenol, can we not react the phenoxide with carboxylic acid to obtain the ester then?

A: No, you can't. Remember, carboxylic acid is a stronger acid than phenol. This would mean that phenoxide ion is a stronger base than a carboxylate ion. Hence, instead of nucleophilic reaction, we would have an acid–base reaction:

$$RCOOH + C_6H_5O^- \longrightarrow C_6H_5OH + RCOO^-.$$

10.6.4.2 *Formation of amides with ammonia and amines — acylation via nucleophilic substitution/condensation*

Reagents:	Excess NH_3, 1° amine or 2° amine
Conditions:	Sealed tube/Room temperature
Options:	—

Tertiary (3°) amines do not undergo acylation because they do not have any replaceable H atom. The products of the acylation consist of an amide and HCl gas. But since ammonia and the amines are basic in nature, they will react with the acidic HCl to produce a salt, which is actually obtained. This would decrease the amount of reactants for the acid chloride. Thus, excess ammonia or amine is required to ensure there are sufficient reactants to react with the acid chloride.

$$NH_3 + HCl \longrightarrow NH_4Cl,$$

$$RNH_2 + HCl \longrightarrow RNH_3Cl,$$

$$R_2NH + HCl \longrightarrow R_2NH_2Cl.$$

Q: Since the nitrogen atom of the amide functional group also possess a lone pair, why can't the HCl react with this nitrogen atom instead?

A: The lone pair of electrons on the nitrogen atom of the amide functional group delocalized into the carbonyl group, hence is less available to be donated as what a base would do. This also explains why an amide in water is neutral.

Q: Can we react carboxylic acid and amine together to form an amide?

A: Yes. But an acid–base reaction would take place first, generating the alkyl ammonium carboxylate salt $(RNH_3^+RCOO^-)$, which under strong heating would give us the amide and water as a side product.

$$RNH_2 + RCOOH \longrightarrow RCOO^-RNH_3^+ \longrightarrow RCONHR + H_2O,$$

$$R_2NH + RCOOH \longrightarrow RCOO^-R_2NH_2^+ \longrightarrow RCONR_2 + H_2O.$$

10.6.4.3 *Formation of carboxylic acid — hydrolysis via nucleophilic acyl substitution/condensation*

$$CH_3-\overset{\overset{\displaystyle O}{\|}}{C}-Cl + H_2O \longrightarrow CH_3-\overset{\overset{\displaystyle O}{\|}}{C}-OH + HCl$$

Acid chlorides hydrolyzed readily upon contact with water. Ethanoyl chloride even reacts rapidly with cold water. This is in contrast to the slow hydrolysis of halogenoalkanes that requires heating.

This gives us the following trend of decreasing ease of hydrolysis:

| RCOCl | RCl | ArCl |

The trend can be explained by the degree of susceptibility of these compounds towards nucleophilic attack, which is attributed to the following two factors:

1. *The electron density on the electron-deficient C atom of the C–Cl bond.*

   Nucleophiles are more likely to attack a highly electron-deficient site. The carbonyl C in RCOCl is the most electron-deficient as its electron density is drawn away by the more electronegative O atom of the carbonyl functional group. Although there is a partial double bond characteristic in the C—Cl bond, this obviously does not decrease its susceptibility to hydrolysis.

   The C atom of the C–Cl bond in RCl is bonded to the electron-donating alkyl group that causes the C–Cl bond to be less polarized.

   As for ArCl, due to the extensive network of overlapping *p*-orbitals, the C–Cl bond acquires some partial double bond character, which also decreases the electron deficiency on the carbon atom, making it less susceptible to nucleophilic attack.

2. *The size and number of substituents bonded to the C atom of the C–Cl bond.*

   Nucleophiles are more likely to attack an electron-deficient site if there is less steric hindrance. The carbonyl C atom in ROCl is sp^2 hybridized, with three substituents attached to it in a trigonal planar geometry. This bonding structure poses less steric hindrance as compared to that of RCl, wherein the sp^3 hybridized C atom of the C–Cl bond is bonded to four substituents in a tetrahedral geometry. Although the carbon atom of the C—Cl of ArCl is also trigonal planar, the steric hindrance is most significant for ArCl as it contains a bulky benzene ring.

The difference in the strength of the C–Cl bond is demonstrated upon the treatment of these compounds with aqueous silver nitrate, as shown by the following observations:
Upon adding $AgNO_3$(aq) to:

CH_3COCl: White precipitate of AgCl(s) is observed **instantly**

CH_3CH_2Cl: White precipitate of $AgCl(s)$ is observed only **after warming** the mixture.

C_6H_5Cl: **No white precipitate** is observed even after prolonged boiling of mixture.

10.6.5 *Chemistry of Esters*

10.6.5.1 *Preparations of esters*

Esters can be prepared using the following methods:

- nucleophilic acyl substitution/condensation of carboxylic acids,
- nucleophilic acyl substitution/condensation of acid chlorides.

10.6.5.2 *Formation of amides with ammonia and amines — nucleophilic acyl substitution/condensation*

Reagents:	Excess NH_3, 1° amine or 2° amine
Conditions:	Heat in a sealed tube
Options:	—

Heating is neceassry as compared to the formation of an amide using an acid chloride as the ester is less reactive to nucleophilic substitution than the acid chloride. The re-reaction of the amide formed with the alcohol to re-form the ester is quite unlikely as the amide itself is not very reactive unless subjected to strong heating.

10.6.5.3 *Formation of carboxylic acid — hydrolysis via nucleophilic acyl substitution/condensation*

Upon adding H_2O,

gains a H

gains a OH

$$R\overset{O}{\underset{\parallel}{C}}OR \longrightarrow R-\overset{O}{\underset{\parallel}{C}}-OH + ROH$$

heterolytic bond
cleavage

Reagents:	HCl(aq) or NaOH(aq)
Conditions:	Heat
Options:	—

Like the hydrolysis of nitriles, the hydrolysis of esters involves the splitting of the molecules into two parts by a water molecule. One part of the molecule gains a proton (H^+) and the other gains a hydroxide ion (OH^-). In a concerted move, the following can be thought to occur:

- the C–O bond of the ester functional group undergoes heterolytic bond cleavage with the O atom acquiring both bonding electrons.
- the fragment containing the –C=O group gains a –OH group, forming RCOOH.
- the other fragment, containing the –OR group, gains a hydrogen atom, forming ROH.

Acids or bases are added as catalysts to speed up the hydrolysis process.

- In **acidic hydrolysis**, the presence of an acid will cause the protonation of basic compounds. In this case, there aren't any, so the products are basically the acid and the alcohol.

$$H-\overset{H}{\underset{H}{\overset{|}{C}}}-\overset{H}{\underset{H}{\overset{|}{C}}}-O\overset{O}{\underset{\parallel}{C}}-H + H_2O \underset{}{\overset{H^+}{\rightleftharpoons}} H-\overset{H}{\underset{H}{\overset{|}{C}}}-\overset{H}{\underset{H}{\overset{|}{C}}}-O-H + H-\overset{O}{\underset{\parallel}{C}}-OH$$

Acidic hydrolysis of an ester is a reversible process, the backward reaction of which is esterification. Mechanism-wise, acid hydrolysis is the opposite of that for esterification (see Chapter 8) — in acidic hydrolysis, the nucleophile is H_2O and the leaving group is the alcohol — these roles are reversed in esterification.

- In **alkaline hydrolysis**, the presence of a base will cause the deprotonation of the acidic RCOOH after it has formed, generating the carboxylate salt.

Unlike acidic hydrolysis, alkaline hydrolysis is essentially irreversible since the carboxylate ion obtained is resonance-stabilized and exhibits little tendency to react with the alcohol to form back the ester. The OH^- nucleophile, which is a stronger nucleophile than H_2O as it is more electron-rich, also helps to promote the hydrolysis of esters.

Q: Why isn't the alcohol deprotonated by the OH^- since it is also a weak acid?

A: Alcohol is a weaker acid than water, which means that the alkoxide ion, RO^-, is a stronger base than OH^-. Thus, it is impossible for the OH^- to deprotonate the ROH, generating H_2O and RO^-. The RO^- forms would be less stable than OH^-.

10.6.5.4 *Formation of alcohols — reduction*

	For Step (i),		For Step (ii),
Reagents:	LiAlH$_4$ in dry ether	Reagents:	Water
Conditions:	Room temperature	Conditions:	Heat
Options:	NaBH$_4$(aq)* or H_2/Pt/Heat$^\#$	Options:	—

*Being a milder reducing agent than LiAlH$_4$, NaBH$_4$(aq) is not able to reduce RCOOH and its derivatives.

$^\#$H$_2$/Pt/Heat cannot be used to reduce an ester, thus catalytic hydrogenation can be used to reduce carbon−carbon unsaturated bonds with the ester functional group present.

10.6.6 Chemistry of Amides

10.6.6.1 Preparations of amides

Amides can be prepared using the following methods:

- nucleophilic acyl substitution/condensation of acid chlorides,
- nucleophilic acyl substitution/condensation of esters.

The best way to get amides is to first convert the carboxylic acid to the more reactive acid chloride and then subject the latter to reaction with ammonia (or amine).

10.6.6.2 Formation of carboxylic acid — hydrolysis via nucleophilic acyl substitution/condensation

As with the other derivatives of carboxylic acids, amides also undergo hydrolysis, and the products consist of carboxylic acid and ammonia/amines (depending on the type of amide).

Hydrolysis of 1° amide,

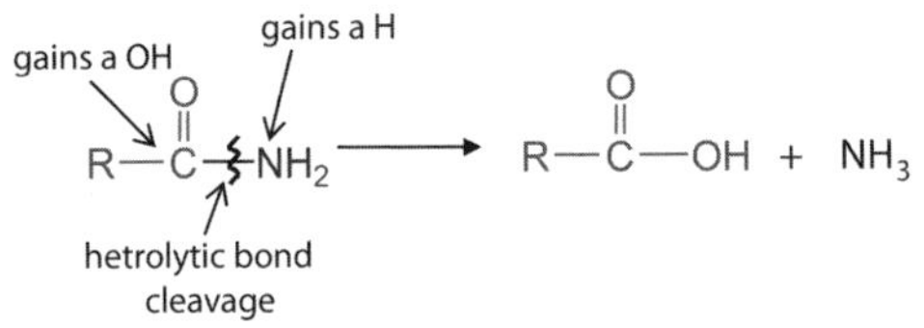

gains a OH
gains a H

$$R-C-NH_2 \longrightarrow R-C-OH + NH_3$$

hetrolytic bond
cleavage

Hydrolysis of 2° amide,

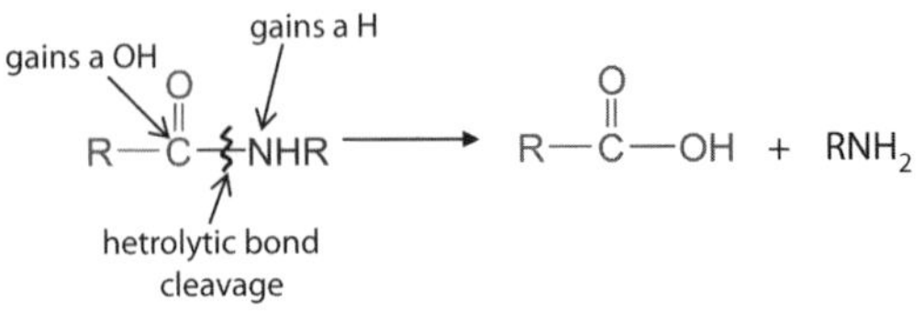

gains a OH
gains a H

$$R-C-NHR \longrightarrow R-C-OH + RNH_2$$

hetrolytic bond
cleavage

Hydrolysis of 3° amide,

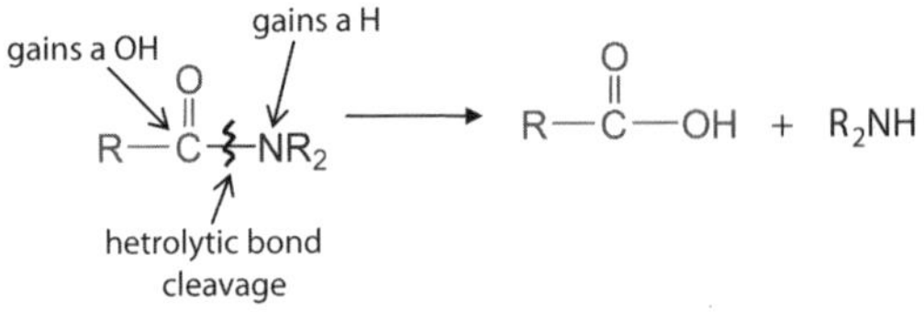

gains a OH
gains a H

$$R-C-NR_2 \longrightarrow R-C-OH + R_2NH$$

hetrolytic bond
cleavage

Depending upon the pH of the medium, one product or the other is actually obtained in the form of its salt.

Reagents:	HCl(aq) or NaOH(aq)
Conditions:	Heat
Options:	—

- In **acidic hydrolysis**, the presence of an acid will cause the protonation of basic compounds such as NH_3 and the amines.

$$R-\overset{\overset{O}{\|}}{C} \ \overset{\overset{H}{|}}{N}-H \ + \ H^+ \ + \ H_2O \ \longrightarrow \ R-\overset{\overset{O}{\|}}{C}-OH \ + \ NH_4^+$$

primary amide

$$R-\overset{\overset{O}{\|}}{C} \ \overset{\overset{H}{|}}{N}-R \ + \ H^+ \ + \ H_2O \ \longrightarrow \ R-\overset{\overset{O}{\|}}{C}-OH \ + \ RNH_3^+$$

secondary amide

$$R-\overset{\overset{O}{\|}}{C} \ \overset{\overset{R}{|}}{N}-R \ + \ H^+ \ + \ H_2O \ \longrightarrow \ R-\overset{\overset{O}{\|}}{C}-OH \ + \ R_2NH_2^+$$

tertiary amide

- In **alkaline hydrolysis**, the presence of a base will cause the deprotonation of the acidic RCOOH, forming instead the carboxylate salt.

$$R-\overset{\overset{O}{\|}}{C} \ \overset{\overset{H}{|}}{N}-H \ + \ OH^- \ \longrightarrow \ R-\overset{\overset{O}{\|}}{C}-O^- + \ NH_3$$

primary amide

$$R-\overset{\overset{O}{\|}}{C} \ \overset{\overset{H}{|}}{N}-R \ + \ OH^- \ \longrightarrow \ R-\overset{\overset{O}{\|}}{C}-O^- + \ RNH_2$$

secondary amide

$$R-\overset{\overset{O}{\|}}{C} \ \overset{\overset{R}{|}}{N}-R \ + \ OH^- \ \longrightarrow \ R-\overset{\overset{O}{\|}}{C}-O^- + \ R_2NH$$

tertiary amide

10.6.6.3 *Formation of amines — reduction*

$$R-\overset{\overset{O}{\|}}{C}-\overset{\overset{H}{|}}{N}-H \ + \ 4[H] \ \longrightarrow \ R-\overset{\overset{H}{|}}{\underset{\underset{H}{|}}{C}}-\overset{\overset{H}{|}}{N}-H \ + \ H_2O$$

primary amide   primary amine

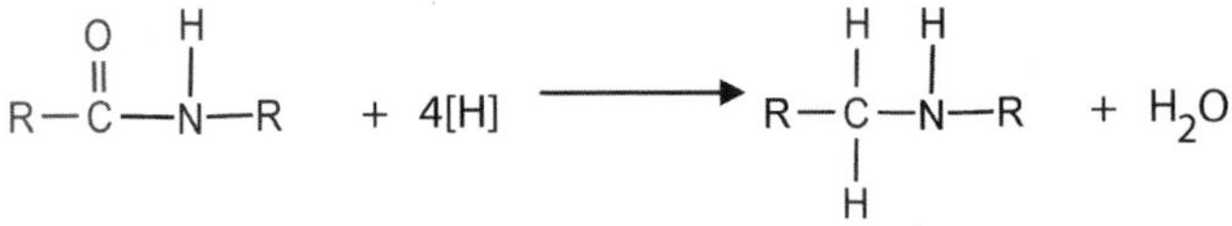

$$R-\overset{\overset{O}{\|}}{C}-\overset{\overset{H}{|}}{N}-R \ + \ 4[H] \ \longrightarrow \ R-\overset{\overset{H}{|}}{\underset{\underset{H}{|}}{C}}-\overset{\overset{H}{|}}{N}-R \ + \ H_2O$$

secondary amide   secondary amine

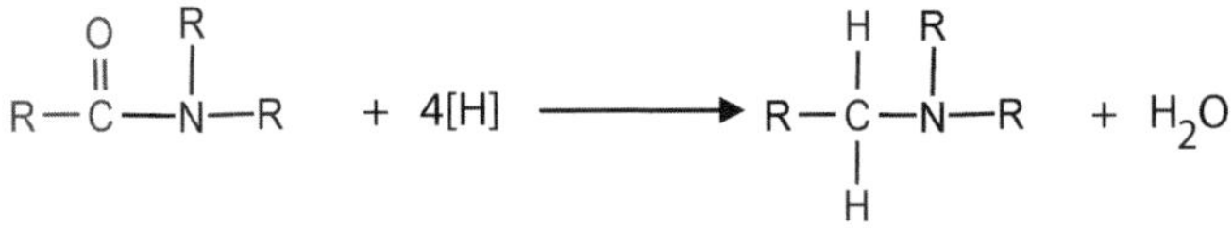

$$R-\overset{\overset{O}{\|}}{C}-\overset{\overset{R}{|}}{N}-R \ + \ 4[H] \ \longrightarrow \ R-\overset{\overset{H}{|}}{\underset{\underset{H}{|}}{C}}-\overset{\overset{R}{|}}{N}-R \ + \ H_2O$$

tertiary amide   tertiary amine

For Step (i),		For Step (ii),	
Reagents:	LiAlH$_4$ in dry ether	Reagents:	Water
Conditions:	Room temperature	Conditions:	Heat
Options:	NaBH$_4$(aq)* or H$_2$/Pt/Heat#	Options:	—

*Being a milder reducing agent than LiAlH$_4$, NaBH$_4$(aq) is not able to reduce RCOOH and its derivatives.
#H$_2$/Pt/Heat cannot be used to reduce an amide, thus catalytic hydrogenation can be used to reduce carbon–carbon unsaturated bonds with the amide functional group present.

The above reactions are important as they allow us to synthesize different types of amine from different amides. One notable point is that a 1° amide would give a 1° amine, etc.

10.7 Summary

The mind map below depicts the reactions of carboxylic acids and their derivatives, focusing on propanoic acid. Propene is used as a starting material so as to integrate the chemistry of carboxylic acid with other functional groups.

My Tutorial

1. Aspirin may be produced from methylbenzene via the following synthetic pathway:

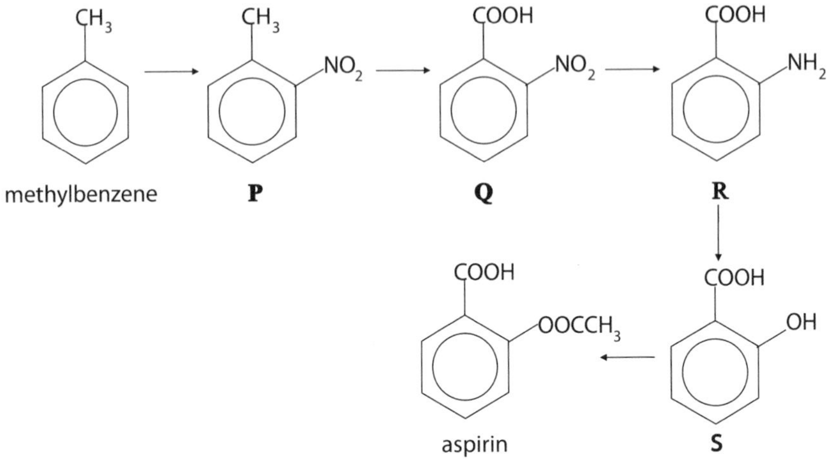

methylbenzene   **P**   **Q**   **R**

aspirin   **S**

(a) Give the reagents and conditions required to convert:

  (i) methylbenzene to **P**
  (ii) **P** to **Q**
  (iii) **R** to **S**
  (iv) **S** to aspirin

(b) (i) Give a mechanistic account for the formation of **P** from methylbenzene.

  (ii) The reaction also produces two other isomers of **P**. Give the displayed formulae of these two isomers. What simple method could be used to show that the purified product is in fact **P**?

  (iii) When the temperature of the reaction is raised above the optimum for the reaction, the overall yield of **P** and its isomers drop. Suggest two reasons for this.

  (iv) Account for why **R** has a higher boiling point than **Q**. Hence, suggest a simple way to separate a mixture of **R** and **Q**.

  (v) The solubility of **Q** in water is low but is enhanced when aqueous sodium hydroxide is added. Explain the observation.

(c) During the esterification process, traces of a polymeric by-product always form. Explain why it is formed and suggest a structure for it.

(d) (i) **P** is a constitutional/structural isomer of **R**. State the type of constitutional/structural isomerism.

(ii) Suggest, with reasons, which of **P** and **R** would have the higher melting point.

(iii) Suggest a simple chemical test to distinguish between **P** and **R**.

(iv) Suggest a different chemical test from (d)(iii) to distinguish between **R** and **S**.

(e) (i) Suggest with reasons, which of **Q** and **R** would have a higher pK_a value.

(ii) Comparing benzoic acid with **Q**, explain why benzoic acid is less acidic than **Q**.

(iii) Suggest, with reasons, how a mixture of **R** and **S** might be separated.

(f) The benzene rings of both **S** and **Q** can be further chlorinated with suitable reagents and conditions. Explain briefly why a higher temperature is required to effect the chlorination of **Q** than **S**. Give the appropriate reagents and conditions.

(g) Ketones and aspirin both have carbonyl groups, but only aspirin is hydrolyzed easily. Explain.

(h) Reaction of **S** to form aspirin does not proceed if CH_3COOH is used. Explain.

2. Lactic acid, $CH_3CH(OH)COOH$, a compound that is produced during cell respiration from pyruvic acid, $CH_3COCOOH$, causes muscle fatigue.

(a) (i) What type of chemical reaction has occurred when pyruvic acid is converted to lactic acid?

(ii) Draw the displayed formulae for both pyruvic acid and lactic acid.

(iii) What type of isomerism is present in lactic acid? Draw the isomers.

(iv) State, with reasons, which of lactic acid or pyruvic acid would be more acidic.

(v) Suggest two simple chemical tests to distinguish between lactic acid and pyruvic acid. Write balanced equations for the positive test.

(vi) Pyruvic acid reacts with hydroxylamine (NH_2OH) to give a condensation product.

(A) Write a balanced equation for the reaction between pyruvic acid and hydroxylamine.

(B) The condensation product exhibits stereoisomerism. Draw the possible stereoisomers and name the type of stereoisomerism.

(b) In the laboratory, lactic acid may be synthesized from ethanol via the following pathway:

$$CH_3CH_2OH \xrightarrow{\textbf{P}} CH_3CHO \xrightarrow{\textbf{Q}} CH_3CH(OH)CN \xrightarrow{\textbf{R}} CH_3CH(OH)COOH$$

(i) Give the reagents and conditions for steps **P**, **Q** and **R**.

(ii) Give a mechanistic account for step **Q**.

(iii) State, with reasons, which of lactic acid or ethanol would be more acidic.

(iv) What is the difference between lactic acid that is synthesized in the laboratory as compared to that formed in the cell?

(v) Suggest a simple chemical test to distinguish between each of the following pairs of compounds:

(A) CH_3CH_2OH and CH_3CHO

(B) CH_3CH_2OH and $CH_3CH(OH)CN$

(C) $CH_3CH(OH)CN$ and $CH_3CH(OH)COOH$

(vi) State, with reasons, which of the following compounds would have a higher boiling point:

(A) CH_3CH_2OH and CH_3CHO

(B) CH_3CH_2OH and $CH_3CH(OH)COOH$

3. An organic compound **Q**, of molecular formula $C_6H_{10}O_3$, reacted with calcium carbonate to give off carbon dioxide. **Q** formed $C_7H_{12}O_3$ with methanol, a sweet-smelling compound, when treated with concentrated sulfuric acid at 170°C. When 2,4-DNPH was added to **Q**, an orange precipitate was formed. **Q** also gave positive result when was treated with iodine in sodium hydroxide, forming a yellow precipitate.

(a) (i) Deduce the constitutional/structural formula of **Q**.

(ii) Write a balanced equation for the reaction of **Q** with:

(A) Calcium carbonate

(B) Concentrated sulfuric acid with methanol

(C) 2,4-DNPH

(D) Iodine in sodium hydroxide

(iii) State the role of the concentrated sulfuric acid.

(b) **R** is an isomer of **Q** with the same molecular formula. **R** gave negative test when treated with iodine in sodium hydroxide but responded positively with other tests. Give a possible structural formula of **R**.

(c) **S** is also isomer of **Q**, containing a chiral carbon. Give a possible constitutional/structural formula of **S** and draw the two stereoisomers.

(d) Write balanced equation when **S** reacts with the following reagents:

   (i) LiAlH$_4$ in dry ether, followed by warming with water.

   (ii) Ethylamine.

   (iii) PCl$_3$.

   (iv) KCN with HCN.

   (v) NH$_2$OH.

   (vi) Sodium phenoxide.

4. (a) Ethanoyl chloride reacts far more rapidly with water than chloroethane. Explain.

  (b) (i) Give the conditions and show the structure of the organic product formed when ammonia reacts with ethanoyl chloride.

    (ii) Thus, give the structure of the organic product formed when an equimolar of $((CH_2)_3COCl)_2$ reacts with $H_2N(CH_2)_6NH_2$.

    (iii) Draw the structure of a possible side product that may be formed in (b)(ii).

    (iv) The product formed in (b)(i) readily hydrolyzes by an acid and a base. Write balanced equations to show the two hydrolysis processes. Explain the different roles played by the acid and the base.

  (c) Carry out the following conversions:

    (i) $RCH_2OH \rightarrow RCON(CH_3)_2$

    (ii) $(CH_3)_3COH \rightarrow (CH_3)_3CCOOH$

    (iii)

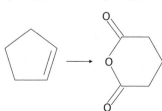

(iv)

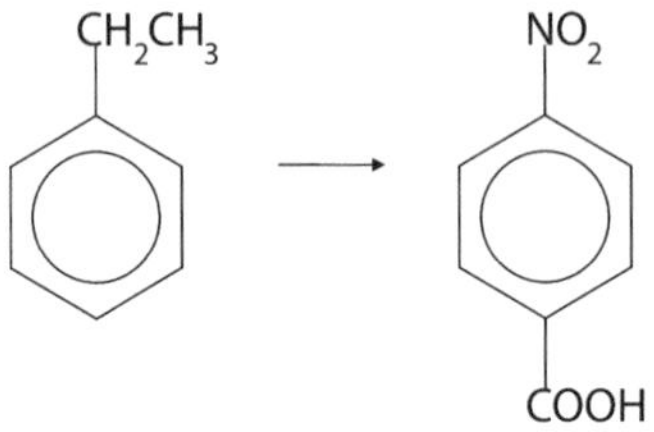

(v)

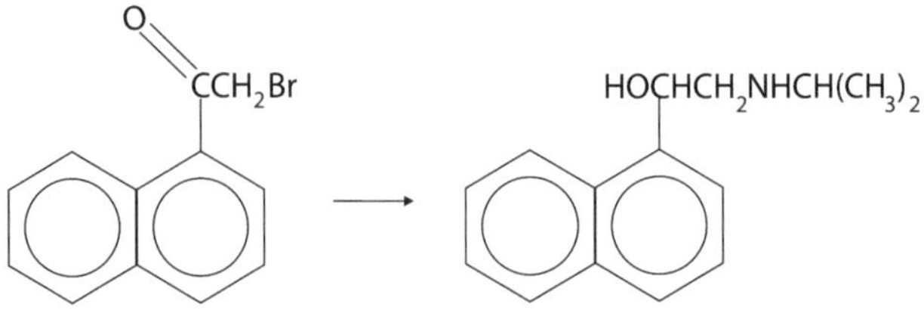

(vi)

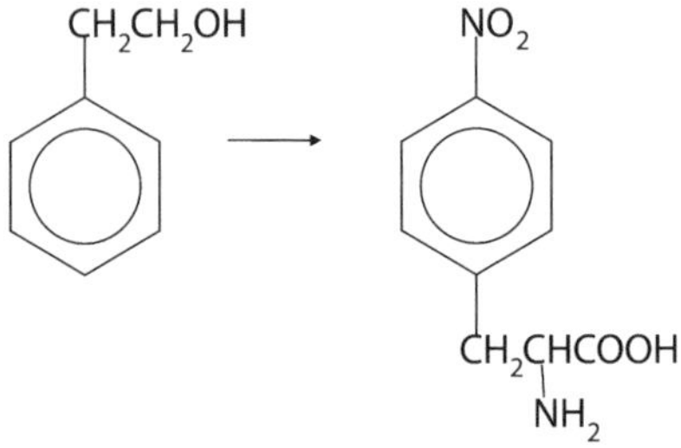

(d) Explain the following observations:

   (i) $CH_3CH_2CO_2H$ is converted into $CH_3CH_2CO_2CH_2CH_3$ by ethanol, in the presence of concentrated sulfuric acid, faster than $H_2NCH_2CO_2H$ is converted to $H_2NCH_2CO_2CH_2CH_3$ under the same conditions.

   (ii) CH_3CO_2H has a pK_a that is smaller than $NCCH_2CO_2H$.

   (iii) $CH_3CH_2NH_2$ is protonated by weak acids whereas CH_3CONH_2 is difficult to protonate.

   (iv) CH_3COCl hydrolyzes much more readily than both CH_3CH_2Cl and C_6H_5Cl whereas CH_3CH_2Cl hydrolyzes more readily than C_6H_5Cl.

   (v) The hydrolysis of a carboxylic ester using acid is reversible, but using alkali is not reversible.

(e) Give the possible organic products formed when **X** reacts with each of the following reagents individually:

$$\mathbf{X}$$

(i) Br_2/CCl_4; Br_2(aq); conc. NH_3/sealed tube; cold alkaline $KMnO_4$; hot acidified $KMnO_4$; cold conc. H_2SO_4; $Cl_2/AlCl_3$; $LiAlH_4$/dry ether; K(s); dilute HNO_3.

(ii) $FeCl_3$(aq); $SOCl_2$; $NaOH$(aq)/heat; HCl(aq); H_2SO_4(aq)/heat; $BaCO_3$(s); Tollens' Reagent; Fehling's Reagent; conc. H_2SO_4/conc. HNO_3.

(iii) Acidified $K_2Cr_2O_7$(aq)/heat; acidified $K_2Cr_2O_7$(aq)/ immediate distillation; $I_2/NaOH$(aq); $HCN/NaOH$; 2,4-DNPH; KCN(aq)/heat; $NaOH$/ethanol/heat; CH_3CH_2Br/ethanol/heat; $CH_3CH_2NH_2$/ethanol/heat; NH_2OH; CH_3COCl; CH_3COOH/H_2SO_4/heat; H_2/Pt/ heat.

(iv) * all the chiral carbons atom in $\mathbf{X}$. How many stereoisomers are there in $\mathbf{X}$?

(f) There are three different straight-chain hydroxybutanoic acids with the molecular formula $C_4H_8O_3$, $\mathbf{P}$, $\mathbf{Q}$ and $\mathbf{R}$. Both $\mathbf{Q}$ and $\mathbf{R}$ can be isolated in optically active forms. When each of these three acids is heated, it loses water and forms a compound: $\mathbf{P}$ gives $\mathbf{S}$ ($C_4H_6O_2$), $\mathbf{Q}$ gives $\mathbf{T}$ ($C_4H_6O_2$) and $\mathbf{R}$ gives $\mathbf{U}$ ($C_8H_{12}O_4$). Compound $\mathbf{T}$ cannot be obtained in the optically pure form even when made from optically active $\mathbf{Q}$, but optically active $\mathbf{R}$ leads to optically active $\mathbf{U}$.

$\mathbf{S}$ reacts with sodium hydroxide to form $C_4H_7O_3Na$, whereas $\mathbf{T}$ forms $C_4H_5O_2Na$. $\mathbf{T}$ decolorizes bromine water and reacts with calcium carbonate whereas **S** and **U** have no effect on these reagents.

(i) Deduce the structures for **P** to **U**.

(ii) Given an unlabeled sample each of **P**, **Q** and **R**, use chemical tests to identify each sample without using the above-mentioned reactions.

(g) Account for why 4-nitrophenyl ethanoate is hydrolyzed faster than phenyl ethanoate, which is in turn hydrolyzed faster than ethyl ethanoate.

(h) The are seven unknown reagent bottles, each containing one of the following compounds:

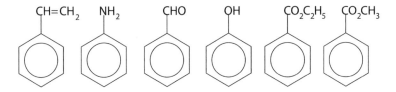

Describe how you would identify each compound using appropriate reagent and conditions. You should not use more than seven positive tests.

CHAPTER 11

Amines

11.1 Introduction

Amines are organic compounds that contain the amino functional group denoted as $-NH_2$. They are derivatives of ammonia as they are formed by replacing one or more hydrogen atoms of the ammonia molecule with either alkyl or aryl groups. Thus, we would expect amines to possess both physical and chemical properties similar to ammonia. There are three classes of amines classified as primary (1°), secondary (2°) or tertiary (3°) amines.

1° amine   2° amine   3° amine

The N atom of the amino group is sp^3 hybridized, and the molecular geometry at this atom is trigonal pyramidal. The lone pair of electrons on the N atom accounts for the chemical properties of amines, as we shall see in later sections.

11.2 Nomenclature

Amines can be named in various ways. Simple amines can be named with the suffix–*amine*. Other substituents are named as prefixes accompanied by the appropriate positional numbers. For higher molecular weight amines, the amino functional group is usually considered as a substituent and is indicated by the prefix "*amino–*".

methylamine   dimethylamine   trimethylamine   methylpropylamine

Another way of naming amines involves indicating the substituents on the N atom with the prefix "*N–*". For instance, the following compound is named as *N, N*-dimethylpentylamine.

Phenylamine is a special name given to the benzene derivative that contains the amino group directly attached to the benzene ring. For substituted phenylamine, the substituents are named as prefixes accompanied by positional numbers.

phenylamine   2,3-dibromophenylamine   2-aminobenzoic acid

11.3 Physical Properties

11.3.1 *Melting and Boiling Points*

For a given class of amines, be it 1°, 2° or 3° amines, melting and boiling points increase with increasing carbon chain length and decrease with increasing degree of branching. Longer carbon chain length leads to an increase in the strength of the instantaneous dipole–induced dipole (id–id) force because of an increase in the number of electrons. The increase in branching would lead to the lone pair of electrons being less available for hydrogen bonding as there would be greater steric hindrance for the electron-deficient hydrogen atom of a neighboring amine molecule to approach the lone pair of electrons. As a result, the intermolecular forces of increasing importance would be of the id–id type between the alkyl groups that are bonded to the nitrogen atom.

But if one is to compare, let's say, the boiling points of isomeric 1°, 2° and 3° amines of similar relative molecular mass, 1° and 2° amines are less volatile (higher boiling point) than 3° amine due

to strong hydrogen bonds that exist between molecules of the 1°
and 2° amines. As molecules of 3° amine lack a H atom bonded to
the electronegative N atom, they can only form weaker permanent
dipole–permanent dipole interactions with other polar molecule. But
if the other molecule has a H atom covalently bonded to the highly
electronegative O, F or N atoms, then the 3° amine can function as
a hydrogen bond acceptor. The lower boiling point of 2° amine than
the 1° amine can be accounted for by the lower availability of the
lone pair of electrons for hydrogen bonding due to steric effect.

Table 11.1

Amine	1° amine	2° amine	3° amine
Structural formula	$CH_3CH_2CH_2-\overset{\displaystyle H}{\underset{\displaystyle \vert}{N}}-H$	$CH_3CH_2-\overset{\displaystyle H}{\underset{\displaystyle \vert}{N}}-CH_3$	$CH_3-\overset{\displaystyle CH_3}{\underset{\displaystyle \vert}{N}}-CH_3$
M_r	59.0	59.0	59.0
Boiling point (°C)	48.0	36.0	2.9

Q: How many hydrogen bonds can each amine molecule form?

A: One NH_3 molecule with one lone pair of electrons and three
hydrogen atoms can form an average of one hydrogen bond per
molecule. Thus, both 1° and 2° amines would each be able to
form an average of one hydrogen bond per molecule too. Hence,
the higher boiling point of methylamine (-6°C) as compared to
ammonia (-33.3°C) can be explained by the additional id–id
interaction between the methylamine molecules.

Q: Since alkyl groups release electron density via inductive effect,
and a 2° amine has more alkyl groups than a 1° amine, shouldn't
the lone pair of electrons of the 2° amine be more available for
hydrogen bond formation than a 1° amine?

A: Yes, indeed the electron-releasing effect of the alkyl groups of the
2° amine would make its lone pair of electrons more available for
hydrogen bond formation than that of a 1° amine. But unfortunately, the contradictory effect of the steric hindrance that arises
due to the greater number of bulky alkyl groups present probably overshadows the electron-releasing effect here. In addition, do
not forget that if the lone pair of electrons on the nitrogen atom
is more available because of the electron-releasing effect of the

alkyl groups, then this same effect would make the nitrogen atom less electron-withdrawing on the H atom. As such, the electron deficiency of the H atom would cause it to form weaker hydrogen bond. And if the electron-rich N atom and electron-deficient H atom effects both cancel out each other, then the possible factor to account for the difference in boiling point would be steric factor.

In general, amines have lower boiling points than alcohols of similar relative molecular mass. An amine molecule on average forms the same number of hydrogen bond per molecule as an alcohol molecule. The hydrogen bonds that are formed between alcohol molecules are much stronger because the $-OH$ group is more polarized than the $-NH_2$ group. This feature arises from the higher electronegativity of the O atom than the N atom, which causes the H atom of the $-OH$ group of alcohol to be more electron-deficient than that of the $-NH_2$ group of an amine.

Table 11.2

	Propylamine	**Propan-1-ol**
Structural formula	$CH_3CH_2CH_2NH_2$	$CH_3CH_2CH_2OH$
M_r	59.0	60.0
Boiling point ($^\circ$C)	48.0	97.1

11.3.2 *Solubility*

The ability to form hydrogen bonds accounts for the high solubility of short-chain amines in water. However, as the length of the carbon-chain increases, solubility in water will decrease due to the increasing size of the hydrophobic alkyl chain. As 3° amines can only function as hydrogen bond acceptors, their solubility is generally lower than that of 1° and 2° amines.

11.3.3 *Basicity*

Based on Brønsted–Lowry theory and the Lewis theory of acids and bases, a base is defined as a proton acceptor and an electron-pair donor, respectively. You will find that these two definitions complement each other, and they highlight the role of a base such as

NH_3 — it accepts a proton by donating a pair of electrons to it and forming a dative covalent bond.

Since amines contain an N atom with a lone pair of electrons available for dative covalent bond formation; they are considered to be bases too. They react with acid to form an alkyl ammonium salt and turn moist red litmus paper blue. However, they are weak bases as they only partially ionize in water. The reversible arrow in the chemical equation indicates two important concepts: (i) the reaction is incomplete, and (ii) the system is in a dynamic equilibrium.

$$RNH_2(l) + H_2O(l) \rightleftharpoons RNH_3^+(aq) + OH^-(aq)$$

Q: Since amine is a weak base that partially ionizes in water to give ions. Does the formation of the ions increase the solubility of the amine molecules?

A: Certainly! The formation of the ions from the partial ionization does increase the solubility of the amine molecules as the ions can form stronger ion-dipole interaction with water. This is also the reason behind why ammonia is very soluble in water. The high solubility of ammonia is not just due to the hydrogen bonding formed with the water molecules but also because of the higher solubility of the NH_4^+ and OH^- ions formed from the hydrolysis of NH_3 molecules in water.

Other examples of weak bases include NH_3 and Na_2CO_3.

$$NH_3(aq) + H_2O(l) \rightleftharpoons NH_4^+(aq) + OH^-(aq)$$
$$CO_3^{2-}(aq) + H_2O(l) \rightleftharpoons HCO_3^-(aq) + OH^-(aq)$$

On the other hand, a **strong** base, such as NaOH and KOH, **completely ionizes** in aqueous solution:

$$NaOH(s) + aq \longrightarrow Na^+(aq) + OH^-(aq)$$

Hence, when comparing the strength of two weak bases, the one that can ionize to a greater extent will be the stronger of the two weak bases. The relative base strength of aromatic and aliphatic amines as compared to ammonia is shown below:

Increasing base strength: arylamines $< NH_3 <$ aliphatic amines.

Generally, this trend can be accounted for by comparing the relative ease of donating the lone pair of electrons on the N atom to form a dative covalent bond with a proton. The greater the ease of donating the lone pair of electrons, the greater the extent of the ionization of the base (indicated by the position of the equilibrium being shifted more towards the right of the reaction) and hence the stronger the base.

Aliphatic amines, such as ethylamine, are more basic than NH_3 because the electron-donating alkyl group(s) increases the electron density on the N atom, making its lone pair of electrons more readily available for dative covalent bond formation with a proton.

Q: If a greater number of electron-donating groups increases the electron density on the N atom, making the compound more basic, why is it that $(CH_3)_3N$, with a pK_b of 4.30, is less basic than $(CH_3)_2NH$, which has a pK_b of 3.29?

A: Now, other than the electron-donating effect of the alkyl groups that would enhance the availability of the lone pair of electrons to act as a base, there is a contradictory factor that would diminish the availability of the lone pair of electrons. This factor is none other than the increased steric hindrance brought about by the presence of a greater number of alkyl groups, which causes the lone pair of electrons to be more "embedded." Thus, the interplay of these two opposing factors results in a less basic $(CH_3)_3N$ than $(CH_3)_2NH$. In addition, do not forget that the solubility of a 3° amine is lower as the amine molecule can't form extensive hydrogen bonds with the water molecules. A lower solubility would naturally lead to a less basic molecule.

Q: I am confused. If we consider a 2° amine against 1° amine of similar molecular weight, the lone pair of electrons of the 2° amine would be more available to be donated as a base than that of the 1° amine due to greater electron-releasing effect by the greater number of alkyl groups. So how come the greater availability of the lone pair doesn't make the 2° amine form a stronger hydrogen bond, since the strength of the hydrogen bond is also dependent on the availability of the lone pair?

A: A proton, when compared to an amine molecule, is much smaller in size. Thus, the steric hindrance faced by amine molecules when trying to form hydrogen bonds with each other is greater than that faced by the smaller proton. The proton can have a closer proximity to the lone pair of electrons than another amine molecule. It is also without doubt that the availability of the lone pair of electrons of a 2° amine is greater than that for a 1° amine. Thus, due to the greater steric effect faced by the amine molecule when trying to form a hydrogen bond with another amine molecule, the strength of the hydrogen bond does not increase because of the availability of the lone pair. Whereas because of lower steric hindrance faced by the smaller proton, there is an increase in the basic strength due to the greater availability of the lone pair caused by the electron-releasing effect of the greater number of alkyl groups. Thus, one needs to recognize that the same factor, availability of the lone pair of electrons, does not necessary lead to the same observations for two different physical phenomena. One needs to take into considerations other factors that are also involved during the interaction, such as solubility, sizes of the species interacting, steric effect, etc.

Arylamines, such as phenylamine, are as a whole the least basic among the three classes of amines because the lone pair of electrons on the N atom partially delocalizes into the benzene ring and is thus less readily available for dative covalent bond formation with a proton.

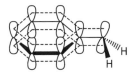

Fig. 11.1. The *p*-orbtial of N atom overlaps partially with that of the adjacent C atom of the benzene ring.

Within the class of arylamines itself, those with electron-donating substituents attached to the benzene ring are generally stronger bases than those containing electron-withdrawing substituents.

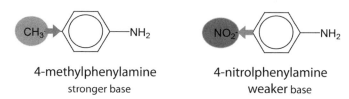

4-methylphenylamine
stronger base

4-nitrolphenylamine
weaker base

For example, 4-methylphenylamine is a stronger base than phenylamine as the electron-donating methyl substituent reduces the extent of delocalization of the lone pair of electrons on the N atom.

Conversely, 4-nitrophenylamine is a weaker base than phenylamine as the electron-withdrawing nitro substituent increases the extent of delocalization of the lone pair of electrons on the N atom.

Q: Amides also contain an N atom with a lone pair of electrons. Why are they not basic?

A: Amides neither turn moist red litmus paper blue nor react with acids to form salts. Their seeming lack of basicity stems from the fact that the lone pair of electrons on the N atom delocalizes into the electron-withdrawing carbonyl group and is thus less readily available for dative covalent bond formation with a proton. In fact, amides are regarded as being neutral. Thus, one can simply use the moist red litmus paper test to differentiate between amides and amines.

Q: How can one explain that there is actually greater delocalization of electrons into a carbonyl group than into a benzene ring, which then accounts for amides to be even less basic than phenylamine?

A: If one looks closely at the various resonance structures below, one would notice that when the lone pair of electrons from the N atom delocalizes into the benzene ring, it is the carbon atoms that hold the negative charge. But in the amide, it is a highly electronegative oxygen atom that holds on to the extra electron.

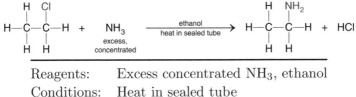

resonance hybrid

Since the carbon atom is less electronegative than an oxygen atom, one would not expect the carbon atoms of the benzene ring to hold the extra electron as "comfortably" as the oxygen atom of the amide. Thus, we would expect the lone pair of electrons of an amide to be more delocalized than that of phenylamine.

11.4 Preparation Methods for Amines

11.4.1 *Nucleophilic Substitution of Halogenoalkanes*

$$H-\overset{\overset{\displaystyle H}{|}}{\underset{\underset{\displaystyle H}{|}}{C}}-\overset{\overset{\displaystyle Cl}{|}}{\underset{\underset{\displaystyle H}{|}}{C}}-H \; + \; NH_3 \xrightarrow[\text{heat in sealed tube}]{\text{ethanol}} H-\overset{\overset{\displaystyle H}{|}}{\underset{\underset{\displaystyle H}{|}}{C}}-\overset{\overset{\displaystyle NH_2}{|}}{\underset{\underset{\displaystyle H}{|}}{C}}-H \; + \; HCl$$

excess, concentrated

Reagents:	Excess concentrated NH_3, ethanol
Conditions:	Heat in sealed tube
Options:	Use of alkylamine to form more highly substituted alkylamines

Excess NH_3 is used to minimize polyalkylation so that the 1° amine will be the major product (refer to Chapter 7 for more details).

Since polyalkylation may compromise the product's yield, reduction of amides can be employed to form the various classes of amines.

Arylamines cannot be prepared using this method as halogenoarenes do not readily undergo nucleophilic substitution reactions. To prepare phenylamine, nitrobenzene is subjected to reduction.

11.4.2 *Reduction of Amides*

$$
\underset{\text{primary amide}}{R-\overset{\overset{\displaystyle O}{\|}}{C}-\overset{\overset{\displaystyle H}{|}}{N}-H} \;+\; 4[H] \;\longrightarrow\; \underset{\text{primary amine}}{R-\overset{\overset{\displaystyle H}{|}}{\underset{\underset{\displaystyle H}{|}}{C}}-\overset{\overset{\displaystyle H}{|}}{N}-H} \;+\; H_2O
$$

$$
\underset{\text{secondary amide}}{R-\overset{\overset{\displaystyle O}{\|}}{C}-\overset{\overset{\displaystyle H}{|}}{N}-R} \;+\; 4[H] \;\longrightarrow\; \underset{\text{secondary amine}}{R-\overset{\overset{\displaystyle H}{|}}{\underset{\underset{\displaystyle H}{|}}{C}}-\overset{\overset{\displaystyle H}{|}}{N}-R} \;+\; H_2O
$$

$$
\underset{\text{tertiary amide}}{R-\overset{\overset{\displaystyle O}{\|}}{C}-\overset{\overset{\displaystyle R}{|}}{N}-R} \;+\; 4[H] \;\longrightarrow\; \underset{\text{tertiary amine}}{R-\overset{\overset{\displaystyle H}{|}}{\underset{\underset{\displaystyle H}{|}}{C}}-\overset{\overset{\displaystyle R}{|}}{N}-R} \;+\; H_2O
$$

For Step (i),		For Step (ii),	
Reagents:	LiAlH$_4$ in dry ether	Reagents:	Water
Conditions:	Room temperature	Conditions:	Heat
Options:	NaBH$_4$(aq)* or H$_2$/Pt/Heat#	Options:	—

*Being a milder reducing agent than LiAlH$_4$, NaBH$_4$(aq) is not able to reduce RCOOH and its derivatives.

#H$_2$/Pt/Heat cannot be used to reduce an amide, thus catalytic hydrogenation can be used to reduce carbon–carbon unsaturated bonds with the amide functional group present.

11.4.3 *Reduction of Nitriles*

$$
\underset{\underset{\displaystyle H}{|}}{H-\overset{\overset{\displaystyle H}{|}}{C}-\overset{\overset{\displaystyle CN}{|}}{\underset{\underset{\displaystyle H}{|}}{C}}-H} \;+\; 4[H] \;\longrightarrow\; \underset{\underset{\displaystyle H}{|}}{H-\overset{\overset{\displaystyle H}{|}}{C}-\overset{\overset{\displaystyle CH_2NH_2}{|}}{\underset{\underset{\displaystyle H}{|}}{C}}-H}
$$

For Step (i),		For Step (ii),	
Reagents:	LiAlH$_4$ in dry ether	Reagents:	Water
Conditions:	Room temperature	Conditions:	Heat
Options:	NaBH$_4$(aq)* or H$_2$/Pt/Heat#	Options:	—

*Being a milder reducing agent than LiAlH$_4$, NaBH$_4$(aq) is not able to reduce RCOOH and its derivatives, plus nitrile.

#H$_2$/Pt/Heat cannot be used to reduce a nitrile, thus catalytic hydrogenation can be used to reduce carbon–carbon unsaturated bonds with the nitrile functional group present.

11.4.4 *Reduction of Nitrobenzene*

For Step (i),			For Step (ii),	
Reagents:	Sn, conc. HCl		Reagents:	NaOH(aq)
Conditions:	Heat		Conditions:	—
Options:	—		Options:	—

The reduction of nitrobenzene using granulated tin and concentrated HCl will result in the formation of a protonated amine salt. To recover the amine from its salt, a subsequent step involves adding NaOH(aq), which will react with the weakly acidic salt. At the same time, the tin would be precipitated out as $Sn(OH)_2$, which makes purification easier.

Q: Can I use $LiAlH_4$ to reduce nitrobenzene to phenylamine?

A: Yes, you can. But take note that if there are other functional groups that can be reduced by $LiAlH_4$, which you would like to preserve, then $LiAlH_4$ is not a good choice.

Q: How is the nitro-group being reduced by the Sn with conc. HCl?

A: There are a few possibilities. One is that Sn, being a metal, reacts with the conc. HCl via the metal-acid reaction to release nascent hydrogen atoms and these hydrogen atoms caused the $-NO_2$ group to be reduced. Another possibility is that the Sn^{2+} that is formed is also responsible for reducing the $-NO_2$ group and itself is being oxidized to form Sn^{4+}. The Sn^{4+} is stabilized by forming a complex, $[SnCl_6]^{2-}$, with the Cl^- ions that is present.

11.5 Chemical Properties

11.5.1 *Acid–Base Reaction with Acids*

$$CH_3CH_2NH_2 + HCl \longrightarrow CH_3CH_2NH_3^+Cl^-$$

Reagents:	$HCl(aq)$ or $H_2SO_4(aq)$
Conditions:	Room temperature
Options:	Can use other weak acids but strong enough to react with the weak bases eg. CH_3COOH

Q: Can we use $HNO_3(aq)$ for the acid–base reaction?

A: You need to be extra careful when using $HNO_3(aq)$, especially if phenylamine is involved. This is because the benzene ring of phenylamine is highly activated by the delocalization of electrons from the N atom. $HNO_3(aq)$ might nitrate the benzene ring by introducing a nitro-group into the ring.

To recover the amine from its salt, all that's needed is to add a strong base such as $NaOH(aq)$.

$$CH_3CH_2NH_3^+Cl^- + OH^- \longrightarrow CH_3CH_2NH_2 + Cl^- + H_2O$$

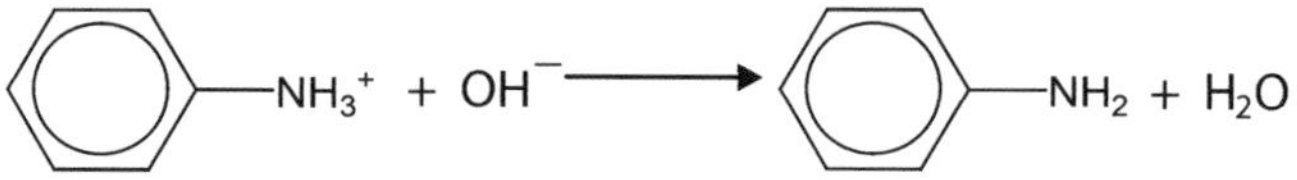

The above reactions are important procedures to separate a mixture of organic amine and acid. One can protonate the organic amine and make it more solubilized in the aqueous medium as the protonated amine is more soluble in water due to its ability to form ion–dipole interaction with water. Both the aqueous and organic layers can then be separated using a separatory funnel. The organic amine is then regenerated by adding a base to neutralize it. One could actually also deprotonate the organic acid (provided it can react with the added base) and convert it into the more soluble anionic form. After the two layers have been separated, the organic acid is regenerated by adding an acid to neutralize it.

Q: Can we use Na_2CO_3 as the base?

A: No, you can't. The existence of $(NH_4)_2CO_3$ just shows that the ammonium, alkyl ammonium and phenylammonium ions are not acidic enough to decompose the carbonate.

11.5.2 Formation of Amides with Acid Chloride — Acylation via Nucleophilic Substitution/Condensation

$$\underset{\substack{\text{O}\\ \| \\ \text{R}-\text{C}-\text{Cl}}}{} \;+\; \underset{\text{primary amine}}{\substack{\text{H}\\ | \\ \text{H}-\text{N}-\text{R}}} \longrightarrow \underset{\text{secondary amide}}{\substack{\text{O}\quad\text{H}\\ \| \quad | \\ \text{R}-\text{C}-\text{N}-\text{R}}} \;+\; \text{HCl}$$

$$\underset{\substack{\text{O}\\ \| \\ \text{R}-\text{C}-\text{Cl}}}{} \;+\; \underset{\text{secondary amine}}{\substack{\text{R}\\ | \\ \text{H}-\text{N}-\text{R}}} \longrightarrow \underset{\text{tertiary amide}}{\substack{\text{O}\quad\text{R}\\ \| \quad | \\ \text{R}-\text{C}-\text{N}-\text{R}}} \;+\; \text{HCl}$$

Reagents:	RCOCl
Conditions:	Sealed tube/Room temperature/Anhydrous
Options:	—

Tertiary (3°) amines do not undergo acylation because they do not have any replaceable H atoms. 1° amides are obtained by reacting acid chloride with ammonia.

The products of the acylation consist of an amide and HCl. But since amines are basic, they will react with the acidic HCl to produce a salt.

$$\text{RNH}_2 + \text{HCl} \longrightarrow \text{RNH}_3\text{Cl}$$

$$\text{R}_2\text{NH} + \text{HCl} \longrightarrow \text{R}_2\text{NH}_2\text{Cl}$$

11.5.3 Formation of Amides with Esters — Nucleophilic Acyl Substitution/Condensation

$$\underset{\substack{\text{O}\\ \| \\ \text{R}-\text{C}-\text{OR}}}{} \;+\; \underset{\text{primary amine}}{\substack{\text{H}\\ | \\ \text{H}-\text{N}-\text{R}}} \longrightarrow \underset{\text{secondary amide}}{\substack{\text{O}\quad\text{H}\\ \| \quad | \\ \text{R}-\text{C}-\text{N}-\text{R}}} \;+\; \text{ROH}$$

$$\underset{\substack{\text{O}\\ \| \\ \text{R}-\text{C}-\text{OR}}}{} \;+\; \underset{\text{secondary amine}}{\substack{\text{R}\\ | \\ \text{H}-\text{N}-\text{R}}} \longrightarrow \underset{\text{tertiary amide}}{\substack{\text{O}\quad\text{R}\\ \| \quad | \\ \text{R}-\text{C}-\text{N}-\text{R}}} \;+\; \text{ROH}$$

Reagents:	Ester
Conditions:	Heat
Options:	—

This reaction is similar in nature to the reaction between 1° and 2° amines with acid chlorides — they both proceed via nucleophilic acyl substitution wherein the amines act as nucleophiles attacking the electron-deficient carbonyl C atom. And as expected, tertiary (3°) amines do not react with esters since they do not have any replaceable H atoms.

Q: Will the alcohol produced further react with the amide since the C atom of the C–N bond is electron deficient?

A: No. This is because an amide is less reactive than an ester to nucleophilic attack (refer to Chapter 10) and in addition, the alcohol molecule is a weaker nucleophile than an amine because the lone pair of electrons for the amine is more available for nucleophilic attack due to the lower in electronegativity of the N atom as compared to the O atom.

Q: Can we get an amide by just reacting an amine with carboxylic acid?

A: No, refer to Chapter 10.

11.5.4 *Formation of Amine — Nucleophilic Substitution*

$$CH_3CH_2Cl + CH_3CH_2\overset{\cdot\cdot}{N}H_2 \longrightarrow CH_3CH_2-\overset{\overset{\displaystyle CH_2CH_3}{|}}{N}-H + HCl$$

primary amine   secondary amine

$$CH_3CH_2Cl + (CH_3CH_2)_2\overset{\cdot\cdot}{N}H \longrightarrow CH_3CH_2-\overset{\overset{\displaystyle CH_2CH_3}{|}}{N}-CH_2CH_3 + HCl$$

secondary amine   tertiary amine

$$CH_3CH_2Cl + (CH_3CH_2)_3\overset{\cdot\cdot}{N} \longrightarrow CH_3CH_2-\overset{\overset{\displaystyle CH_2CH_3}{\overset{+|}{}}}{\underset{\underset{\displaystyle CH_2CH_3}{|}}{N}}-CH_2CH_3 + Cl^-$$

tertiary amine   quaternary ammonium salt

Reagents:	Limited RX, ethanol
Conditions:	Heat in sealed tube
Options:	—

In this reaction, the amine acts as the nucleophile attacking the electron-deficient C atom bonded to the Cl atom in the halogenoalkane.

A nucleophile's role is that of an electron-pair donor and its reactivity depends on the ease of donating a lone pair of electrons to the electrophile. Hence, the reaction proceeds most readily for a 3° amine, followed by a 2° amine and then a 1° amine based on the electronic factor. However, the degree of "bulkiness" around the lone pair increases as the number of alkyl groups increases. Thus, due to the steric factor, a 3° amine would be less reactive than a 2° amine, and so on. The two factors, electronic and steric, would likely interact, and the reactivity outcome would depend on the nature and number of alkyl groups.

To prevent the amine product formed from further reacting with the halogenoalkane reactant to form poly-substituted products, the amine reactant must be introduced in excess relative to the amount of halogenoalkane.

Q: Is phenylamine nucleophilic enough to take part in nucleophilic substitution reactions since its lone pair of electrons is partially delocalized into the benzene ring?

A: Yes, phenylamine does react with halogenoalkane to give nucleophilic substituted products. The reaction is essentially similar to that of aliphatic amines. But the amide that is formed totally cannot function as a nucleophile as the lone pair is even less available than the lone pair on the N atom of phenylamine. This is because the lone pair of electrons can now delocalize both into the carbonyl (C=O) and phenyl functional groups.

11.5.5 *Formation of Halogenophenylamine — Electrophilic Substitution*

Apart from nucleophilic reactions involving the amino group, phenylamine can also undergo electrophilic substitution. Reactions of this nature proceed more readily since the benzene ring is more electron-rich due to the electron-donating effect of the amino group in phenylamine. The more reactive nature of phenylamine is observed in its reaction with aqueous bromine, which generates a poly substituted product — similar to that obtained in the halogenation of phenol.

2,4,6-tribromophenylamine
(a white ppt)

Reagents:	$Br_2(aq)$
Conditions:	Room temperature
Options:	—

Brown $Br_2(aq)$ is decolorized, and a white precipitate of 2,4,6-tribromophenylamine is obtained due to the low solubility of the organic product that is formed.

Q: Will halogenation of phenylamine occur when it is carried out using Br_2 in a non-polar solvent such as CCl_4? Will monosubstituted products be obtained as in the case of phenol?

A: Multiple substitution would still take place. So in order to get only the mono-substituted product, it is a common practice to react the phenylamine with ethanoyl chloride to form phenyl acetamide, which is less activating than the amino group of phenylamine. After halogenation of the benzene ring, the amide "cap" is removed by hydrolysis using a base.

Q: Why is the benzene ring of phenylamine much more electron rich than that of phenol?

A: As the electronegativity of the N atom in phenylamine is lower than that of the O atom in phenol, the lone pair of electrons on the N atom of phenylamine is more delocalized into the benzene ring than for phenol, causing the benzene ring of phenylamine to be more electron-rich.

11.5.6 *Reaction of Primary Amine with Nitrous Acid*

A primary amine or phenylamine would react with nitrous acid (HNO_2) to yield nitrogen gas. This typical reaction is useful to test for the presence of a primary amine as both 2° and 3° amines would not react to give off N_2 gas.

$$RNH_2 + 2HNO_2 \rightarrow ROH + N_2 + H_2O$$

11.6 Summary

The mind map below depicts the reactions of propylamine, with propene as the starting material so as to integrate the chemistry of propylamine with other functional groups.

The mind map below depicts the reactions of phenylamine, with benzene as the starting material so as to integrate the chemistry of phenylamine with other functional groups.

My Tutorial

1. Compounds **X**, **Y** and **Z** are isomers with the molecular formula $C_5H_{11}NO$. **X**, **Y** and **Z** contain a maximum of two functional groups each. Both **Y** and **Z** give negative test results when warmed with iodine in sodium hydroxide, whereas **X** gives a yellow precipitate. All three compounds give a negative result when tested with sodium nitrate and hydrochloric acid at 0°C.

   (a) (i) Identify the yellow precipitate that is formed.
      (ii) Name and give the constitutional/structural formula of the group identified with iodine in sodium hydroxide.

   (b) (i) Write a formula for compound **X**.
      (ii) Write the constitutional/structural formula of the organic product formed when **X** reacts with sodium nitrate and hydrochloric acid at 0°C.
      (iii) Write the formula of a constitutional/structural isomer of **X** having the same functional groups, which would give a negative result with iodine in sodium hydroxide and a positive test result with sodium nitrate and hydrochloric acid at 0°C.

   (c) **Y** is an isomer of **X** containing the functional group of a primary amide, $-CONH_2$. **Y** is not chiral.
      (i) State the type of isomerism that **X** and **Y** exhibit.
      (ii) State the physical state of **Y** and suggest reasons for it.
      (iii) Suggest a structure for **Z**, which is a chiral isomer of **Y**, and mark the chiral center with asterisk.
      (iv) Describe briefly how isomers **X** and **Z** might be distinguished.

   (d) (i) Describe a simple chemical test which would supply further evidence for the presence of *one* of the functional groups in **X**.
      (ii) Write the formula of the major organic product of the suggested test.
      (iii) Explain how this test might be extended to allow the positive identification of compound **X**.

   (e) Give the constitutional/structural formula of another isomer having the same molecular formula, which shows the same results as **X** for the iodine in sodium hydroxide and sodium nitrate and hydrochloric acid at $0°C$, but which would also exhibits *cis-trans* (geometric) isomerism.

2. When the neutral compound **P**, $C_{10}H_{13}NO$, was refluxed with dilute acid, it formed two products **Q**, C_2H_7N, and **R**. On analysis, **R** was found to contain 70.59% carbon, 23.53% oxygen and 5.88% hydrogen by mass. The relative molecular mass of **R** was found to be 136 via mass spectrometry. On reaction with alkaline potassium manganate (VII) solution, **R** was oxidized to **S**, $C_8H_6O_4$.

   S, which was acidic, was readily dehydrated to the neutral compound **T**, $C_8H_4O_3$. **Q** reacted with gaseous hydrogen bromide to form an ionic solid, **U**, C_2H_8NBr. When **U** was dissolved in dilute hydrochloric acid and sodium nitrate (III) solution, a yellow oil, **V**, was formed, and no effervescence occurred.

   (a) Determine the empirical and molecular formulae of **R**.
   (b) Deduce the constitutional/structural formulae for compounds **P** to **V**.
   (c) Give the IUPAC names of compounds **Q**, **R** and **G**.

3. (a) Explain why phenylamine is a weaker base than ethylamine.
   (b) Predict, with reasons, whether ethanol will be a weaker or stronger base than ethylamine.
   (c) Both nitrobenzene and phenylamine contain a nitrogen atom, yet phenylamine is basic while nitrobenzene is not. Explain.
   (d) Describe briefly how a sample of phenylamine can be prepared by reducing nitrobenzene using a metal–acid reaction. Describe how you can isolate the phenylamine and give equations for the reactions.

CHAPTER 12

Amino Acids

12.1 Introduction

Amino acids are organic compounds that contain both an acidic carboxyl group (–COOH) and a basic amine group (–NH$_2$). α-amino acids have both these functional groups attached to the same C atom, known as the α-carbon, and their general formula is RCH(NH$_2$)COOH.

$$H_2N - \overset{\overset{\displaystyle H}{|}}{\underset{\underset{\displaystyle R}{|}}{C}}{}^{\alpha} - COOH$$

Q: What is an α-carbon?

A: α-carbon refers to the first carbon atom that is bonded to a functional group, such as –COOH, –OH, –Cl, –NH$_2$, –CHO, –COOR, etc. In the case of an amino acid, there are two functional groups, namely the –COOH and –NH$_2$. Since –COOH has a higher priority than –NH$_2$, the parent is an acid. The amine group being considered as a substituent is given the prefix, amino. The next carbon atom that is bonded to the α-carbon is known as β-carbon and the following one is the γ-carbon, so on and so forth.

Accordingly, the differences among these α-amino acids lie with the type of side chain (R-group) that is attached to the α-carbon. When the R group is an H atom, we have the simplest α-amino acid known as glycine.

Apart from glycine, an important feature of the α-carbon of the other amino acids is that the α-carbon is chiral in nature, and this causes them to be optically active and thus exist as a pair of enantiomers.

mirror plane

Just like the use of Lego building blocks in creating structures of unique features, different combinations of amino acids in linear chains result in various types of proteins and polypeptides. The proteins and polypeptides in the human body are essentially derived from twenty α-amino acids which are classified according to the properties of the R-group. Depending on the nature of the R-group, some of these render the amino acids to be either weakly acidic or weakly basic while others render the amino acids to be either hydrophilic or hydrophobic. And because of these special natures of the different R-groups, different proteins consisting of different amino acids would have different physical properties, such as different solubility or interaction with different types of substrate molecules.

- Hydrophobic, neutral α-amino acids

These amino acids contain an R-group that is non-polar. Under this group of amino acids, we can have both aliphatic and cyclic hydrocarbons and aromatic rings. Thus, protein molecules possessing these types of amino acids are likely to be less soluble in water medium or more likely to interact with non-polar molecules.

Glycine (Gly)　　Alanine (Ala)　　Valine (val)　　Isoleucine (Ile)　　Leucine (Leu)

Phenylalanine (Phe)　　Tryptophan (Trp)　　Methionine (Met)　　Proline (Pro)

If you have noticed the structure of proline, its side chain is linked to the amine group at the α-carbon, making it a secondary amine, but it is still counted as one of the α-amino acids. As for methionine, the electronegativity difference between C (2.55) and S (2.58) is so small that the C–S bond is considered non polar.

Q: Why is tryptophan considered hydrophobic non-polar when its side chain has a secondary amine that is capable of hydrogen bonding and there is also a lone pair of electrons on the N-atom that is capable of being donated for hydrogen bond formation?

A: As mentioned in earlier chapters, the solubility of polar organic compounds can be compromised by the presence of large hydrophobic groups. In order for a base to act as a base and for an acid to behave as such, it must first dissolve. If solubility is low, then we would expect the corresponding acidity or basicity to be low too. In addition, the lone pair of electrons on the N-atom can delocalize into the benzene ring, rendering it less likely to be donated, hence a weaker base. At the same time, the lone pair of electrons is less likely to be used for hydrogen bond formation.

Q: Why is proline not basic in nature?

A: As for proline, the side chain is part of the amine group at the α-carbon, and we shall see in Section 12.2 that both these amine and carboxyl functional groups on the α-carbon do undergo what is known as an intramolecular acid–base reaction that produces a species that is better known as a zwitterion. This would result in the zwitterions behaving both as a basic and acidic species.

• Hydrophilic, neutral α-amino acids

These amino acids have a neutral polar R-group. Side chains of this nature contain the polar–OH and –SH groups as well as the amide –$CONH_2$ group, all of which are capable of forming hydrogen bonds with water molecules. Thus, protein molecules possessing these types of amino acids are likely to be more soluble in water medium or more likely to interact with more polar molecules.

Serine (Ser)   Cysteine (Cys)   Threonine (Thr)

Tyrosine (Tyr)   Asparagine (Asn)   Glutamine (Gln)

• Weakly acidic α-amino acids

The R-group of these amino acids contains the weakly acidic carboxylic acid. At physiological pH, which refers to pH 7.4 of blood, the carboxyl group will undergo an acid–base reaction and be deprotonated. Hence, at physiological pH, these amino acids would have negatively-charged R-groups. Therefore, protein molecules possessing these types of amino acids are likely to be more soluble in water medium or more likely to interact with more polar molecules.

Aspartic acid (Asp)   Glutamic acid (Glu)

- Weakly basic α-amino acids

  The R-group of these amino acids contains the weakly basic amine group (this does not refer to the amine group that is bonded to the α-carbon). At physiological pH, the amine group will undergo an acid–base reaction and be protonated. Hence, at physiological pH, these amino acids have positively-charged R groups. Therefore, protein molecules possessing these types of amino acids are likely to be more soluble in water medium or more likely to interact with more polar molecules.

Lysine (Lys)   Arginine (Arg)   Histidine (His)

Q: The physiological pH is about 7.4, which is a basic pH, so how can the alkaline physiological pH protonate the –NH_2 of the R-group, which is basic in nature?

A: The pK_b of NH_3 is about 9.26, which means that when ammonia gas dissociates in pure water at 25°C, the pH of the solution would be around 9.26 (we say "around" because the actual pH of the solution would depend on the initial concentration of NH_3). Thus, at the physiological pH, which is more acidic than the pK_b

of NH_3, certainly NH_3 would be protonated. If we treat the $-NH_2$ in the R-group as being similar to a NH_3 in terms of basic strength, then it is logical to expect the more basic $-NH_2$ group to be protonated too at the relatively more acidic physiological pH.

Q: Is there a way to show that if the pK_b of NH_3 is about 9.26, then the pH would also be around that value?

A: From Chapter 8, pH $\approx \frac{1}{2} pK_a$ only for a weak acid of concentration $1\,mol\,dm^{-3}$. For a weak base of concentration $1\,mol\,dm^{-3}$, then pOH $\approx \frac{1}{2} pK_b$. Therefore, for NH_3, pH $\approx 14 - \frac{1}{2} pK_b = 9.37$. Hence, if one knows the pK_a or pK_b value of a $-NH_2$ group, one can predict whether it would be more protonated or less protonated under a specific pH. For example, the pK_a value of the $-NH_2$ group of glutamic acid has a pK_a value of about 9.5, which indicates that it is an extremely weak acid. The conjugate base of the $-NH_2$ group of glutamic acid, which is $-NH^-$, would have a pK_b of about 4.5. This would mean that if the pH of the solution is less than the pK_a value of 9.5, the $-NH_2$ group would be protonated and vice versa.

Q: Why is there a pK_a value for the $-NH_2$ group? Do you mean that the $-NH_2$ group can act as an acid too? Shouldn't it just be basic?

A: Refer to Chapter 8; the $-NH_2$ group can act as an acid because the N–H bond is polar, hence in the presence of a very strong base or a very reactive metal such as Na, the $-NH_2$ group can act as an acid. For example, $Na(s) + NH_3(l) \rightarrow NaNH_2(s) + 1/2H_2(g)$. But as we discussed in Chapter 8, if the pK_b of the base is small, then its pK_a would be large, unless this species is equally strong in its acidic and basic properties.

Example: Given that there are three pK_a values associated with aspartic acid: 2.3, 4.0 and 9.3, suggest the major species present in solutions of aspartic acid with the pH values (i) 1, (ii) 3, (iii) 7 and (iv) 12.

Solution: The pK_a value of 2.3 is associated with the $-COOH$ group attached to the α–carbon, and it is more acidic than the $-COOH$ group in the R-group, which has a pK_a value of 4.0. The higher acidity of the former is due to the –COOH group being closer to an electron-withdrawing –NH$_2$ group, which is attached to the same α–carbon. The pK_a value of 9.3 corresponds to the –NH$_3^+$ group, indicating it is an extremely weak acid. Take note that for such a case here, the amino acid is always taken as the most protonated form. Meaning? If the amino acid has –COOH groups, these groups are not deprotonated. And if the amino acid contains –NH$_2$ groups, they are considered in the protonated –NH$_3^+$ form.

(i) At pH $= 1$, which is lower than all the pK_a values, the cationic species is the dominant one:

$$
\begin{array}{c}
\text{H} \\
| \\
\text{H}_3\overset{+}{\text{N}}-\text{C}-\text{COOH} \\
| \\
\text{CH}_2 \\
| \\
\text{COOH}
\end{array}
$$

(ii) At pH $= 3$, the –COOH group attached to the α–carbon would be deprotonated as it is the strongest acidic group amongst all, generating a zwitterion:

$$
\begin{array}{c}
\text{H} \\
| \\
\text{H}_3\overset{+}{\text{N}}-\text{C}-\text{COO}^- \\
| \\
\text{CH}_2 \\
| \\
\text{COOH}
\end{array}
$$

(iii) At pH $= 7$, both the –COOH groups would be deprotonated, generating the singly charged anionic species as the dominant one:

$$
\begin{array}{c}
\text{H} \\
| \\
\text{H}_3\overset{+}{\text{N}}-\text{C}-\text{COO}^- \\
| \\
\text{CH}_2 \\
| \\
\text{COO}^-
\end{array}
$$

(iv) At pH $= 12$, the doubly charged anionic species is the dominant one:

$$
\begin{array}{c}
\text{H} \\
| \\
\text{H}_2\text{N}-\text{C}-\text{COO}^- \\
| \\
\text{CH}_2 \\
| \\
\text{COO}^-
\end{array}
$$

12.2 The Formation of Zwitterions — An Intramolecular Acid–Base Reaction

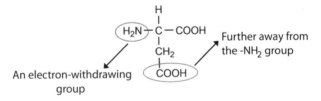

The intramolecular acid–base reaction is to be expected since the amine and carboxyl functional groups are in very close proximity to each other. The zwitterion that is produced is ionic, and it is the predominant form for amino acids molecule to exist in. It got its name from the German word *zwitter*, which means *hybrid*, as it is a dipolar ion. However, overall, the zwitterion is considered an electrically neutral species even though it carries formal charges within the molecule itself. The quantity of the formal charge is greater than the partial electric dipole on, let's say, a polar H–Cl molecule. Hence, we would expect the electrostatic interaction between the formal charges to be stronger than that between the partial dipoles. Due to this dipolar nature of zwitterions, amino acids exhibit physical properties rather unusual for covalent molecules but similar to those of ionic compounds.

Q: Taking the example of aspartic acid, can an intramolecular acid–base reaction occur between the amine group and the side chain carboxyl group instead?

A: Not likely. Both the –COOH groups on the α-carbon and the R-group have different acidic strength. We would expect the –COOH group on the α-carbon to be more acidic than the one on the R-group due to the presence of an electron-withdrawing amine group that is also bonded to the α-carbon. This electron-withdrawing group would enhance the acidity of the –COOH group by stabilizing the conjugate base that forms (refer to

Chapter 8). Thus, being more acidic, the –COOH group on the α-carbon would be more likely to dissociate and hence suppress the dissociation of the less acidic –COOH on the R-group. In addition, the proximity of both groups on the α-carbon makes the protonation more favorable as the close proximity of the two oppositely charged groups that form, the $-NH_3^+$ and $-COO^-$ groups, allows intramolecular ionic interaction to take place.

12.3 Physical Properties of Zwitterions

12.3.1 *Melting and Boiling Points*

One evidential fact that amino acid exists as zwitterions is their high melting points, and this, for the simplest amino acid glycine, is $233°C$. Under normal conditions, you will find them to be colorless crystalline solids.

The high melting point can only be explained if we consider amino acids to be in the zwitterionic form with strong ionic bonds between oppositely-charged groups that need to be overcome for the phase change to occur. Undeniably, the idea of amino acids existing just as simple discrete covalent molecules could not account for the high melting point since the weak intermolecular attractive forces can be easily overcome.

12.3.2 *Solubility*

As zwitterions are charged species, we would expect the amino acids to be capable of forming strong ion–dipole interactions with water molecules and therefore dissolve readily in water. But in actual fact, the solubility of amino acids is not very high in water. This is because the close proximity of the two oppositely charged groups restricts the formation of the hydration sphere. As a result, the amount of energy that is released during the hydration process is insufficient to compensate for the energy that is required to overcome the strong ionic bond in the solid state. Being in the zwitterionic form, the amino acids are insoluble in non-polar solvents.

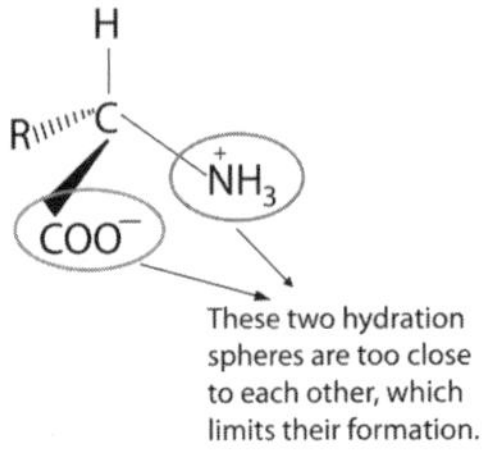

These two hydration
spheres are too close
to each other, which
limits their formation.

12.4 Chemical Properties of Zwitterions

12.4.1 *Acid–Base Reaction with Acids or Bases*

An aqueous solution of an amino acid can be perceived as a system in dynamic equilbrium consisting of two species, as shown below:

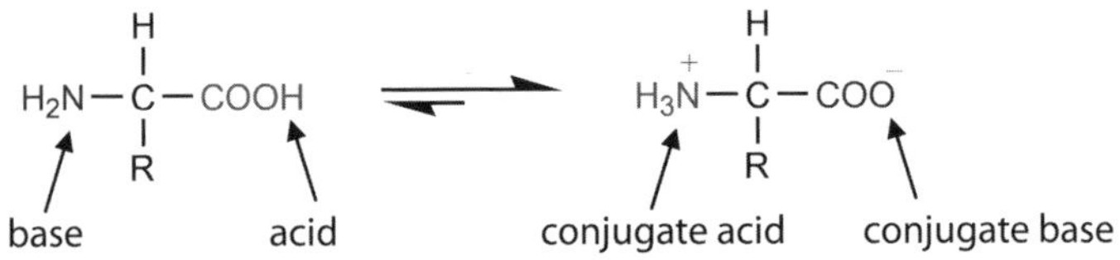

base   acid   conjugate acid   conjugate base

Q: Why is the double-headed arrow being used, and why is it not of the same length?

A: The usage of the double-headed arrow is to indicate that when the amino acid is dissolved in water, some of the zwitterions would be converted back to the non-zwitterionic form. But the predominant species is still the zwitterionic form, which is thus indicated by a longer arrow towards the right. In fact, both the cationic and anionic forms of the amino acid molecules do exist in water too. The quantity of each of the species, i.e. non-zwitterionic, zwitterion, anionic and cationic forms, would depend very much on the pH of the solution!

Regardless of the type of R group present, all amino acids are amphoteric.

- The carboxylate group ($-COO^-$) is the conjugate base of the weakly acidic carboxyl group, and it is thus the species in the zwitterions that reacts with acids.
- The protonated amine group ($-NH_3^+$) is the conjugate acid of the weakly basic amine group, and it is thus the species in the zwitterion that reacts with bases.

$$\underset{\substack{| \\ R}}{\overset{\substack{H \\ |}}{H_3\overset{+}{N}-C-COO^-}} + H^+ \longrightarrow \underset{\substack{| \\ R}}{\overset{\substack{H \\ |}}{H_3\overset{+}{N}-C-COOH}}$$

Cationic product

$$\underset{\substack{| \\ R}}{\overset{\substack{H \\ |}}{H_3\overset{+}{N}-C-COO^-}} + OH^- \longrightarrow \underset{\substack{| \\ R}}{\overset{\substack{H \\ |}}{H_2N-C-COO^-}} + H_2O$$

Anionic product

Although the zwitterionic form of the amino acids is amphoteric, they are generally less acidic than carboxylic acids (using the K_a of CH_3COOH, which is $\sim 1.8 \times 10^{-5}$ as a guide) and less basic than most amines (using the K_b of NH_3, which is $\sim 1.8 \times 10^{-5}$, and $CH_3CH_2NH_2$, which is $\sim 4.7 \times 10^{-4}$, as a guide). Why is this so? It is simply because $-NH_3^+$ is a weaker acid (K_a of $CH_3CH_2NH_3^+$ is $\sim 2.1 \times 10^{-11}$) than a $-COOH$ group due to the stronger attractive force between the lone pair of electrons on the N atom and the H^+ ion, even though the $-NH_3^+$ is positively charged and thus more likely to be attracted by a base. The strong attractive force between the lone pair of electrons and the H^+ ion makes it less likely to be extracted away, unless a stronger base such as OH^-, H^-, NH_2^-, etc. is used.

Similarly, the basicity of the conjugate base, $-COO^-$ (using the K_b of CH_3COO^-, which is $\sim 5.6 \times 10^{-10}$, as a guide), is not as high as that of a $-NH_2$ group even though it contains a negative charge. This is because the negative charge is being dispersed into an electron-withdrawing carbonyl group through resonance (Chapter 9). Hence, for an amino acid with an R-group that contains the $-COOH$ group, for example aspartic acid, the resultant acidity of the amino acid that has arisen is due to the more acidic $-COOH$ group on the R-group and not due to $-NH_3^+$ group. In addition, $-COOH$ being a stronger acid than the $-NH_3^+$ group, the dissociation of the $-COOH$ would suppress the dissociation of $-NH_3^+$. In a similar vein, for an amino acid with a basic $-NH_2$ group in the R-group, the resultant basicity that has arisen is from the $-NH_2$ in the R-group and not the $-COO^-$ on the α-carbon. Anyway, just think logically: The fact that the zwitterions can be formed would mean that $-COOH$ is a stronger acid than $-NH_3^+$, or else how could the $-NH_2$ group be protonated to create $-NH_3^+$? In a similar argument, if $-COO^-$ is a stronger base than $-NH_2$ group, then it would be able to deprotonate the $-NH_3^+$ and exist as both $-NH_2$ and $-COOH$ instead, but it did not!

All these explanations can thus be used to account for the existence of $(NH_4)_2CO_3$, but not when carboxylic acid and Na_2CO_3 are mixed. Usually, if we have two weak acids, HX and HY, and if HX is a stronger weak acid than HY, it would be indicated by a higher K_a value for HX. Then the conjugate base of HX, which is X^-, would be a weaker base than Y^-, which is indicated by a higher K_b value for Y^-. Remember when the K_a of HX is multiplied by the K_b of X^-, it is equal to the K_w, which has a value of $1.0 \times 10^{-14} \, mol^2 \, dm^{-6}$ at 25°C?

If a solid amino acid of the zwitterionic form is dissolved in water, the following equilibria would be established:

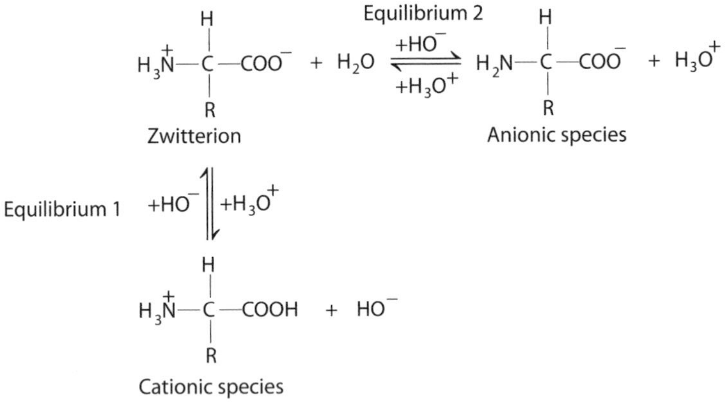

If the K_b of the $-COO^-$ is greater than the K_a of the $-NH_3^+$ group, then the predominant species that are present in the system would be both the zwitterions and the cationic species. Hence, an alkaline solution would result as the hydrolysis of the $-COO^-$ group to form $-COOH$ is more rampant than the dissociation of the $-NH_3^+$ group to form $-NH_2$, as shown by Equilibrium 1. An addition of H_3O^+ to Equilibrium 1 would "encourage" the formation of the cationic species in accordance to the prediction by Le Chatelier's principle of equilibrium shift. Oppositely, the addition of OH^- to Equilibrium 1 results in the formation of the zwitterion. "Over addition" of OH^- to Equilibrium 1 would result in the formation of the anionic species, i.e., shifting the position of equilibrium 2 to the right.

If the K_b of the $-COO^-$ is smaller than the K_a of the $-NH_3^+$ group, then the predominant species that are present in the system would be both the zwitterions and the anionic species, with an acidic solution being formed. This is because the dissociation of the $-NH_3^+$ group to form $-NH_2$ is more rampant than the hydrolysis of the $-COO^-$ group to form $-COOH$. An addition of OH^- to Equilibrium 2 would "encourage" the formation of the anionic species in accordance to the prediction by Le Chatelier's principle of equilibrium shift. Oppositely, the addition of H_3O^+ to Equilibrium 2 results in the formation of the zwitterion. "Over addition" of H_3O^+ to Equilibrium 2 would result in the formation of the cationic species, i.e. shifting the position of equilibrium 1 "downwardly".

Hence, different types of amino acid would give rise to solutions of different pH. One should take note that in the solid state, only the zwitterionic form exists! But in the solution, there are the zwitterions, the cationic species and the anionic species, each existing in different concentrations, and these concentrations would in turn depend on the nature of the amino acid itself and the pH of the solution.

Q: Is it necessary for the cationic species to form the zwitterions before forming the anionic species? Or vice versa?

A: It is not necessary. If a cationic species collides with two OH^- ions at the same time, then the anionic species can be formed without the formation of the zwitterions. But obviously, it is statistically less probable for three species to collide at one go than for two species to do so. In addition, for the reaction to be fruitful, it is also necessary for an appropriate molecular orientation to be in place when three particles collide.

Q: Is it possible to have a solution where only the zwitterions exist?

A: Yes, amino acids exist predominantly as the zwitterions only at a certain pH, which we call the **isoelectric point (pI)**. Thus, pI is actually a pH value needed to create the zwitterions only.

If **the pH of a medium is lower than the pI** value of an amino acid, its weakly basic $-COO^-$ group (along with the basic R-group, if any) will react with the H^+ ions in the medium and the **cationic form predominates** under this type of acidic condition.

If the pH of medium is higher than the pI value of an amino acid, its weakly acidic $-NH_3^+$ group (along with the acidic R group, if any) will react with the OH^- ions in the medium and the **anionic form predominates** under this type of basic condition.

Q: Since amino acids are in the zwitterionic form when they are in the solid state, if we now dissolve two amino acids which do not have any other additional acidic or basic R-groups, say glycine and cysteine, in water, would these two amino acids give the same amount of zwitterions at the same pH?

A: The amount of zwitterions for glycine would be different from that of cysteine. In addition, the pH of the solution would also be different. Why is it so? Although both glycine and cysteine do not have additional acidic or basic groups on the R-groups, but the very fact that these two amino acids have different R-groups would mean that the acidity of the $-NH_3^+$ group and the basicity of the $-COO^-$ group would be different for each of the zwitterions. Thus, we would expect the pH of the solution of glycine to be different from cysteine because a different amount of zwitterions would be present for each amino acid.

Zwitterion of Glycine — Equilibrium 1:
$$H_3\overset{+}{N}-\underset{H}{\overset{H}{C}}-COO^- + H_2O \rightleftharpoons H_2N-\underset{H}{\overset{H}{C}}-COO^- + H_3O^+$$

Zwitterion of Cysteine — Equilibrium 2:
$$H_3\overset{+}{N}-\underset{CH_2SH}{\overset{H}{C}}-COO^- + H_2O \rightleftharpoons H_2N-\underset{CH_2SH}{\overset{H}{C}}-COO^- + H_3O^+$$

Concentration of the H_3O^+ higher than in Equilibrium 1, hence needs a lower pH to shift Equilibrium 2 to the left. This results in a lower pI value.

Electron withdrawing group stabilises the conjugate base, making Cysteine a stronger acid than Glycine.

Indeed, the pI value of glycine is 5.97 whereas the value for cysteine is 5.07. What is the key significance of the different pI values? A lower pI value means that we need to create a more acidic pH (a higher concentration of H^+) to protonate the anionic form so as to convert all species into the zwitterionic form. This lower pI value can only happen if comparatively to glycine, most of the zwitterions of cysteine have already been dissociated to give the anionic species and H$^+$. Which means that the pH of cysteine in pure water must be more acidic than glycine (Equilibrium 2 is more to the right than Equilibrium 1). Thus, we need a more acidic pH to "suppress" the dissociation of the zwitterions of cysteine as compared to for glycine. In a nutshell, the smaller the pI value of an amino acid, the more acidic would be its zwitterions when placed in pure water!

Recall that when the pH of a buffer solution is equal to the pK_a value of the weak acid that makes up the buffer solution, the concentration of the conjugate acid (HA) is equal to the concentration of the conjugate base (A$^-$). This pH = pK_a concept is exactly the same as the pI concept of an amino acid.

Q: If we place aspartic acid in a basic medium, will both acidic groups, that is the –COOH and –NH$_3^+$ groups, in the zwitterion react with the OH$^-$ ions?

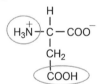

A: They would, BUT it depends on the pH of the medium. Why is this so? The –NH$_3^+$ group is less acidic than the –COOH group, and thus under a limited supply of OH$^-$ (indicated by a lower pH value), the OH$^-$ will more likely react with the stronger of the two acids, which is the –COOH group here. However, if [OH$^-$] is high, deprotonation of the –NH$_3^+$ group will also occur, since there is now excess OH$^-$ after all the –COOH groups have been reacted. The same can be said for a basic amino acid such as lysine wherein the side chain amine (–NH$_2$) group is more basic than the carboxylate (–COO$^-$) group. Thus, one would expect the –NH$_2$ group to be protonated first before protonating the –COO$^-$. In short, one would expect the quantity of charge and the overall sign of the charge of an amino acid molecule to vary with respect to pH when it is placed in water.

Q: If the –COO⁻ group carries a negative charge, shouldn't it be more likely to be protonated first than the neutral –NH_2 group?

A: Good question! The negative charge (which means an extra electron) of the –COO⁻ group is actually delocalized to the highly electron-withdrawing carbonyl group, and the negative charge is equally dispersed among two highly electronegative O atoms. As a result, the basicity of the –COO⁻ group is lower than that of the neutral –NH_2 group. The K_b of CH_3COO^- is ~5.6×10^{-10}, while the K_b of NH_3 is ~1.8×10^{-5}. This is a good gauge for estimating the basicities of the –COO⁻ and –NH_2 groups in the zwitterions. In conclusion, a negative charged species may not necessarily be more basic than a neutral one. For example, Cl^- is negatively charged, but it is a very weak conjugate base of HCl.

Q: Can we ever get back the non-ionized form of the amino acid $RCH(NH_2)COOH$?

A: Theoretically, yes. We can adjust the pH of the solution in such a way that the –NH_3^+ group of the zwitterions is deprotonated while protonating the –COO⁻ group at the same time. In actual fact, an aqueous solution of an amino acid consists of a mixture of species ranging from the electrically neutral to the cationic and the anionic forms. The predominant composition would solely depend on the pH of the solution and the nature of the amino acid.

Each of the amino acids has a characteristic isoelectric point which reflects the type of R group that we had earlier used to classify the amino acids (see Table 12.1). Amino acids with neutral side chain have pI values ranging from 5.0 to 6.3. Both aspartic acid and glutamic acid have isoelectric points in the lower acidic pH range, whereas the basic amino acids have isoelectric points at higher pH values.

The differences in isoelectric points can be used to separate a mixture of amino acids and establish their identity in an analytical procedure known as electrophoresis (see Chapter 18).

A simple electrophoresis procedure involves placing a mixture of amino acids in the center of a strip of filter paper soaked in a buffer solution

Table 12.1   Isoelectric Points of the 20 α-Amino Acids.

Neutral amino acids					
• Glycine	5.97	• Phenylalanine	5.48	• Cysteine	5.07
• Alanine	6.00	• Tryptophan	5.89	• Threonine	5.60
• Valine	5.96	• Methionine	5.74	• Tyrosine	5.66
• Isoleucine	6.02	• Proline	6.30	• Asparagine	5.41
• Leucine	5.98	• Serine	5.68	• Glutamine	5.65
Acidic amino acids					
• Aspartic acid	2.77	• Glutamic acid	3.22		
Basic amino acids					
• Lysine	9.74	• Arginine	10.76	• Histidine	7.59

with a pre-determined pH. Depending on the pI of the amino acid relative to the pH of the solvent, we can have the neutral, cationic or anionic species being formed. These species are then separated by applying an electric field across the paper strip. Depending on their charge and molecular weight, these species will have different migration patterns under the influence of the applied electric field:

- If **pH of solvent > pI of amino acid**, the **anionic form** of the amino acid predominates, and it migrates towards the **positive electrode**(anode). This is because higher pH value signifies higher concentration of OH^-, hence more deprotonation.

- If **pH of solvent < pI of amino acid**, the **cationic form** of the amino acid predominates, and it migrates towards the **negative electrode**(cathode). This is because lower pH value signifies higher concentration of H^+, hence more protonation.

- If **pH of solvent = pI of amino acid**, the amino acid exists predominantly as the **neutral zwitterion**, which **does not migrate**.

After electrophoresis, the positions of the amino acids on the filter paper can be identified by the use of ninhydrin, which forms a violet-colored product with the amino acids (except for proline and hydroxyproline).

Example: What is observed when a mixture of glycine (Gly), glutamic acid (Glu), aspartic acid (Asp) and lysine (Lys) are subjected to electrophoresis at a pH of about 5.9?

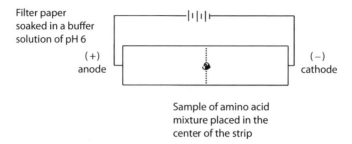

Filter paper
soaked in a buffer
solution of pH 6

(+)
anode

(−)
cathode

Sample of amino acid
mixture placed in the
center of the strip

Solution:

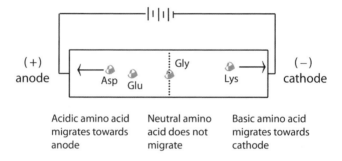

(+)
anode

Asp   Glu

Gly

Lys

(−)
cathode

Acidic amino acid   Neutral amino   Basic amino acid
migrates towards   acid does not   migrates towards
anode   migrate   cathode

- Since glycine's pI of 5.97 is very close to the pH of the buffer, glycine exists predominantly as the neutral zwitterion, which does not migrate under the effect of an electric field.
- Since lysine's pI of 9.74 is greater that the pH of the buffer, it is said to behave as a base in the "acidic" medium and is protonated. The resultant cationic species are attracted to the negatively charged cathode and migrate towards it.

$$H_3\overset{+}{N}-\underset{\underset{\underset{NH_2}{|}}{\overset{|}{(CH_2)_4}}}{\overset{\overset{H}{|}}{C}}-COO^- + 2H^+ \longrightarrow H_3\overset{+}{N}-\underset{\underset{\underset{\overset{+}{NH_3}}{|}}{\overset{|}{(CH_2)_4}}}{\overset{\overset{H}{|}}{C}}-COOH$$

Lysine (Lys)   Cationic product

- Since both aspartic acid and glutamic acid have pI values less than that the pH of the buffer, these are said to behave as acids in the relatively "basic" medium and are deprotonated. The resultant anionic species are attracted to the positively charged anode and migrate towards it.

$$H_3\overset{+}{N}-\underset{\underset{\underset{COOH}{|}}{\underset{CH_2}{|}}}{\overset{\overset{H}{|}}{C}}-COO^- \quad + \quad OH^- \quad \longrightarrow \quad H_3\overset{+}{N}-\underset{\underset{\underset{COO^-}{|}}{\underset{CH_2}{|}}}{\overset{\overset{H}{|}}{C}}-COO^- \quad + \quad H_2O$$

Aspartic acid (Asp)　　　　　　　　Anionic product

$$H_3\overset{+}{N}-\underset{\underset{\underset{\underset{COOH}{|}}{\underset{CH_2}{|}}}{\underset{CH_2}{|}}}{\overset{\overset{H}{|}}{C}}-COO^- \quad + \quad OH^- \quad \longrightarrow \quad H_3\overset{+}{N}-\underset{\underset{\underset{\underset{COO^-}{|}}{\underset{CH_2}{|}}}{\underset{CH_2}{|}}}{\overset{\overset{H}{|}}{C}}-COO^- \quad + \quad H_2O$$

Glutamic acid (Glu)　　　　　　　　Anionic product

- As the rate of migration $\propto \frac{\text{charge}}{\text{mass}}$, the anionic form of aspartic acid, which has a lower molecular weight, will be located nearer to the anode than that of glutamic acid.

Q: Can we use electrophoresis to separate a mixture of amino acids with neutral side chains?

A: Yes. As has been mentioned before, the degree of deprotonation of the $-NH_3^+$ and protonation of $-COO^-$ in water depends on the value of the K_a and K_b, respectively. And the K_a and K_b values would in turn also be affected by the nature of the R-group. The R-groups of, let's say, three amino acids may all be neutral, but there may still be differences in their K_a and K_b. Even if the K_a and K_b values were the same, the molecular weight of these amino acids would certainly be different. Hence, the overall charge/mass ratio would be different at a particular pH. Thus, separation could be effected.

12.4.2 *Formation of Amides*

$$R-\overset{\overset{O}{\|}}{C}-Cl \quad + \quad H-\underset{\underset{R}{|}}{\overset{\overset{H}{|}}{N}}-\overset{\overset{H}{|}}{C}-COOH \quad \longrightarrow \quad R-\overset{\overset{O}{\|}}{C}-\overset{\overset{H}{|}}{N}-\underset{\underset{R}{|}}{\overset{\overset{H}{|}}{C}}-COOH \quad + \quad HCl$$

$$R-\overset{\overset{O}{\|}}{C}-OR \quad + \quad H-\underset{\underset{R}{|}}{\overset{\overset{H}{|}}{N}}-\overset{\overset{H}{|}}{C}-COOH \quad \longrightarrow \quad R-\overset{\overset{O}{\|}}{C}-\overset{\overset{H}{|}}{N}-\underset{\underset{R}{|}}{\overset{\overset{H}{|}}{C}}-COOH \quad + \quad ROH$$

Reagents:	Acid chloride
Conditions:	Room temperature/Anhydrous
Options:	Ester/Heat

Q: In the acid–base reaction of amino acids, we consider these to be the reactions of zwitterions. Why is it that in the formation of amides, we revert back to the non-ionized form?

A: Zwitterions are not nucleophilic enough to attack the electron-deficient C atom of the acid chloride. It is the amino group of the non-ionized form of the amino acid that has nucleophilic properties. In fact, even though amino acids exist predominantly as zwitterions, an equilibrium exists between these two forms when it is being placed in a solvent, with the majority being the zwitterionic form.

$$H_2\ddot{N}-\underset{\underset{R}{|}}{\overset{\overset{H}{|}}{C}}-COOH \;\rightleftharpoons\; \overset{+}{H_3N}-\underset{\underset{R}{|}}{\overset{\overset{H}{|}}{C}}-COO^-$$

nucleophilic site

Thus, when the non-ionized form "hits" the acid chloride, a reaction occurs and the non-ionized form is consumed; its concentration will drop and, according to Le Chatelier's principle, the backward reaction will be favored in an attempt to increase the concentration of the non-ionized form. Hence, just as how the continuous addition of a strong base will cause further dissociation of the weak acid, addition of RCOCl will cause more of the zwitterions to form back the non-ionized form, which then reacts with the added RCOCl.

12.4.3 *Polymerization — Reaction between Amino Acids*

Peptides are polymers made up of numerous amino acid units linked together by strong peptide bonds or amide bonds as commonly known by chemists (–CONH–). Proteins are polypeptides that consist of 100 or more amino acid residues.

Formation of the peptide linkages (or amide linkages) requires the amino group of one amino acid to react with the carboxyl group of another amino acid with the elimination of a water molecule. Such a polymerization process is known as condensation polymerization. The amino acid that is linked to the polymeric chain is commonly known as a residue.

With the 20 common amino acids, combinations of these in any number and type give rise to a whole wide range of polymers.

When one molecule of Glycine (Gly) combines with that of Alanine (Ala), we get not one but two different types of dipeptides:

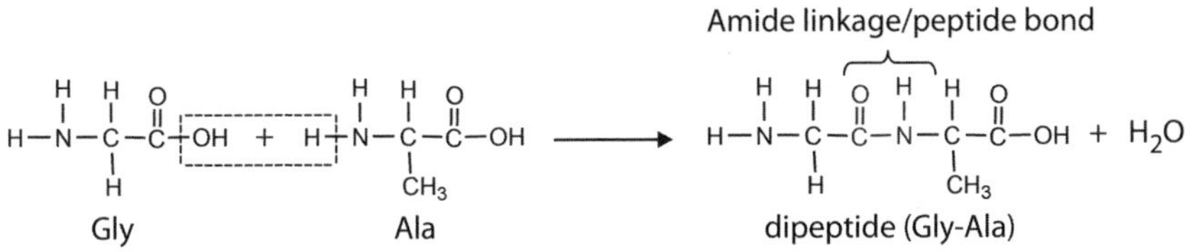

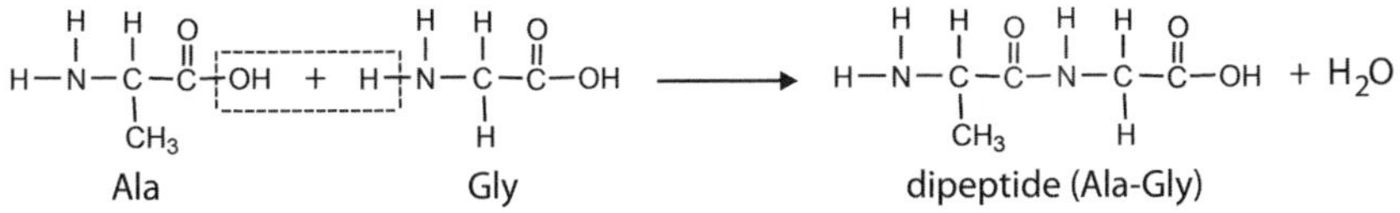

Going by the 3-letter code, the dipeptides Gly-Ala and Ala-Gly are two different compounds. Even though they contain the same amino acid residues, their specific bonding pattern differs. In Gly-Ala, the Gly amino acid residue has a free amino group and is called the N-terminal amino acid residue. Ala with the free carboxyl group is known as the C-terminal residue. The reverse is true for Ala-Gly.

By convention, the formula of peptides starts with the N-terminal amino acid residue written on the left and ends with the C-terminal amino acid residue on the right:

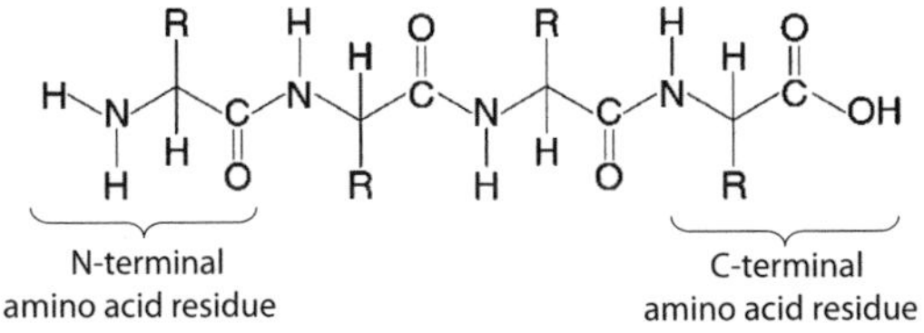

Exercise: What are the possible tripeptides that can be formed from combining one each of Glutamic acid (Glu), Serine (Ser) and Tyrosine (Tyr). Use the 3-letter code to name them.

Solution: There are a total of six possible tripeptides, namely, Glu-Ser-Tyr, Glu-Tyr-Ser, Ser-Glu-Tyr, Ser-Tyr-Glu, Tyr-Glu-Ser and Tyr-Ser-Glu.

My Tutorial

1. (a) An organic compound has the following composition by mass: 61.3% carbon; 5.1% hydrogen; 10.2% nitrogen and 23.4% oxygen. Determine the empirical formula of the compound.
   (b) Given the relative molecular mass of the compound in (a) is 137, determine the molecular formula of the compound.
   (c) The compound may be one of the following three isomers:

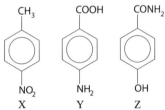

   (i) Explain which isomer, **X** or **Y**, would have the higher boiling point.
   (ii) With appropriate reagents and conditions, describe a simple chemical test to differentiate the following pair of compounds:

   (A) **X** and **Y**
   (B) **X** and **Z**
   (C) **Z** and **Y**

   (iii) Compound **X** can be converted to compound **Y** via a two-step process. Give suitable reagents and conversions for each of the two steps.

   (d) Compound **Y** belongs to a class of compounds known as amino acids.

   (i) Give the structures of **Y** which is present predominantly in each of the solution:

   (A) pH 2

(B) pH 7

(C) pH 10

(ii) What important property does the presence of the amino acid confer on the solution? Suggest an important application of this solution.

(e) (i) Rank the three compounds trimethylamine, aminoethanoic acid and ethanamide in order of increasing melting point, give your reasons.

(ii) Predict with reasons, whether the ethyl ester of aminoethanoic acid will have a higher or lower melting point than that of aminoethanoic acid itself.

(f) Phenylalanine, $C_6H_5CH_2CH(NH_2)COOH$, is an amino acid that is commonly found in nature.

(i) Explain why the solubility of phenylalanine is low in water but the solubility increases when a base or an acid is added.

(ii) In a solution where the pH is about 7, there is very little movement of phenylalanine toward both the positive and negative electrodes. Draw the structure of the species that is present at this pH.

(iii) Draw a polymeric section of poly(phenylalanine), showing three repeat units. Name the functional group that holds these monomers together.

CHAPTER 13

Polymers

13.1 Introduction

Polymers are macromolecules made up of many smaller molecules combined together to form really long chains. The number of molecules can easily be in the hundreds of thousands, or even millions. Polymers can be classified as either natural or synthetic polymers. The former includes rubber and polysaccharides such as starch and cellulose. We also have biopolymers such as DNA and proteins, which play a key role in biological processes and from which silk and wool are made. Synthetic polymers, the man-made ones, include synthetic rubber, nylon, silicone and a broad range of plastics such as poly(styrene) and poly(vinylchloride) (commonly abbreviated as PVC).

Plastics is a term that aptly describes these polymeric materials' ability to be molded and shaped into many forms — tubes, fibers, films, boxes and even the irregular shaped ones, such as toys. These plastics can be classified as either of two types — *thermoplastics* and *thermosets* — based on the different effects that heat has on them. Thermoplastics soften when heated and harden again on cooling. They can be subjected to repeated cycles of the molding process, which promotes recycling, but there is an eventual limit because of the chemical degradation that happens during the heating process. Examples of thermoplastics include nylon, polyester, acrylic, poly(styrene) and PVC. Such materials can be melted and molded or reshaped into various useful products such as toys, food packaging, beverage bottles and advertising signage.

On the other hand, thermosets can be melted to take shape permanently. They harden when heated and do not soften again. Examples of thermosets include melamine, used in dinnerware; epoxy resin, used in fiberglass and as a strong adhesive that can be used on metal, glass and stone; and Bakelite, made from phenol and methanal, which is used in plastic ware and as an electrical insulator. As thermosets do not melt when heated, recycling them poses a problem. Yet, they are valued for their very nature — they are generally stronger than thermoplastics and they are suitable materials for use in situations where strong heat (below decomposition temperature) and abrasion are involved.

Q: Why does heat have a different effect on thermoplastics and thermosets, causing them to exhibit different physical properties and hence uses?

A: Polymers are different because of differences in their compositions and molecular structures. These differences in the compositions and molecular structures cause the bonding between the polymeric molecules to be different, hence results in different physical and chemical properties.

Not all molecules can come together to form polymers. Those that can do so are termed monomers, and a polymer is made up of many repeating units of monomers. When subjected to the polymerization process, new covalent bonds form between the monomers, linking them up to form either long linear chains or an extensive three-dimensional structural network.

When linear polymeric chains just stack on top of each other, we have what is called a linear polymer. These linear chains pack well together, with extensive van der Waals forces of attraction between them giving rise to high density, high melting point and high tensile strength. High-density poly(ethene) is one such polymer.

Q: What is tensile strength?

A: Tensile strength refers to the maximum stress that a material can withstand before it shows significant narrowing of its cross-sectional area or deformation of its body shape. The stress is measured by acting a force on a unit area of the material. It has a unit of force per area, which is $N\,m^{-2}$, same as for pressure.

If one is to trace the backbone of a polymeric chain and find that there are branches extending from the main chain, the polymer is known as a branched-chain polymer. These branched chains tend to pack in an irregular manner, and this causes branched-chain polymers, such as low-density poly(ethene), to have lower density, melting point and tensile strength than their linear counterparts.

Q: Are the branches in branched-chain polymers basically referring to the side-chains of the monomer molecules?

A: The branches that we are discussing here do not originate from the monomer molecules. These branches are created during the polymerization process and are formed due to the nature of the polymerization, which is free radical in nature. Thus, if we use ethene, which does not have any side chain, as the monomer, we would still be able to get branched-chain poly(ethene). Just like a "parent river" (linear chain of molecules) containing "distributaries" (branching of molecules).

Q: Why does branched-chain poly(ethene) have lower density, melting point and tensile strength?

A: The branches of branched-chain poly(ethene) prevent maximum surface area of contact between the polymeric molecules. As a result, the intermolecular forces between the molecules are not extensive, leading to lower melting point and also greater distance of separation between the molecules. The greater distance of separation results in lower density as the mass per unit volume is smaller. The lower tensile strength arises because of the weaker intermolecular forces. So take note that generally branched-chain polymers would have a lower melting point, density and tensile strength than linear chain polymers of the same type.

Bakelite – phenol-methanal plastic

A method known as cross-linking introduces strong covalent or ionic bonds between individual polymer chains. The polymers formed in this way are known as cross-linked polymers. The extent of cross-linking affects the freedom of movement of individual chains and hence changes the polymer's mechanical properties. A high degree of cross-linking causes materials, such as phenol-methanal plastics, to be hard and rigid, while a lower degree of cross-linking can produce materials with elastomeric properties.

Thermosets are typically cross-linked polymers. When heated, these plastics harden as new covalent bonds are formed between different polymeric chains, and this increase the degree of cross-linking. Re-heating in an effort to melt the thermoset is not possible as the energy consumed in breaking the cross-linkages is sufficient to decompose the polymer itself.

Q: I thought that once the polymer is formed, whatever covalent bonds that should be formed would already have been formed? How does heating cause more covalent bonds to be formed?

A: For thermoset polymers, when the polymer is synthesized, cross-linking is not completed, which means there are still cross-linking sites available, but not within close proximity. When we heat the polymer again, the heating causes the molecules to be more mobile due to the greater amount of vibrational and rotational energies. This allows the molecules to re-orientate themselves, resulting in the cross-linking sites coming in close proximity to each other.

On the other hand, linear polymers are good thermoplastics since there are only weak van der Waals forces of attraction and only a few cross-links formed between the polymeric chains. The temperature needed to overcome these weak intermolecular forces of attraction is below the decomposition temperature of the thermoplastic. Thus, the melting process only breaks existing weak intermolecular attractive forces between chains, and in the re-molding process, when the chains are forced to take on a new arrangement, new intermolecular attractive forces are formed that hold the polymeric chains together in their new conformation.

Q: What characteristic features must molecules have in order to function as monomers?

A: There are two general ways that monomers bond together and form polymers. Monomers that contain unsaturated carbon–carbon double or triple bonds bind together by using pi (π) electrons to form new sigma (σ) bonds between each other. The polymer formed in this manner is called an **addition polymer**. Obviously, this polymerization process is known as **addition polymerization**. The other way of getting a polymer is through a condensation reaction in which monomers bind together with the elimination of small molecules such as water or ammonia. In this case, there must be monomers that behave as nucleophiles (electron-rich species), attacking those that contain electron-deficient center(s) — two inherent features for a condensation reaction. The polymer formed is called a **condensation polymer** and this process of forming the polymer is known as **condensation polymerization**.

At this point, we have another useful classification scheme for polymers. Addition polymerization and condensation polymerization are named according to the type of polymer produced. Some examples under this classification scheme are shown below:

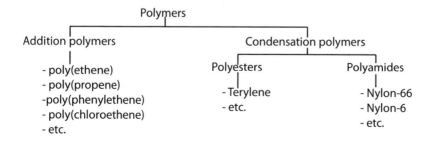

13.2   Addition Polymers

If you notice the names of addition polymers, they have the word "poly" placed before the name of an alkene such as ethene and phenylethene. Addition polymers are usually formed from monomers that contain C=C bonds. When these monomers combine to form the polymer, there is no loss of any segment of the monomeric unit, giving rise to the term "*addition*" polymer.

The notion of "addition" is the same as that covered in the chemistry of alkenes — a π-bond in an alkene cleaves with the two π-electrons used to form two σ-bonds with atoms from other species. Addition polymerization can proceed via different methods. A common method is free radical polymerization, which bears similarities to the free radical substitution of alkanes.

13.2.1   *Free Radical Polymerization of Alkene*

The following example shows the formation of poly(ethene) via a chain reaction comprising three stages: *initiation, propagation* and *termination.*

Step 1: Chain initiation

A substance, called an initiator, is added to start off the chain reaction. An example is benzoyl peroxide. At high temperature and pressure, homolytic bond cleavage occurs and the benzoyl peroxide molecules break up forming free radicals.

$$C_6H_5-\overset{\overset{O}{\|}}{C}-O-O-\overset{\overset{O}{\|}}{C}-C_6H_5 \xrightarrow{\text{heat}} 2\,C_6H_5-\overset{\overset{O}{\|}}{C}-O\cdot \longrightarrow C_6H_5^{\bullet} + CO_2$$

The phenyl initiator radical then proceeds to attack an ethene molecule to start the polymer chain. In the process, the π-bond cleaves homolytically with one of the bonding electrons used to form a bond with the initiator radical, and a new radical, with a longer alkyl chain, is thus generated.

Step 2: Chain propagation

The chain lengthens with each successive attack of the growing polymer chain radical on a monomer.

$$C_6H_5CH_2CH_2{}^\bullet + \underset{H}{\overset{H}{C}}=\underset{H}{\overset{H}{C}} \longrightarrow C_6H_5CH_2CH_2CH_2CH_2{}^\bullet$$

Chain propagation ends when there are no more individual monomers or when chain termination occurs.

$$C_6H_5(CH_2CH_2)_xCH_2CH_2{}^\bullet + \underset{H}{\overset{H}{C}}=\underset{H}{\overset{H}{C}} \longrightarrow C_6H_5(CH_2CH_2)_{x+1}CH_2CH_2{}^\bullet$$

Step 3: Chain termination

When two polymer chain radicals react with each other, the resultant polymer chain is unreactive and termination occurs with no further lengthening of the chain.

$$C_6H_5(CH_2CH_2)_xCH_2CH_2{}^\bullet + {}^\bullet CH_2CH_2(CH_2CH_2)_yC_6H_5 \longrightarrow C_6H_5(CH_2CH_2)_{x+y+2}C_6H_5$$

Q: In the free radical substitution of alkanes, the halogen radical extracts a H atom from the alkane molecule, cleaving the C−H bond homolytically and generating an alkyl free radical. In the free radical polymerization of alkenes, can alkyl radicals also be formed by cleaving C−H bonds?

A: Yes. It is also possible that the polymer chain radical extracts a H atom bonded to a carbon atom that is part of its backbone or from another polymer chain.

$$^\bullet CH_2(CH_2)_y\overset{H}{C}HCH_2CH_2 \text{-----} \longrightarrow CH_3(CH_2)_y\overset{\bullet}{C}HCH_2CH_2 \text{-----}$$

This radical can proceed to attack a monomer molecule or combine with another polymer chain radical.

$$\begin{array}{l} ^\bullet CH_2(CH_2)_yCH_2CH_2\text{----} \\ CH_3(CH_2)_x\overset{\bullet}{C}HCH_2CH_2\text{-----} \end{array} \longrightarrow \begin{array}{l} CH_2(CH_2)_yCH_2CH_2\text{----} \\ | \\ CH_3(CH_2)_x\overset{\bullet}{C}HCH_2CH_2\text{-----} \end{array}$$

What is subsequently produced are two polymeric chains linked at a juncture. But it doesn't stop here — through this "back-biting" or chain-transfer mechanism, branches multiply in numbers and eventually branched-chain polymers are formed. Branching usually occurs at high temperatures as there is a greater degree of dexterity of the chains due to a greater amount of vibrational and rotational energies. So, take note that the proposed three-stages of addition polymerization — *initiation, propagation* and *termination* — are an ideal mechanism used to explain how the addition polymer is being formed. The actual polymerization process is of much complexity, with a range of products being formed.

Just as the ratio of the amount of halogen to alkane would affect the degree of polyhalogenation of an alkane molecule (refer to Chapter 4), the ratio of the amount of initiator to the monomer would also affect the molecular weight of the polymer obtained. If more initiator is used, there would be a greater amount of the free radical monomer units being created in the beginning, which would lead to a greater number of monomers being added to the large number of different radical sites per unit time. This would cause a high depletion rate of the monomers, resulting in shorter polymer chains being formed.

In addition, from the above mechanism, one can see that if the time for the polymerization process is lengthened, one would tend to get polymers of higher molecular weight. This is because the longer the time, the higher the chances for the polymer radicals to meet each other and undergo termination.

Q: If an alkene can undergo electrophilic addition reaction, can addition polymerization be initiated by a cation?

A: Yes. Other than free radical initiation, an alkene monomer can undergo cationic and anionic polymerization. An example of cationic polymerization is the usage of a Lewis acid such as $AlCl_3$ in the Ziegler–Natta catalyst used for the formation of poly(alkene). Similarly, for anionic polymerization, a likely initiator would be the NH_2^- ion. Using different types of initiator would lead to polymers with different physical and chemical properties.

13.2.2 *Types of Addition Polymers and Their Properties*

Poly(ethene) is an example of a homopolymer — a polymer that is made up of only one kind of monomer (i.e. X). The constitutional/structural formula of a polymer is written as $[X]_n$, where n is a large integer and enclosed in square brackets is the repeat unit of the polymer:

$$n\,CH_2 = CH_2 \longrightarrow -[-CH_2-CH_2-]_n-$$

Another example of a homopolymer is poly(vinyl chloride) (PVC), which is made from chloroethene(or vinyl chloride) monomers:

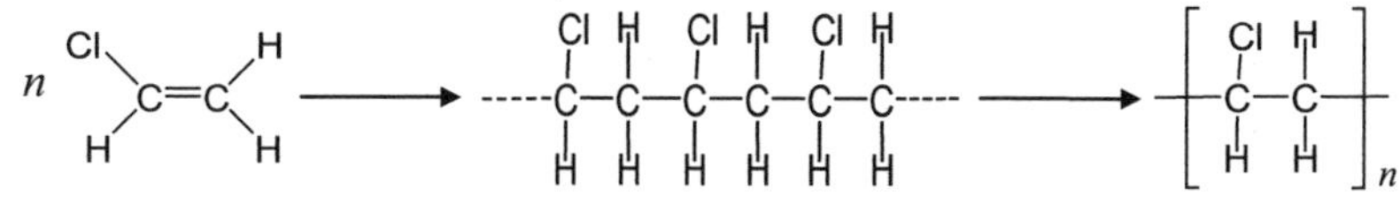

Two or more types of monomer (i.e. X and Y) chemically combine to give a copolymer.

$$n X + n Y \rightarrow -X-Y-X-Y-X-Y-X-Y-X-Y- \longrightarrow -[X-Y]-_n$$

Q: What is the difference between a repeat unit and the monomer for an addition polymer?

A: Both the repeat unit and monomer share the same molecular formula. The main difference lies in the number and type of bond. A monomer is more unsaturated than a repeat unit as, after addition polymerization, the number of π-bonds in the repeat unit would be less than that in the monomer. But the number of σ-bonds has increased for the repeat unit.

Exercise 1: Draw two repeat units of the polymer formed from the monomer(s) given.

(a) $CH_3CH=CHCl$
(b) $CH_3CH=CH_2$ and $CH_2=CHCN$
(c) $CH_2=CH-CH=CH_2$

Solution:

(a)

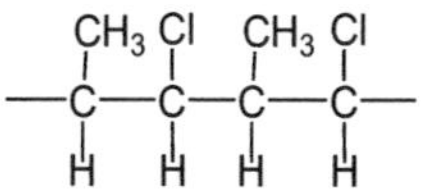

(b)

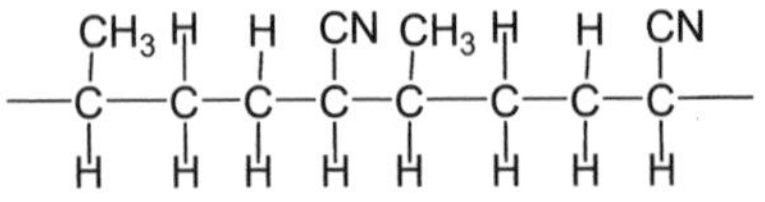

(c)

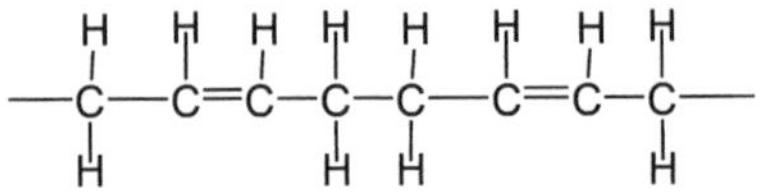

The monomer $CH_2 = CH–CH = CH_2$ has two $C=C$ bonds. When a polymer chain radical attacks this molecule, the following electron transfer occurs:

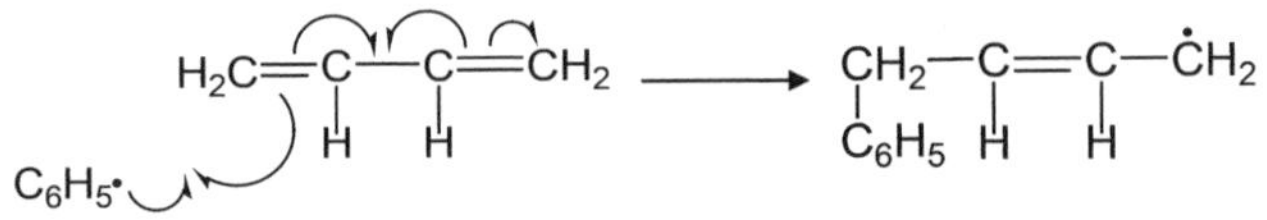

The radical that is formed is resonance stabilized, as shown below:

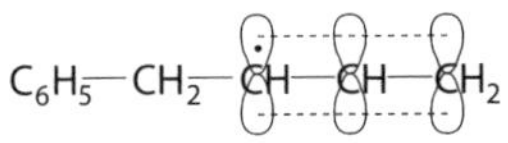

The double bonds that are present in the polymer chains allow different polymer chains to cross-link to each other. For example, heating natural rubber with sulfur allows the polymer chains to cross-link to each other through the formation of S—S bridges.

Q: Can the polymer that is formed from the $CH_3CH = CHCl$ monomer be of the following configuration?

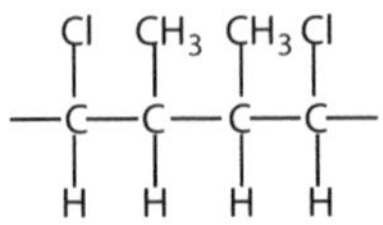

A: Yes. When the initiating radical attacks the monomer, two likely possibilities would take place:

Pathway I is favored by the delocalization of the lone electron into the Cl group but destabilized by the electron-withdrawing effect (via inductive effect) of the more electronegative Cl group. The latter effect would cause the carbon atom with the lone electron to be even more electron deficient. But if this pathway is favored, then formation of the head-to-tail configuration would be favored. And this would show that the delocalization effect has predominated over the electron-withdrawing effect.

Pathway II is favored by the stabilization of the radical through the electron-releasing (via inductive effect) of the CH_3—group; this would decrease the electron deficiency of the carbon atom, leading to the formation of the tail-to-tail configuration. The formation of the radical via pathway II is not as favorable as pathway I, possibly due to the fact that the carbon atom that is bonded to the Cl group is actually less electron-rich than the other carbon atom. This would diminish the attack of the initiating radical or subsequent radical for the monomer. The dominant type of polymer, i.e. head-to-tail or tail-to-tail, that is formed would really depend on which of the stabilization factors dominate. Sometimes, steric factor can also play a part. In the polymerization of chloroethene, the head-to-tail configuration is favored because the steric factor plus the delocalization of the lone electron into the Cl group favor its formation. Hence, for the poly(chloroethene), it would be unlikely to get the tail-to-head polymer.

In the actual polymerization, take note that due to the statistical factor, the following type of polymer may also be formed, or may be found in a section of the polymeric chain:

head-to-head

$$-\overset{\overset{\displaystyle CH_3}{|}}{\underset{\underset{\displaystyle H}{|}}{C}}-\overset{\overset{\displaystyle Cl}{|}}{\underset{\underset{\displaystyle H}{|}}{C}}-\overset{\overset{\displaystyle Cl}{|}}{\underset{\underset{\displaystyle H}{|}}{C}}-\overset{\overset{\displaystyle CH_3}{|}}{\underset{\underset{\displaystyle H}{|}}{C}}-\overset{\overset{\displaystyle CH_3}{|}}{\underset{\underset{\displaystyle H}{|}}{C}}-\overset{\overset{\displaystyle Cl}{|}}{\underset{\underset{\displaystyle H}{|}}{C}}-\overset{\overset{\displaystyle Cl}{|}}{\underset{\underset{\displaystyle H}{|}}{C}}-\overset{\overset{\displaystyle CH_3}{|}}{\underset{\underset{\displaystyle H}{|}}{C}}-$$

tail-to-tail

Thus, in order to get the desired polymer with particular physical and chemical properties, it is very important to take note that the types of initiator, monomer, ratio of initiator to monomer and reaction conditions would all affect the polymerization process. There would never be a polymer that consists of purely a specific type of molecule with a particular molecular weight and configuration. Thus, we always talk about the average molecular weight of a polymer.

Even if the polymer consists of purely the head-to-tail molecules of the same chain length, the physical properties might still vary because the spatial orientation of the side groups may differ. For example, poly(propene):

isotactic (the CH_3 groups are on the same side of the carbon skeleton)

syndiotactic (the CH_3 groups alternate on different sides of the carbon skeleton)

atactic (the CH_3 groups are randomly distributed on both sides of the carbon skeleton)

The different types of poly(propene) have different melting points: Isotactic poly(propene) melts at about 165°C, syndiotactic at about 128°C while atactic melts below 0°C. The higher melting point of the isotactice polymer arises because the electron cloud is more symmetrically distributed than in the other two types of polymer. This allows maximum surface area of contact between the isotactic polymer molecules. Different melting points give rise to different physical properties, hence different applications.

Q: What is the difference between a head-to-tail and a tail-to-head configuration? Aren't they the same in terms of the sequential arrangement of the repeating units?

A: The main difference lies in where the initiator is bonded. In the head-to-tail configuration, the initiator is bonded to the tail end of the first repeat unit, whereas in the tail-to-head, it is bonded to the head end. This is exactly the same as for polypeptides, for example, Arg-Lys is not the same as Lys-Arg (refer to Chapter 12).

Q: How would one know which is the head and which is the tail, like in 1-chloropropene?

A: Well, you look at which carbon atom the functional group is bonded to. For 1-chloropropene, the carbon atom possessing the Cl group is the head, but in propene, the carbon atom with the CH_3-group is the head. So basically, the head of a monomer is defined by the priority of the group of atom/s in the monomer.

Q: Can the copolymer that is formed from the monomers $CH_3CH = CH_2$ and $CH_2 = CHCN$ have the following repeating units too?

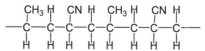

A: Yes. What type of polymer is formed depends very much on the stability of the radical that is formed and, of course, on the ratio of the two monomers used during the reaction, plus the reaction conditions that are being imposed on the system. Generally, the polymer that is formed from two monomers would not only have an alternating bonding pattern of monomers (i.e. –A–B–A–B–A–B–) but also other bonding patterns such as –A–A–B–B–A–A–B–B– and –A–B–B–B–A–A–B–. Of course, the situation would be made more complicated if there were more than two monomers being used.

Exercise 2: Shown below are segments of different polymers. Identify the monomer(s) from which these are formed and hence give the constitutional/structural formula of each polymer.

(a)

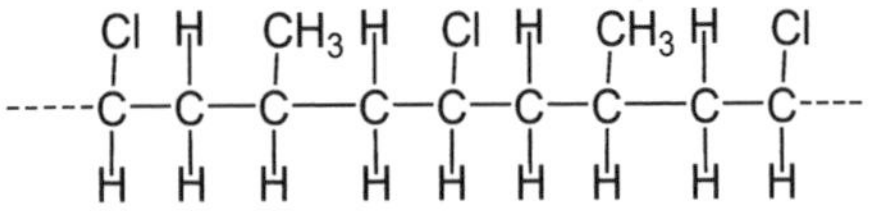

(b)

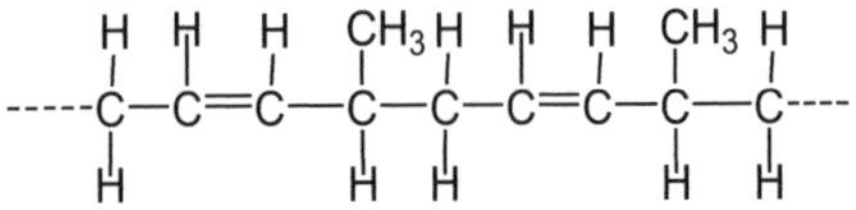

Solution:

(a) The monomers which make up this copolymer are:

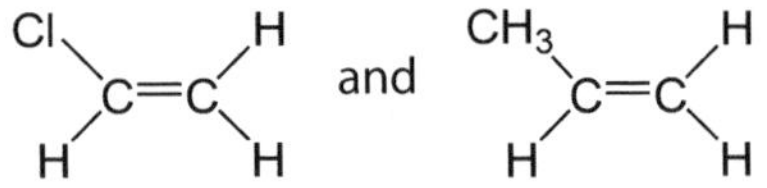

Constitutional/structural formula of copolymer:

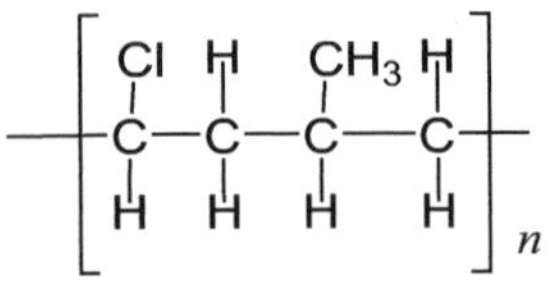

(b) Monomer: $CH_2{=}CH{-}CH{=}CHCH_3$

Constitutional/structural formula of homopolymer:

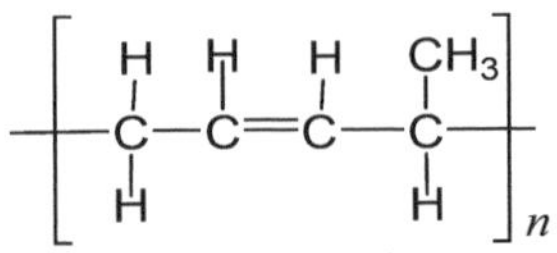

Q: How can one identify the monomer/s that make up a polymer?
A: It is very simple. The polymer is usually presented in the head-to-tail or tail-to-head configuration. All you need to do is to move along the carbon backbone (from left to right) and identify the next repeat unit. The position where the head is bonded to the tail of the two repeat units would be the positions of the double bonds in the monomers. If there is systematic occurrence of a double bond in the polymer (like in Exercise 2(b)), then the original monomer is likely to contain a 1,3-dienes group (–C=C–C=C–). If there are two double bonds that occur systematically in the polymer, then the monomer contains the 1,3,5-trienes group (–C=C–C=C–C=C–).

Working with particular monomers and introducing special methods generates polymers of different physical properties which can be made into a great range of products. For instance, one property we tend to promptly associate rubber with is its elasticity. Yet, we can alter this property of rubber through the chemical process called *vulcanization*, which increases cross-linking by forming disulfide bonds between individual chains. Vulcanized rubber is hard and more durable, with superior mechanical properties. Products made from vulcanized rubber include car tires, shoe soles and even bowling balls.

Another example is poly(ethene), made from ethene monomers. It can be made into a highly branched polymer known as low-density poly(ethene), commonly abbreviated as LDPE, or transformed into the linear polymer high-density poly(ethene), abbreviated as HDPE. LDPE has high flexibility and chemical resistance. It is widely used in manufacturing a broad range of packaging applications, which include plastic bags, containers and dispensing bottles.

Since there is little branching in HDPE, it has higher tensile strength, and is more resistant to heat and corrosion than LDPE. It, too, has a wide variety of applications, which include containers, folding tables, chairs and pipes. Table 13.1 lists a few other addition polymers and their properties.

Table 13.1   Examples of Addition Polymers and Their Properties

Monomer	Polymer	Properties	Examples of of Uses

propene
 	
CH_3 H
  C=C
H   H
	Poly(propene)
	
	$[CH_2-CH]_n$
	      CH_3
		— tough and flexible
		— resistant to chemicals
		— resistant to heat
			— packaging
			— ropes
			— polymer banknotes
			— stationery

chloroethene (vinyl chloride)
 	
Cl   H
  C=C
H   H
	Poly (chloroethene) or poly (vinyl chloride)
	
	$[CH_2-CH]_n$
	      Cl
		— resistant to chemicals
		— relatively unstable to heat and light.
			— flexible hoses
			— electrical cable insulation
			— pipes
			— waterproof clothing
			— shoes and bags
			— signage

phenylethene (styrene) $CH_2=CH-$⬡	Poly (phenylethene) or poly (styrene) $[CH_2-CH]_n$ ⬡	— resistant to chemicals — hard with limited flexibility — transparent	— disposable cutlery — foam drink cups — CD cases — insulation materials
tetrafluoroethene F F C=C F F	Poly (tetrafluoroethene) (brand name $[CF_2-CF_2]_n$	Its properties stem from the aggregate effect of strong C–F bonds: e.g. — resistant to heat and chemicals — "anti-stick" properties — low friction	— non-stick coating for cook ware — lubricant — laboratory apparatus

Most of the polymers listed in Table 13.1 are resistant to chemical attack because of the lack of electron-rich or electron-deficient sites on the polymer chain that would "invite" an attack by electrophiles or nucleophiles, respectively. But polymers such as PVC can actually be hydrolyzed by a strong base under extreme conditions. This behavior is very similar to the reaction of chloroalkane with NaOH. Poly(phenylethene) is hard with limited flexibility because of the presence of the bulky benzene ring along the polymer chain. The benzene rings lock the polymer molecules in a particular spatial configuration, preventing the molecules from "wriggling." As for the hard property of poly(phenylethene), it arises from the strong inter-molecular forces of the instantaneous dipole–induced dipole type because of the massive electron cloud from the benzene rings. In short, take note that the physical properties of the polymer, and hence its usage, are highly related to the types and strength of the chemical bonds present both intramolecularly and intermolecularly!

13.2.3 *Pollution Problems Associated with Poly(Alkenes)*

In addition polymerization, the unsaturated monomers are converted to long saturated chains which give the polymer a high degree of inertness since there are no reactive sites. This in turn accounts for the chemical resistance of all poly(alkenes) in general when exposed to reagents such as acids and alkalis, and hence their non-biodegradable nature.

One disposal method for the non-biodegradable poly(alkenes) is to burn them, but chances are that this adds to air pollution. Incomplete combustion produces poisonous carbon monoxide. Those polymers that contain benzene (e.g. poly(styrene)) do not only pro-duce large amounts of soot but they also release poisonous hydro-carbons. When burnt, PVC and other chlorinated polymers produce poisonous hydrogen chloride gas. These toxins and many others that arise from the burning process are detrimental to both human health and the environment. The best, albeit dire, disposal method is basi-cally dumping them into landfills, which pose a grim yet stark reality of our society's growing consumption of disposable plastics which are made to last. While scientists and companies are churning out bio-degradable plastic products, the least we could do is to make a con-scious effort in practicing the three Rs — Reduce, Reuse and Recycle.

13.3 Condensation Polymers

Up to this chapter, we have seen a few condensation reactions, such as the formation of esters and amides. Both of these reactions require an electron-deficient carbonyl carbon center and a nucleophile that attacks it to form either the ester or amide bond. Condensation polymerization is nothing new. In order to form a long chain consisting of numerous ester or amide linkages, there needs to be a monomer that contains twice the number of electron-deficient sites and another monomer that contains twice the number of nucleophilic sites. Other than this, condensation polymers can also be formed from bifunctional monomers such as amino acids, which contain both an electrophilic and nucleophilic site within each molecule.

When these monomers combine to form the polymer, there is a loss of one small molecule (e.g. water, hydrogen chloride, ammonia) for each covalent bond formed between a pair of monomers. Hence, for condensation polymer, the molecular formula of the repeat unit is not the sum of those of the monomers from which it is made. This is in contrast with addition polymers, whose repeat unit has the exact molecular formula as the monomer.

13.3.1 *Polyesters*

Two methods to form polyesters are as follows:

- Reaction between a diacid halide and diol

- Reaction between a dicarboxylic acid and diol

Q: The reaction between a carboxylic acid and an alcohol to yield an ester is a reversible one. Why is it not the case for the polymerization reaction between a dicarboxylic acid and a diol?

A: In the industry, when polyester is made by reacting a dicarboxylic acid and a diol, the water that is produced is continuously being distilled away from the reactor. Thus, the diminishing amount of water impedes the backward reaction from occurring. In addition, the amount of water that is collected during the process can be used to estimate the extent of the polymerization process and hence the molecular weight of the polymer that is formed.

Q: Is it better to produce polyester from diacid halide and diol than using diacid and diol?

A: If you need to produce a phenolic ester, then using diacid halide is better (refer to Chapter 10). This is because the HCl fumes that are produced are corrosive in nature. It would corrode the metal reactor.

Poly(ethylene terephthalate) (commonly abbreviated as PET or PETE) is an example of a polyester made from benzene-1,4-dicarboxylic acid and ethane-1,2-diol. Its name is derived from the old names of the respective monomers — terephthalic acid and ethylene glycol. It is also known by the brand name Terylene. This polyester is used in synthetic fibers. Its use in the textile industry accounts for the majority of the world's production of the polyester. Its other major use is in plastic bottle production. The term "PET" is often used when talking about its use in packaging, whereas the term "polyester" is associated with its role as a synthetic fiber.

ethane-1,2-diol   benzene-1,4-dicarboxylic acid   Terylene
                                                                    (one type of polyester)

Unlike addition polymers, condensation polymers are unstable in the presence of acids and alkalis since hydrolysis will occur. Prolonged exposure to these reagents may cause degradation of the polyester since it is broken down to smaller molecules.

13.3.2 *Polyamides*

Polyamides consist of both natural and synthetic ones. The former includes proteins such as silk and wool while the latter includes nylons. Nylon is a generic term for a class of synthetic polyamides. Nylon is used in many applications. Nylon fibers are used to make fabrics, musical strings and ropes, to name a few. Solid nylon is used in auto and machine parts such as gears and bearings.

Belonging to this family is Nylon-6,6, whose name is based on the number of carbon atoms in the main chains of both the dicarboxylic acid and diamine monomers, i.e. hexanedioic acid and hexane-1,6-diamine.

hexane-1,6-diamine   hexanedioic acid   nylon-6,6

Q: How can the diacid, which is acidic, reacts with the basic diamine to give nylon? Wouldn't an acid–base reaction take place rather than nucleophilic substitution?

A: You are right that an acid–base reaction between the diacid and diamine indeed can happen. But do not forget that the protonated diamine can also be deprotonated by the carboxylate ion that forms. So in the reactor, the system would be at equilibrium consisting of a mixture of the protonated and unprotonated diamines. As time goes by, the unprotonated diamine would get a chance to make a nucleophilic attack on the diacid, thus forming the nylon. This would cause more protonated diamine to be deprotonated via shifting of the position of equilibrium.

Nylon-6,6 can be produced in the laboratory by reacting hexanedioyl chloride and hexane-1,6-diamine at room temperature:

hexanedioyl chloride   hexane-1,6-diamine   nylon-6,6

Another widely used member of the nylon family is Nylon-6. It is a homopolymer which is made from the bifunctional monomer 6-aminohexanoic acid (this explains the single digit in its name).

6-aminohexanoic acid   nylon-6

Caprolactam (6-aminohexanoic acid lactam) is a cyclic amide that is another precursor to Nylon-6. Polymerization involves ring-opening using hydrolysis and subsequent removal of water at high temperatures to form the polymer.

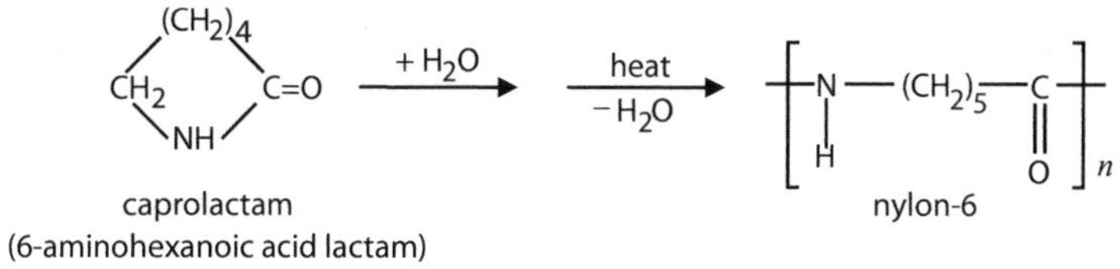

caprolactam
(6-aminohexanoic acid lactam)

nylon-6

In close proximity, a H atom (bonded to the highly electronegative N atom) can form a hydrogen bond with the lone pair of electrons on the carbonyl O atom of another chain. This leads to extensive intermolecular hydrogen bonding between individual polymer chains, which explains why Nylon-6,6 and other nylons have high tensile strength.

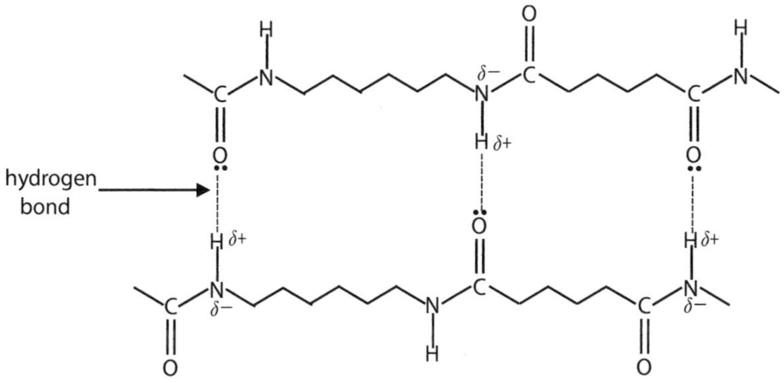

Just like polyesters, polyamides are susceptible to hydrolysis in the presence of acids and alkalis.

Example: Shown below are portions of two polymers. For each of them, identify the products formed when the polymer is subjected to (i) acidic hydrolysis and (ii) alkaline hydrolysis.

(a) $-COOCH_2CH(CH_3)OCOCH_2CH_2COOCH_2CH(CH_3)OCO-$

(b) $-OCOCH_2CH(CH_3)COOCH_2CH_2OCOCH_2CH(CH_3)COO-$

(c) $-CONHCH_2CH_2NHCOCH_2C(CH_3)_2CONHCH_2CH_2NHCO-$

Solution:

It is assumed that all ester and amide linkages are broken. Locate these functional groups at which bond cleavage occurs. Apart from hydrolysis, we also have to consider the occurrence of acid–base reaction. Carboxylic acids are unstable in an alkaline medium and become deprotonated. Amines are basic and they are protonated in an acidic medium. Alcohols are neutral and thus unaffected by both acids and bases.

(a) (i) $HOCH_2CH(CH_3)OH$ and $HOOCCH_2CH_2COOH$
   (ii) $HOCH_2CH(CH_3)OH$ and $^-OOCCH_2CH_2COO^-$

(b) (i) $HOOCH_2CH(CH_3)COOH$ and $HOCH_2CH_2OH$
   (ii) $^-OOCCH_2CH(CH_3)COO^-$ and $HOCH_2CH_2OH$

(c) (i) $H_3N^+-CH_2CH_2-NH_3^+$ and $HOOCCH_2C(CH_3)_2COOH$
   (ii) $H_2NCH_2CH_2NH_2$ and $^-OOCCH_2C(CH_3)_2COO^-$

Note that the polyesters in (a) and (b) are different, and this is noticeable when you draw out the structure starting from the left-hand side of the chain, i.e.

• Polymer in (a) has the following bonding pattern:

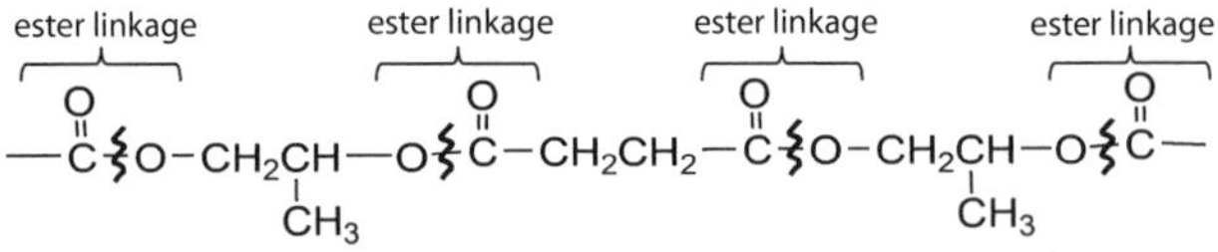

Polymer in (b) has the following bonding pattern:

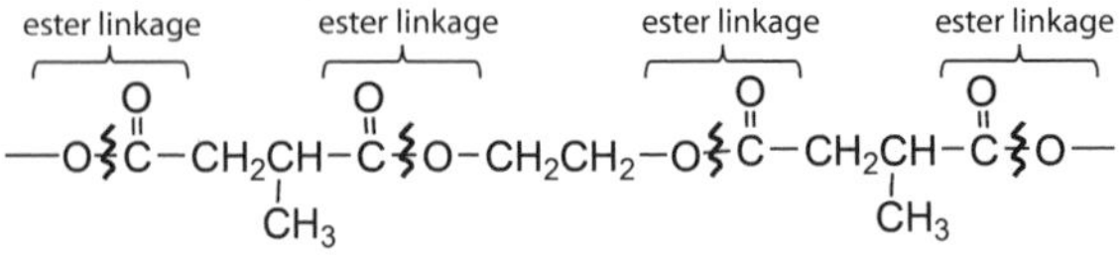

13.4 Biological Polymers — Polypeptides and Proteins

In this section, we will focus our discussions on the biological polyamides formed from amino acid monomers (refer to Chapter 12). The formation of peptide linkages (or amide linkages) requires the amino group of one amino acid to react with the carboxyl group of another amino acid with the elimination of a water molecule. This type of linkage is known as a head-to-tail linkage (the carboxyl group has a higher priority than the amino group, hence the end of the molecule possessing the carboxyl group is considered the head while the end with the amino group is the tail) of peptide bonds and it gives rise to linear polypeptides and proteins.

amide linkage/peptide bond

$$ n \; \underset{\underset{R}{|}}{\overset{\overset{H}{|}\;\overset{H}{|}\;\overset{O}{\|}}{H-N-C-C}}-OH \;+\; n \; \underset{\underset{R'}{|}}{\overset{\overset{H}{|}\;\overset{H}{|}\;\overset{O}{\|}}{H-N-C-C}}-OH \longrightarrow \left[\underset{\underset{R}{|}}{\overset{\overset{H}{|}\;\overset{H}{|}\;\overset{O}{\|}}{N-C-C}} \; \underset{\underset{R'}{|}}{\overset{\overset{H}{|}\;\overset{H}{|}\;\overset{O}{\|}}{N-C-C}} \right]_n + (2n-1)H_2O $$

Polypeptides and proteins are susceptible to hydrolysis in the presence of acids and alkalis. Upon hydrolysis, the amino acids produced can be separated and identified by chromatographic or electrophoretic methods. This is, in fact, how the structure of polypeptide is elucidated. Polypetides contain long chains of amino acids, and there are countless ways in which monomers can combine. Determining the structure of a particular polypeptide chain requires identifying the *type* of amino acid residues, the *quantity* of each and the *sequence* in which these are linked together in the chain.

Before research and analysis can be done on a specific protein, purification is necessary to isolate the protein from a complex mixture that is typical of biological samples. After a series of purification procedures, sequencing of amino acids in the protein can be carried out. Complete hydrolysis can cause all the peptide bonds to cleave and give the various types and relative quantities of amino acids present. But it does not provide information on the exact bonding pattern of the polymeric chain.

Sequencing of amino acids is usually accomplished by a combination of analytical methods, one of which is partial hydrolysis. Partial hydrolysis produces fragments of short-chain peptides. We can then analyze the fragments and look for overlaps of sequences to deduce the overall amino acid sequence of the protein. The piecing together of the fragments can be likened to working on a jigsaw puzzle. *N*-terminal amino acid analysis and *C*-terminal amino acid analysis are carried out to determine the amino acid that form the *N*-terminus and *C*-terminus of a peptide chain, respectively. This helps in ordering the sequences of individual fragments.

Partial hydrolysis can be achieved through acidic hydrolysis or enzymatic hydrolysis. Acidic hydrolysis is not selective, and it produces a mixture of fragments from random cleaving of peptide linkages.

On the other hand, enzymatic hydrolysis is selective as it utilizes enzymes that target specific sites at which cleavage occurs. For instance, the enzyme trypsin cleaves peptide bonds specifically at the carboxyl end of the basic amino acid residues arginine and lysine.

Met —Asp —Tyr — Ala —Lys — Gly — Pro — Phe —Ser

Trypsin cleaves this bond at the carboxyl end of lysine

Take note that, by convention, the formula of peptides starts with the N-terminal amino acid residue written on the left, and it ends with the C-terminal amino acid residue on the right.

If the enzyme chymotrypsin is used, cleavage of peptide bonds occurs at the carboxyl end of the aromatic amino acid residues phenylalanine, tryptophan and tyrosine.

Chymotrypsin cleaves these bonds only

Met —Asp —Tyr — Ala —Lys — Gly — Pro — Phe — Ser

The bonds at the amino end of Tyr and Phe are not cleaved

Example: Given below are three fragments of a hexapeptide obtained from its partial acidic hydrolysis. Deduce the amino acid sequence of the hexapeptide.

Pro-Gly-Leu, Arg-Pro, Leu-Asp-Val

Solution:
Matching the overlapping segments of the fragments gives the structure of the hexapeptide as Arg-Pro-Gly-Leu-Asp-Val.

Example: A polypeptide is subjected to enzymatic hydrolysis using trypsin, and the following fragments are obtained:

Trp-Val-Asp-Lys,   Gly-Ser-Phe-Ala,   Met-Leu-Arg

When the same polypeptide is hydrolyzed using chymotrypsin, the following fragments are produced:

Ala,   Met-Leu-Arg-Trp,   Val-Asp-Lys-Gly-Ser-Phe

From the information given, deduce the amino acid sequence of the polypeptide.

Solution:
Matching the overlapping segments of the fragments produced in both reactions gives the following amino acid sequence:

Met-Leu-Arg-Trp-Val-Asp-Lys-Gly-Ser-Phe-Ala

To check whether the answer is correct, we just need to cleave the deduced structure according to each type of enzymatic hydrolysis and check that the fragments obtained match those given in the question.

13.4.1 *The Different Levels of Protein Structure*

There are four levels of protein structure: the primary structure, the secondary structure, the tertiary structure and the quaternary structure.

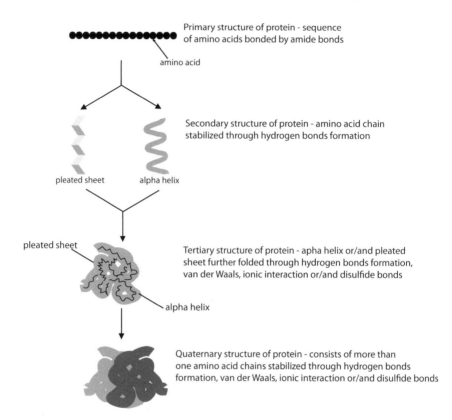

- ## Primary Structure

  The primary structure of a protein is the sequence of amino acid residues in the linear protein chain. The primary structure is held together by strong covalent bonds (peptide bonds) between adjacent amino acid residues.

- ## Secondary Structure

  When we say that the protein chain is linear, it does not mean that the chain is rigid and all amino acid residues are lined up in a straight line. There is free rotation about the single bonds (C_α–C and C_α–N) at the α-carbon, which gives a certain dexterity to the linear chain that leads to interactions formed between particular groups on the protein chain.

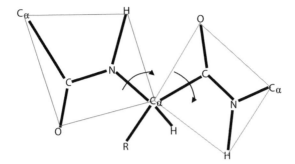

Free rotation about C_α—N and C_α—C bonds

Q: Is there free rotation about the peptide bond, which is also a single bond?

A: The peptide bond is not really a single bond, rather, it has partial double bond character — the p-orbital of the N atom interacts with the π-electron cloud of the carbonyl group resulting in the delocalization of electrons. There is restricted rotation (i.e. rigidity) about the C–N bond in the peptide group as maximum overlap of p-orbitals is achieved only when these orbitals are on the same plane.

The secondary structure refers to the spatial orientation of the residues of the amino acids bought about by the hydrogen bonds that are formed between the peptide groups within the primary structure of the protein. These hydrogen bonds cause the backbone to adopt a highly regular pattern. There are two types of stable secondary structures, namely the α-helix and the β-pleated sheet structures.

(a) *The α-helix*

Features of the α-helix

  o The alpha helix contains 3.6 amino acid residues per complete turn with the N–H group pointing upwards and the C=O groups pointing downward.

  o Hydrogen bonding occurs between the –C=O group of the nth amino acid and the –NH group of the $(n+4)^{th}$ amino acid which is in the adjacent turn. These hydrogen bonds help to link the turns, making the helix structure stable.

  o The side chain R-groups extend outwards from the center of the helix structure.

○ Types of proteins with the α-helix structure include the fibrous α-keratin present in hair and wool and myoglobin.

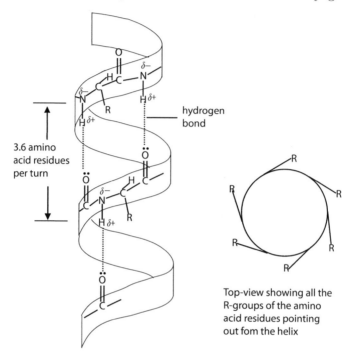

Fig. 13.1.

(b) *β-pleated sheet*

Features of the β-pleated sheet

○ When two different strands of the protein molecules lie side by side, intermolecular hydrogen bonding occurs between the –C=O group of an amino acid in one strand and the –NH group of an amino acid in an adjacent strand. If the two strands are aligned in such away that the *N*-terminal of one strand faces the *C*-terminal of the other, then the β-pleated sheet that forms is *anti-parallel*. If both the *N*-terminal faces each other, it is known as a *parallel* β-pleated sheet. Intermolecular hydrogen bonding would be responsible for holding the different strands of protein molecules together.

○ Other than between different strands of protein molecules, if the protein molecule is long enough, it can also coil in such a way to form either the parallel or the anti-parallel β-pleated sheet structure. Then the hydrogen bonding that holds the secondary structure together would be intramolecular in nature.

- Due to the sp^2 hybridization nature of the carbonyl carbon atom of the peptide bond, the shape about the carbonyl carbon is trigonal planar. As a result, the zigzag appearance gives rise to the pleated sheet structure.

- The residue side chains of the various amino acids are perpendicular to the plane of the pleated sheet and are either above or below the plane.

- Types of proteins with the β-pleated sheet structure include the fibroin β-keratin present in silk fibers, muscle fibers, collagen, nails and feathers.

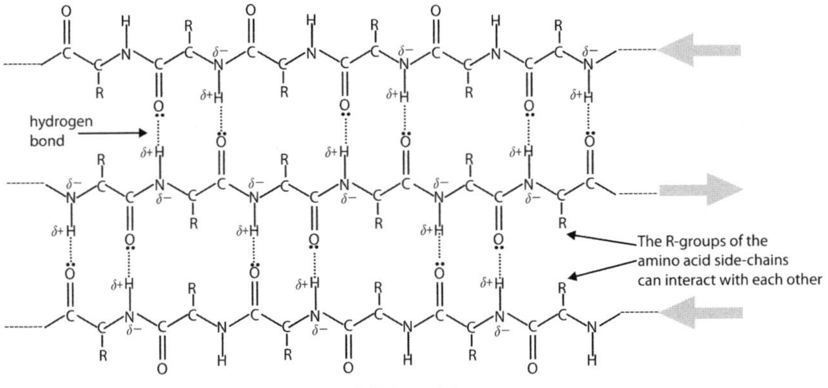

Anti-parallel pleated sheets

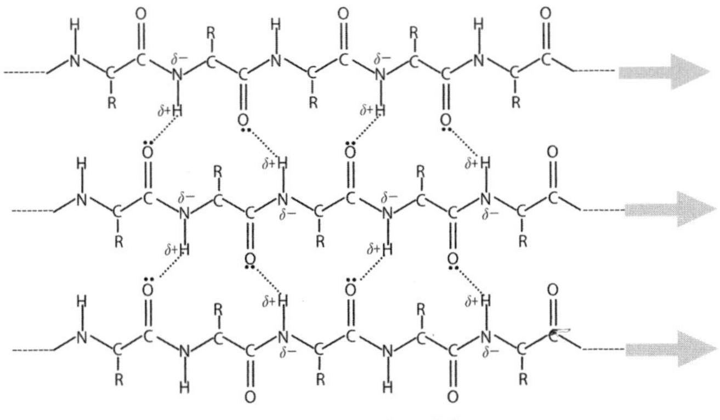

Parallel pleated sheets

A particular protein molecule may not necessarily have the same secondary structure throughout the whole strand. Some parts of the molecule may be coiled in the α-helix form while other parts are lined up as the β-pleated sheet. It would not be surprising to find that neither type of secondary structure exists for certain sections within the protein molecule. The absence of the secondary structure is termed random coil, which separates segments of the α-helix or β-pleated structures. The random coil serves the function of allowing the protein molecule to fold into a three-dimensional globular structure.

- **Tertiary Structure**

    The tertiary structure of a protein refers to the three-dimensional structure of a single protein chain that arises from the folding of the secondary structures, i.e. the α-helix or β-pleated sheet, as a result of the <u>interactions between the side chain R-groups</u>. There are four types of R group interactions that would stabilize the tertiary structure:

    (i) *ionic bonds between oppositely charged groups*

    Charged groups are formed when side chains containing acidic or basic groups ionize in water, e.g.

    ○ The carboxylic acid side chain of aspartic acid deprotonates in water or a basic medium to form a negatively charged group:

$$-----N-CH-C----- + H_2O \rightleftharpoons -----N-CH-C----- + H_3O^+$$

with side groups $\overset{H}{|}$ N, CH_2, $COOH$ on the left and CH_2, COO^- on the right.

    ○ The amino side chain of lysine protonates in water or acidic medium to form a positively charged group:

$$-----N-CH-C----- + H_2O \rightleftharpoons -----N-CH-C----- + OH^-$$

with side groups $(CH_2)_4NH_2$ on the left and $(CH_2)_4NH_3^+$ on the right.

    (ii) *hydrogen bonding between polar groups*

    Side chains of this nature include the polar $-OH$, $-NH$, $=NR$ and $-COOH$ groups as well as the amide $-CONH_2$ group, all of which are capable of hydrogen bonding. The protein with the α-helix or/and β-pleated structures would fold in such a way that the R-groups that are capable of assuming intermolecular hydrogen bonding would be facing each other.

(iii) *van der Waals forces of attraction between uncharged groups*

Also known as hydrophobic interactions, these interactions constitute both the permanent dipole–permanent dipole and instantaneous dipole–induced dipole interactions. Nature is so intelligent that the protein molecule would curl in such a way that the polar R groups, such as carbonyl and amide, would face each other, allowing pd–pd interaction to take place. Non-polar R-groups, such as the alkyl group, would be aggregated within the center of the tertiary structure so that they are kept away from the hydrophilic medium that is commonly present in Nature. Instantaneous dipole–induced dipole interaction is responsible for attracting these non-polar R-groups together, hence maintaining a thermodynamically stable tertiary structure.

(iv) *disulfide bonds between the cysteine residues*

One of the amino acids, cysteine, has sulfur in its side chain. If α-helix or β-pleated structures fold in such a way that two cysteine residues in the polypeptide chain are brought in close proximity, a redox reaction can occur between two sulfur atoms resulting in the formation of a disulfide covalent (S—S) bond. Such a bond is strong, but not unbreakable by using a suitable reagent such as ammonia, which provides an alkaline medium. This is commonly used in the perming and straightening of hair in a hair studio.

Knowing what constitutes the tertiary structure of a protein and how it is formed is important for us to understand why different types of proteins behave differently in our biological medium. The way that protein molecules fold results in two main groups of proteins, namely the globular proteins and fibrous proteins.

(a) Globular protein

The name has clearly spelt out how the protein molecule would look, i.e. in globule form. The shape is the result of compact folding of the protein molecule with most of the hydrophobic R-groups clustered within the core of the globule. The hydrophilic R-groups are exposed to the hydrophilic aqueous medium. As a result, such protein molecules are soluble in water. Most enzymes, antibodies and hormones are of the globular nature.

(b) Fibrous protein

After the secondary structure has been formed, fibrous protein molecules would fold in such a way parallel to one another along a single axis to give long fibers, such as the twisted helices, or pleated sheets. Most of the R-groups of such proteins are directed outwards of the chain and, as a result, such proteins are generally insoluble in water. Fibrous proteins serve the function for support or as structural materials, such as keratin of hair and the collagens of tendons.

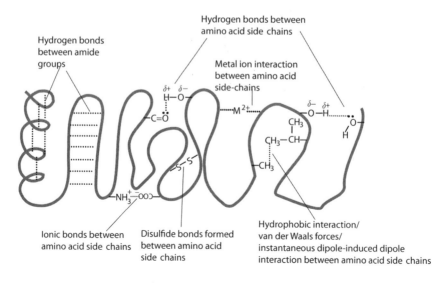

- **Quaternary Structure**

The quaternary structure of a protein refers to the three-dimensional structure whereby several subunits of the protein cluster together forming a final specific shape. The types of interactions that hold these subunits together can be ionic interactions, hydrogen bonding, disulfide bridges and van der Waals forces. If there are more than two subunits present, such a protein is known as an oligomeric protein. A good example to discuss is haemoglobin.

Haemoglobin is a macro-biological molecule in the red blood cell that is responsible for transporting oxygen molecules to the other cells during respiration. The quaternary structure of the haemoglobin consists of four globular protein subunits — two α-subunits and two β-subunits — non-covalently bonded together via the interactions between the R-groups from the different subunits. Each of the subunits is strongly associated with a non-protein haem group, which contains an iron (II) ion at the center of the haem group. The main difference between the two types of subunits is in the number of amino acid residue in the primary structure. The secondary structure of each subunit has the helix structure in some of the segments of the protein chain separated by random coil in between.

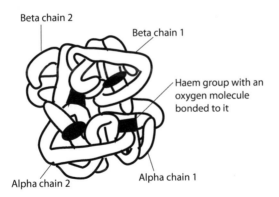

Beta chain 2

Beta chain 1

Haem group with an oxygen molecule bonded to it

Alpha chain 2

Alpha chain 1

13.4.2 Denaturation of Proteins

Protein molecules have a specific biological function because of their unique three-dimensional conformation. This conformation is stabilized via the various types of interactions of the R-groups — ionic interactions, hydrogen bonding, disulfide bridges and van der Waals forces. Hence, if such interactions are disrupted, the protein molecule will lose its biological function, and we say that the protein has been denatured.

Denaturation of protein is the result of the disruption or possible destruction of both the secondary and tertiary structures. To destroy the primary structure, that is, the peptide bonds, more drastic conditions need to be applied. The following section focuses on how various factors such as heat, pH, heavy metal ions, alcohol, detergent and reducing agents can bring about the denaturation of protein.

- **Heat**

When a protein molecule is subjected to heating, the increase in vibrational and rotational kinetic energies disrupts the various interactions between the R-groups. The result is the unfolding of the molecule, causing a change in the tertiary structure. If heating is not drastic and is ceased, the protein molecule may resume its original tertiary structure. But if heating continues, more serious unfolding takes place as the stabilization forces — hydrogen bonding — of the helical structure will be destroyed. The surrounding water molecules can now interact with the peptide bond via hydrogen bonding. In addition, the originally embedded hydrophobic R-groups are now being exposed to the hydrophilic medium. As a result, these hydrophobic groups would prefer to interact with each other, causing coagulation of the protein. A good example to quote is how when the egg white of the egg is fried, it turns white! The high heat during frying causes the albumin in the egg white to denature as the tertiary globular structure is totally lost. The protein chains are not able to fold back into the original compact globular structure. This decreases the solubility of the protein and hence it becomes a white solid mass.

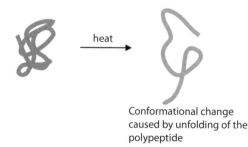

Conformational change
caused by unfolding of the
polypeptide

Q: Is there any advantage that we can harness knowing that heating can denature protein?

A: Yes. Knowing that heating causes denaturation of protein, we can employ such idea in sterilization techniques, since heating can also denature the enzymes in bacteria, and in so doing, kill the bacteria. Remember, enzymes are actually protein molecules. Other than heating, sometimes ultraviolet radiation (UV) is also used in sterilization. This is because the highly energetic UV radiation destroys the nuclei acids in the microorganisms such that their DNA is disrupted by the UV radiation. This removes their reproductive capabilities and kills them.

- *pH*

The R-groups of a protein may contain acidic or/and alkaline groups such as the polar –OH, –NH, =NR and –COOH. Under acidic conditions, the basic R-groups would be protonated, creating positively charged groups. Under alkaline conditions, the acidic groups could be deprotonated, creating anionic groups. The similarly charged groups that are being created would repel each other, causing the protein to unfold, hence destroying its tertiary and secondary structures. Serious unfolding would cause the hydrophobic groups to be exposed to the hydrophilic medium, resulting in coagulation of the protein. If strong heating is applied together, the primary structure could be destroyed, resulting in hydrolysis of the peptide bonds giving free amino acids.

Q: Why does milk coagulate when it becomes sour?

A: When milk turns sour, it is because bacteria such as *Lactobacilli* convert the lactose that is present in milk to lactic acid, which is acidic. The drop in pH causes the protein, which is mostly casein, to denature and hence lose its globular tertiary structure. This decreases the solubility of the protein, and hence it precipitates out as flocculent curds.

- *Heavy metal ions*

Heavy metal ions such as Ag^+, Cu^{2+} or Hg^{2+} can compete with other positively charged R-groups of the protein molecule for attraction to the negatively charged groups such as COO^-. As a result, the original ionic interactions that are present to hold the protein in the native three-dimensional conformation would be disrupted, causing unfolding to take place. In addition, the Hg^{2+} ion is also capable of forming bonds with the –SH groups. Sometimes, the heavy metal ions may actually displace the original residential metal ion that is present. In the worst scenario, serious unfolding causes the protein to precipitate out of the solution as an insoluble metal–protein salt. This phenomenon has an advantage as it makes some of the heavy metal salts suitable for use as antiseptics. For example, silver nitrate is dropped into the eyes of newborn babies to prevent gonorrhoea chlamydia infections from their mothers, and mercuric chloride, another antiseptic, acts to precipitate the proteins in infectious bacteria.

During the making of *tofu* (bean curd), calcium sulfate or calcium chloride is added to soy milk. The Ca^{2+} ions interact with the protein molecules and, through this interaction, the conformation of the protein molecules changes, causing unfolding to take place. As a result, the solubility of the protein molecules decreases, resulting in coagulation.

• *Alcohol*

Alcohol is a common disinfectant for the skin because of its ability to denature protein in bacteria. One of the common interactions between the R-groups of the protein molecule in the secondary and tertiary structures is hydrogen bonding. The –OH group of the alcohol molecule is capable of hydrogen bonding with some of the R-groups of the protein. This disrupts the original interactions and causes unfolding to take place. As a result, the native conformation of the protein is gone, and with it goes its biological function.

• *Detergent*

Detergents consist of a special type of molecule that is amphiphilic in nature, i.e. there is both a hydrophobic and a hydrophilic part. When this amphiphilic molecule comes in contact with a protein molecule, the hydrophobic part of the molecule can interact with the non-polar R-groups of the protein, whereas the hydrophilic part can interact with the polar R-groups. As a result, the original interactions that are responsible for stabilizing the secondary and tertiary structures would be disrupted. This causes unfolding and hence affects the native conformation of the protein molecule. The consequences associated with the unfolding are a change in the solubility of the protein and loss of its biological function.

- **Reducing agents**

As discussed in Section 13.4.1, the disulfide bridges that are present in the secondary and tertiary structures of protein are formed by the oxidation of the –SH groups on cysteine molecules. The presence of a reducing agent can cause the S—S bridges to break. As a result, this would cause unfolding and hence this affect the conformation of the protein.

My Tutorial

1. (a) (i) Explain the meaning of the terms *addition polymerization* and *condensation polymerization*, illustrating your answer with one example of each, of industrial importance.

   (ii) Draw the structure of the repeating unit in the polymer made from each of the following:

   (A) $CH_2CHCOOCH_3$
   (B) H_2NCH_2COOH

   (iii) Name the type of bond fission that is involved in both (a)(ii)(A) and (a)(ii)(B).

   (b) Propene, $CH_3CH{=}CH_2$, a valuable chemical that is produced during the cracking of petroleum, is principally used for the manufacturing of poly(propene).

   (i) Write an equation to represent the polymerization of propene, representing the polymer with four repeat units.

   (ii) Give the name of the catalyst that is used in the industrial polymerization of propene.

   (iii) Explain how the use of this catalyst influences the structure of the polymer form.

   (iv) Draw three different types of poly(propene), explaining the differences between them.

   (v) Poly(propene) is non-biodegradable. With reference to two different uses of poly(propene), suggest how this can be both an advantage and a disadvantage.

(vi) Poly(propene) is a popular material for the making of carpets. Suggest two advantages of using poly(propene) for carpets.

(vii) When poly(propene) is heated, it softens over a range of temperatures rather than melting sharply at a particular temperature. Suggest reasons for the observed phenomenon.

(c) The structures of the repeating units of two polymers **P** and **Q** are given below:

$$\left[CF_2 \right]_n \qquad \left[CH_2OOC \underset{}{\bigcirc} COOCH_2 \right]_n$$

P   **Q**

(i) Draw the constitutional/structural formulae of the monomers from which **P** and **Q** are synthesized.

(ii) Classify the type of polymerization and name the other compound produced when **Q** is made as in (i) above.

(iii) Clothes that are made from polymer **Q** feel more comfortable than those made from poly(propene) but are not as comfortable as those that are made from cotton, which is actually poly(glucose). Explain the observed phenomenon.

poly(glucose)

(d) **R** and **S** represent the repeating unit of a synthetic polymer and a section of a natural material, respectively.

(i) State the type of compound represented by **R** and the type of compound represented by **S**.

(ii) Name the polymerization process that results in **R** and **S**.

(iii) What is the name given to the unit that makes up **S**?

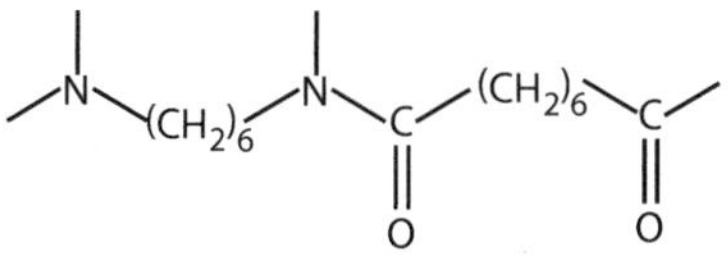

R

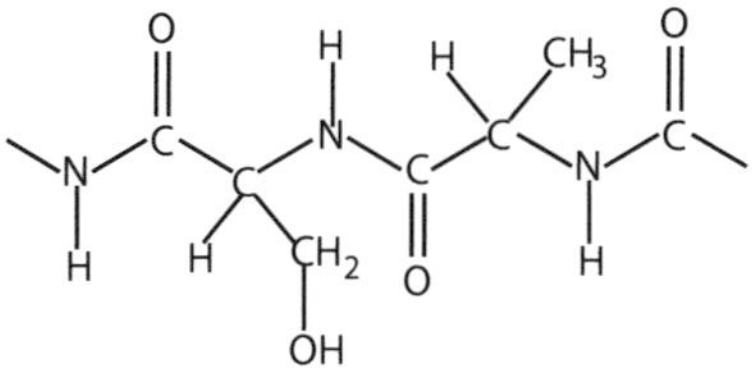

S

(iv) Suggest two similarities and two differences between **R** and **S**.

(v) Both the synthetic polymer **R** and the natural material **S** can be dyed to give uniform color.

(A) Why do you think that the molecules of the two polymers might behave in the same way with the molecules of a dyestuff?

(B) What type of bonding will be involved between the polymer and molecules of the dyestuff.

(C) Suggest a disadvantage of making carpets with such a polymer.

(vi) Polymer **R** is known as nylon-6,6 in the industry, and it is made by reacting a diacid chloride rather than the corresponding dicarboxylic acid with a diol. Suggest a reason why the dicarboxylic is not preferred.

(vii) Nylon-6,6 is almost twice as strong as poly(propene). Suggest possible explanations for the observed phenomenon.

(e) Nylon-6 is produced from caprolactam:

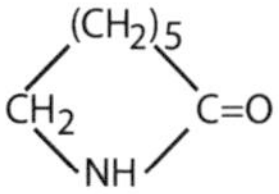

(i) Draw a section of the polymer showing three repeat units.

(ii) Name the functional group that is present in the polymer.

(iii) Suggest what conditions might be required for the polymerization of caprolactam.

(iv) Give a name for the mechanism responsible for the polymerization.

(v) Suggest two uses of nylon-6.

(vi) Explain, with the aid of a diagram, why nylon-6 has a high tensile strength when drawn into fibers.

(vii) Although nylon is a good electrical insulator, there are polymers which are good electrical conductors, such as the following:

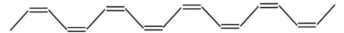

(A) Identify the monomer.

(B) What type of stereoisomerism is exhibited by the above polymer? Draw a diagram of a section of the above polymer in the other stereoisomeric form.

(C) Give a reason why the above polymer is able to conduct electricity and predict what will happen when the polymer is heated with buta-1,2-diene.

(D) What similar property does the polymer have with graphite?

2. (a) Compound **T**, $C_5H_8O_2$, reacts with acidified potassium manganate (VII) to give **U** of molecular formula $C_5H_8O_4$. Compound **T** also reacts with $LiAlH_4$ to form compound **V**, $C_5H_{12}O_2$. Both **U** and **V** react together with a catalyst to form a macromolecule.

(i) Deduce the constitutional/structural formulae of **T** to **V**.

(ii) Write a repeat unit of the macromolecule formed.

(iii) Suggest a suitable catalyst for the synthesis of the macromolecule.

(iv) The macromolecular compound is not suitable to be made into a container to hold alkaline solutions. Suggest a reason for this.

(b) A polymer represented by the following formula is obtained from a monomer **W**:

$$\left[\begin{array}{c} CH - CH_2 \\ | \\ COOR \end{array}\right]_n$$

(i) What type of polymerization is taking place?

(ii) Draw the constitutional/structural formula of monomer **W**.

(iii) The polymerization process is initiated by a radical $X^\bullet$. Write equations to show how the radical initiates the reaction, followed by another reaction with a second monomer.

(iv) If the R-group in the monomer is a phenyl group, the polymer formed is less flexible, with higher melting point than another similar polymer, where the R-group is a CH_3–group. Explain the differences in the physical properties observed.

(c) Muconic acid, $HOOC-CH=CH-CH=CH-COOH$, is capable of undergoing polymerization to produce a polymer with a high melting point.

(i) Draw a section of the polymer showing three repeat units.

(ii) Suggest a reason why the melting point of the polymer is high.

(iii) The polymer is likely to disintegrate under alkaline conditions. Suggest a reason for the phenomenon.

(iv) The polymer can be further hardened through heating with sulfur. Explain why it is possible for the hardening to take place.

(d) A bottle containing phenyl(ethene) turns cloudy with some white suspension when exposed to air. Account for the formation of the white solid.

(e) Explain how ethene can produce both low-density and high-density polymers.

(f) Explain the differences between thermosetting and thermoplastic materials and give an example of each.

CHAPTER 14

Mass Spectrometry

14.1 Introduction

Mass spectrometry is an analytical technique that enables us to determine the chemical composition of a sample and elucidate the structures of compounds based on the mass-to-charge ratio of charged particles.

The fundamental principle lies in the fact that charged particles of different mass and charge behave differently in an electric or magnetic field. If we have a beam of positively charged particles moving across an electric field, it will deviate from its original path and get deflected towards the negative plate. If the beam is comprised of negatively charged particles, it will deflect towards the positive plate. All this is in accordance with the concept of *like charges repel and unlike charges attract*.

An atom is electrically neutral since it contains equal numbers of protons and electrons. A beam of atoms moving across an electric field will continue in its path and not be deflected at all. Molecules made up of atoms are also neutral species, so how can we use this technique to identify molecules? We can actually transform these neutral species into positively charged particles through the process known as *ionization*. Electrons are forcefully removed from atoms (even those which tend to form anions), which makes ionization an energetically demanding but possible process.

If a sample containing different molecules is ionized into positively charged species, how do we identify them if they deflect towards the same negative plate? These like charges are differentiated by the amount of deviation from their trajectory, defined as the angle of deflection (θ):

$$\text{angle of deflection} \propto \frac{\text{charge}}{\text{mass}}.$$

The positively charged species with the higher charge experiences greater attractive force exerted by the negative plate and deviates more from its trajectory. For two ions of the same unit charge but different masses travelling at the same speed, the heavier particle has a greater kinetic energy (K.E. $= 1/2\, mv^2$), which means that more energy must be exerted on it for the heavier ion to deflect from its original trajectory. Since the applied electric field exerts the same amount of force on particles that carry the same charge, the heavier ion deflects to a smaller extent.

The amount of deviation can also be measured in terms of the radius of deflection:

$$\text{radius of deflection} \propto \frac{\text{mass}}{\text{charge}}.$$

This is possible if we imagine that after deflection, the particle moves in a circular path. Hence, the factors that affect the radius of deflection are reciprocal to that for the angle of deflection. For two positively charged particles of the same charge but different masses, the heavier particle would experience a smaller angle of deflection and hence a larger radius of deflection when these two particles travel through the same electric field.

Similar to the deflection in an electric field, when a positively charged particle moves through an applied magnetic field, the particle would also experience deflective force. This is because a moving positively charged particle, which is considered to be a positive current, would itself generate a magnetic field, and this magnetic field is aligned in the direction of the applied field. As a result, there will be a deflection which can be predicted by using the Fleming's left-hand rule.

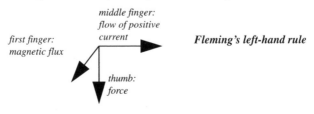

middle finger:
flow of positive
current

first finger:
magnetic flux

Fleming's left-hand rule

thumb:
force

Q: What would happen if the charged particle were negatively charged instead?

A: The direction of movement of an electron is always considered to be opposite to that of positively charged particles, i.e. current. So you can still apply Fleming's left-hand rule to determine the direction of deflection by using the middle finger to point to the opposite direction of the electron's movement. The magnetic field that is generated by a moving negatively charged particle is in opposition to the applied field. The following diagram depicts the trajectories of different types of particles passing through a magnetic field.

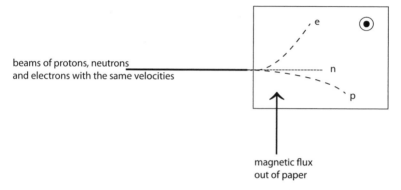

For a positively charged particle of mass m, velocity v and charge q passing through a magnetic field with field strength B, the radius of deflection r follows the following equation:

$$r = \frac{mv}{Bq}.$$

Hence, one can fix the radius of deflection of different particles that have different masses, quantities of charge and velocities, simply by varying the strength of the applied magnetic field acting on these particles. A more massive and/or higher velocity particle would need a stronger magnetic field strength to bend it whereas higher charged particle would need weaker magnetic field strength to bend it to the same radius of deflection.

14.2 Basic Features of a Mass Spectrometer

The mass spectrometer is the instrument that performs the above processes. It is operated under an ultra-high vacuum (UHV) to exclude the presence of air molecules which would interfere with the trajectory of the ions. Mass spectrometry comprises six basic processes:

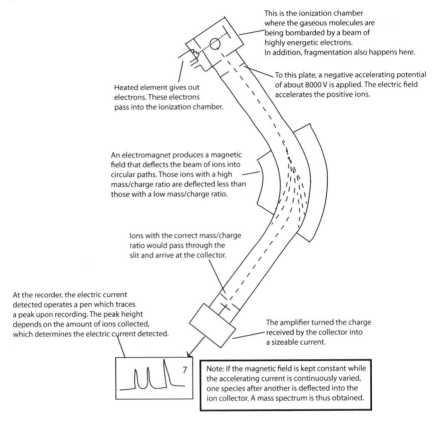

This is the ionization chamber where the gaseous molecules are being bombarded by a beam of highly energetic electrons. In addition, fragmentation also happens here.

To this plate, a negative accelerating potential of about 8000 V is applied. The electric field accelerates the positive ions.

Heated element gives out electrons. These electrons pass into the ionization chamber.

An electromagnet produces a magnetic field that deflects the beam of ions into circular paths. Those ions with a high mass/charge ratio are deflected less than those with a low mass/charge ratio.

Ions with the correct mass/charge ratio would pass through the slit and arrive at the collector.

At the recorder, the electric current detected operates a pen which traces a peak upon recording. The peak height depends on the amount of ions collected, which determines the electric current detected.

The amplifier turned the charge received by the collector into a sizeable current.

Note: If the magnetic field is kept constant while the accelerating current is continuously varied, one species after another is deflected into the ion collector. A mass spectrum is thus obtained.

1. The compound has to be converted into the gaseous form via heating before it enters the spectrometer, especially if the original physical state is a non-gaseous one. The compound must not break down after vaporization. This is to ensure that the molecule as a whole is still intact, which would thus enable us to determine the relative molecular mass of the compound.

2. The gaseous molecules now pass through an ionization chamber where they are "hit" by a beam of high-energy electrons. As the energetic electron passes through a molecule, it transfers some of its kinetic energy to the molecule. As a result, the molecule is energized and becomes highly unstable. Thus, the energized molecule needs to "lose" some of its energy. To lower its energy

content, the molecule can either "eject" an electron and becomes a positively charged ion or undergo fragmentation through bond breaking. These two processes are endothermic in nature.

$$M \rightarrow M^+ + e^-.$$

The positively charged ions now pass through an electric field with strength of about 5–10 kV. This is to accelerate all ions to roughly the same speed such that later, during the deflection stage, the angle of deflection would solely be dependent on the mass and charge of the particle and not on the speed.

3. To further ensure the independency of the angle of deflection on the speed of the ions, the stream of ions passes through an electro-static analyzer, which selects ions of similar kinetic energy within a narrow range. As a result, the amount of "noise" (unnecessary signals) diminishes.

4. The highly energetic ions pass through the poles of an electro-magnet, which causes ions of different masses and charges to deflect differently. The magnetic field strength of the electromagnet varies from low-field strength to high-field strength so as to allow these ions to be captured by the detector after they have moved through a circular pathway of a fixed radius. The equation governing the radius of deflection (directly proportional to $\sqrt{m/e}$) of ions under a particular applied magnetic field strength follows the following equation:

$$r = \sqrt{\frac{2\,mV}{eB^2}}$$

where
r = radius of circular path in the magnetic field
m = mass of ion
V = accelerating voltage
e = electrical charge on the ion
B = strength of magnetic field
$B \propto \sqrt{m/e}$ when V, e and r are kept constant.

5. The ion beam of a particular m/e ratio value is focused and hits the collector plate, causing a current to flow. This current is amplified and recorded. An output known as a mass spectrum is produced. The mass spectrum is a plot of the relative abundance (measured by the amount of current detected) against the m/e ratio.

Q: What happen to the ions that are not collected by the detector?

A: Well, they eventually get sucked away by the vacuum pump that creates the ultra-high vacuum condition.

14.3 Mass Spectrum Analysis

The vertical axis, labeled as signal intensity or peak height, indicates the relative abundance of the ions. An ion of greater abundance will produce a taller peak due to its greater stability under the conditions in the spectrometer. The tallest peak is known as the base peak, and its intensity is assigned an arbitrary value of 100. All other intensities are measured and calculated relative to the base peak.

The horizontal axis shows the mass/charge ratio (m/e) of the ion responsible for a given peak. For singly charged ions, the m/e ratio is numerically equivalent to its mass, but this is not true for ions of higher charges. Nonetheless, most of the ions formed in the mass spectrometer are singly charged and are thus more abundant than ions of higher charges. For instance, with ^{208}Pb ions, you will have a peak at $m/e = 208$ that corresponds to $^{208}Pb^+$, which has much higher signal intensity than the peak at $m/e = 104$, which is produced by $^{208}Pb^{2+}$.

Q: Why is it not very likely for there to be ions of high positive charges?

A: The idea is the same as in successive ionization. More energy is required to remove an electron from a particle that is already positively charged due to stronger net electrostatic force of attraction exerted on the remaining electrons. The conditions in the ionization chamber are conducive for the production of singly charged particles which would directly give us the mass of the particle.

Q: Is it possible to have a case whereby a peak could be due to either A^+ or B^{n+}, both of which have the same m/e ratio? If yes, how can the ambiguity be resolved?

A: Yes, it is possible to have more than two different ions with the same m/e value being detected at the same time. Well, you can't resolve the ambiguity. What one can do is to exhaustively extract as much information from other peaks as possible. The most important information from the mass spectrum is the highest m/e value, which corresponds to the molecular ion. It gives us the relative molecular mass of the compound. Other important information would be to determine the relative isotopic mass and hence the relative atomic mass of an element after taking into consideration the relative abundances of the various isotopes from the heights of the peaks.

The peaks shown on a mass spectrum can be classified into distinct sets that are useful in analysis:

- Molecular ion peak
- Peaks due to isotopes
- Peaks due to fragmentation of molecular ion

When a molecule is ionized, it loses an electron and forms a unipositively charged molecular ion (given the symbol M^+).

$$M + e^- \rightarrow M^+ + 2e^-$$

Usually, the molecular ion (or parent ion) peak is the one with the largest m/e ratio that corresponds to its relative molecular mass. However, since the molecular ion may be unstable and undergo fragmentation before reaching the detector, the molecular ion peak may not surfaced, and thus the peak at the greatest m/e ratio need not necessary give us information on the M_r of the molecule.

Q: So a base peak need not necessary be due to the molecular ion?

A: The base peak simply indicates the most "common" ion that is present in the ion beam. It may be the most stable ion, or it may be the fragment ion that is most frequently formed from the various ways of breaking up the molecular ion. It may not be the molecular ion peak!

Peaks observed at m/e ratios larger than that of the molecular ion are due to isotopic distribution. This stems from the fact that the molecule may contain atomic isotopes which are atoms of the same element with different number of neutrons and hence different masses. Thus, a $(M + 1)$ peak is observed if a fraction of the molecules contains one higher isotope such as ^2H and ^{13}C which are isotopes of ^1H and ^{12}C, respectively. A $(M + 2)$ peak can be attributed to two higher isotopes. If we are dealing with a halogen-containing compound, the $(M + 2)$ peak is attributed to the higher isotope of the halogens and a $(M + 4)$ peak indicates the presence of two such higher isotopes.

The molecular ion can be fragmented in many ways which depend on which bond in the molecule is the weakest or whether the fragment that is formed is stable enough to "survive" under the conditions of the spectrometer. If a more energetic electron beam is used to ionize the sample, this would mean that more energy can be transferred to the molecule resulting in a more energized molecule. Thus, other than multiple successive ionization to form more highly charged ions, fragmentation would be a common phenomenon too. One needs to note that different operating conditions in a mass spectrometer, such as energy of the electron beam, pressure, etc., would affect the fragmentation pattern obtained.

The peaks produced by these fragments are called daughter ion peaks. Likewise in this set, we have peaks that are due to isotopic distribution. By identifying the various fragments produced, we can piece these together to obtain the structure of the parent molecule.

Q: So is fragmentation valued in the mass spectrometer?

A: Fragmentation is valued because it provides us with the possible fragments that may be formed from a molecule. But if all the parent molecules undergo fragmentation, then we lose information regarding the relative molecular mass of the parent molecule, which is also a highly valued piece of information. By piecing these fragments together, we can get an overall idea of the consitutional/structural formula of the molecule. Each molecule has its own unique fragmentation pattern under a particular set of operating conditions, and thus the fragmentation pattern is also known as the fingerprint of the compound. Sometimes, other than fragmentation, processes such as rearrangement may take place. For instance, if we analyze the mass spectrum of butanal, we would see a peak at m/e 44. It could be due to either $[CH_3CH_2CH_3]^+$ or ethanal $[CH_3CHO]^+$, but close analysis points to the enol (presence of both an alkene and alcohol functional groups) form of ethanal that is produced from the following unimolecular arrangement:

butanal (m/e 72)   (m/e 28)   (m/e 44)

The relative signal intensities can help to identify the species responsible for the peak. The low natural abundance of higher isotopes of elements such as carbon and hydrogen will translate to isotope peaks that are low in intensity. It is quite the opposite for the halogens such as chlorine and bromine, whose isotopes have high natural abundance. High intensity peaks are associated with these and in such specific proportions that it serves as a characteristic blueprint to identify the presence of the halogens.

In fact, it is through mass spectrometry that stable isotopes of elements were discovered. Before their discovery, scientists were baffled by the non-integral values of atomic mass obtained empirically. For instance, the A_r of chlorine of 35.5 is now understood to be the weighted average of one atom of the element, taking into account the relative abundance of its isotopes and their masses. Table 14.1 shows some of these naturally occurring isotopes.

Table 14.1   Examples of Naturally Occurring Isotopes.

Element	Stable Isotope	Relative Isotopic Mass	Relative Abundance (%)	Relative Atomic Mass
Hydrogen	H-1 (protium)	1.0078	99.99	1.008
	H-2 (deuterium)	2.0141	0.01	
Carbon	C-12	12.0000	98.93	12.011
	C-13	13.0034	1.07	
Oxygen	O-16	15.9949	99.76	15.999
	O-17	16.9991	0.04	
	O-18	17.9991	0.20	
Chlorine	Cl-35	34.9689	75.76	35.453
	Cl-37	36.9659	24.24	
Bromine	Br-79	78.9183	50.69	79.903
	Br-81	80.9162	49.31	

Each molecule produces characteristic fragmentation patterns. Thus, with a list of such patterns recorded for molecules, we have a database that allows us to identify and distinguish different molecules.

In the following sections, we will study the mass spectrum of elements and simple organic compounds. We will then use what we have learnt to elucidate the structure of unknown compounds based on their mass spectra.

14.4   Analyzing the Mass Spectrum of an Element

Example: Analyzing the mass spectra of chlorine

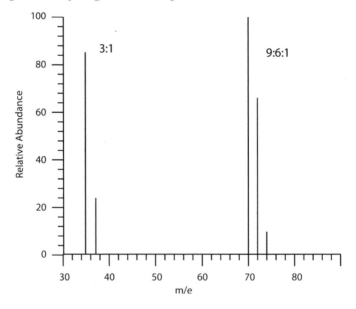

From the mass spectrum of chlorine, we have two distinct groups of peaks:

- *Daughter ion peaks and isotope peaks*

  Fragmenting the diatomic chlorine molecule breaks it into monatomic species, one which is positively charged and the other, an uncharged atom, i.e. with the Cl–Cl bond cleaved:

$$Cl_2^+ \rightarrow Cl + Cl^+$$

The Cl atom produced is an uncharged free radical (i.e. it contains an unpaired electron) that will just bump around in the spectrometer and be eventually removed. The Cl^+ ions are detected, and they are responsible for the peaks at m/e ratio of 35 and 37.

The values of 35 and 37 inform us that there are two types of chlorine isotopes, namely chlorine-35 and chlorine-37.

- *Molecular ion peak and isotope peaks*

The molecular ion (M) peak, at $m/e = 70$, corresponds to the molecular ion $[^{35}Cl^{35}Cl]^+$.

At $m/e = 72$, we have the so-called (M + 2) peak. The molecular ion responsible for this peak is $[^{35}Cl^{37}Cl]^+$, in which one higher chlorine-37 isotope is present.

When a chlorine molecule is comprised of two chlorine-37 atoms, the (M + 4) peak at $m/e = 74$ is observed.

The relative signal intensity (relative abundance) of the two daughter ion peaks is in the ratio of 3:1. This means that for every four chlorine atoms, three of them are chlorine-35 and one of them is chlorine-37. This information, along with the isotopic mass obtained from the m/e value of the daughter ion peaks, allow us to calculate the A_r of chlorine as follows:

$$A_r \text{ of chlorine} = \frac{3(35) + 1(37)}{(3 + 1)} = 35.5.$$

The relative abundance can also be expressed as a percentage, i.e. 75% for chlorine-35 and 25% for chlorine-37. The A_r of chlorine is thus calculated as

$$A_r \text{ of chlorine} = \frac{75(35) + 25(37)}{(75 + 25)} = 35.5.$$

Based on the relative abundance of the chlorine isotopes, we can account for the relative intensity of the three molecular ion peaks using statistical probability. The relative abundance of the molecular ion is calculated as the product of the relative abundance of the atomic isotopes.

m/e value	70	72	74
Molecular ion	$[^{35}Cl^{35}Cl]^+$	$[^{35}Cl^{37}Cl]^+$	$[^{37}Cl^{37}Cl]^+$
Relative abundance	$\frac{3}{4} \times \frac{3}{4} = \frac{9}{16}$:	$2 \times \frac{3}{4} \times \frac{1}{4} = \frac{6}{16}$:	$\frac{1}{4} \times \frac{1}{4} = \frac{1}{16}$
	9 :	6 :	1

Q: In tabulating the relative abundance of $[^{35}Cl^{37}Cl]^+$, why is it necessary to multiply it by a factor of 2?

A: Based on probability, there are two ways to arrange the atoms in the molecule, and we get, across from left to right, either ^{35}Cl–^{37}Cl or ^{37}Cl–^{35}Cl, which is the same molecule after all.

Q: Why did you use a square bracket to "round up" the ion?

A: After a species has lost an electron, you do not know specifically which atom in the species has lost this electron, so the positive charge is placed as a superscript outside the square bracket to represent the delocalization of the charge throughout the ion.

Q: Would the relative abundances of the isotopes be preserved in the daughter ions after fragmentation from the parent molecule?

A: Yes, it would indeed.

Example: Analyzing the mass spectrum of bromine

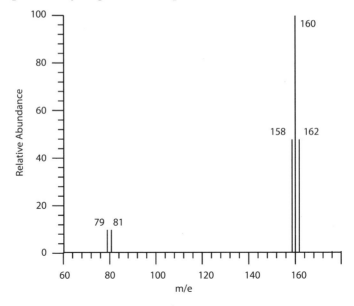

We see here again two distinct groups of peaks:

- *Daughter ion peaks and isotope peaks*

  The values of 79 and 81 inform us that there are two types of bromine isotopes. The equal intensity of these peaks indicates that bromine-79 and bromine-81 are present in equal proportions.

- *Molecular ion peak and isotope peaks*

  The molecular ion (M) peak, at m/e ratio $= 158$, corresponds to the molecular ion $[^{79}Br^{79}Br]^+$.

  At $m/e = 160$, we have the (M + 2) peak attributed to the presence of one higher isotope of bromine, i.e. it is due to the $[^{79}Br^{81}Br]^+$ ion.

  At $m/e = 162$, we have the $[^{81}Br^{81}Br]^+$ ion responsible for this (M + 4) peak.

  Based on statistical probability, the relative abundance of the molecular ion is calculated taking into account the equal relative abundance of the atomic isotopes:

m/e value	158	160	162
Molecular ion	$[^{79}Br_2]^+$	$[^{79}Br^{81}Br]^+$	$[^{81}Br_2]^+$
Relative abundance	$\frac{1}{2} \times \frac{1}{2} = \frac{1}{4}$:	$2 \times \frac{1}{2} \times \frac{1}{2} = \frac{2}{4}$:	$\frac{1}{2} \times \frac{1}{2} = \frac{1}{4}$
	1 :	2 :	1

The relative abundance of the molecular ions can be inferred from the relative height of the peaks in the mass spectrum. However, take note that there is no relationship between the relative intensities of the molecular ion peaks and those of the daughter ion peaks. This is because the daughter ions are formed from molecular fragmentation which depends on the energy of the ionizing electrons and stability of the daughter ions that formed.

14.5 Analyzing the Mass Spectrum of Alkanes

Example: Analyzing the mass spectrum of methane

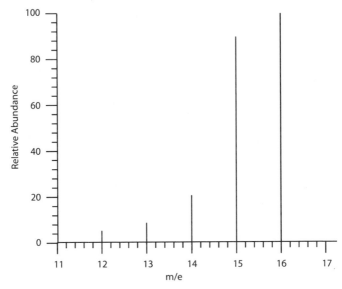

- The mass spectrum of methane shows the molecular ion peak at $m/e = 16$. This value corresponds to the M_r of $^{12}CH_4$.
- Given the very low relative abundance of deuterium at 0.01%, we can rule out its contribution to any isotope peaks. Thus the $(M+1)$ peak of around 1% is due to a small fraction of $^{13}CH_4$.
- The peak at $m/e = 15$ is due to the $[^{13}CH_3]^+$ ion formed from the cleavage of a C–H bond. Further fragmentation of the molecular ion produces the following fragments:

m/e ratio	Ions responsible for the peak observed
15	$[^{12}CH_3]^+$, $[^{13}CH_2]^+$
14	$[^{12}CH_2]^+$, $[^{13}CH]^+$
13	$[^{12}CH]^+$, $[^{13}C]^+$
12	$[^{12}C]^+$

Notice that a peak can represent more than one ion. Even the peak at $m/e = 16$ can be attributed to a small fraction of $[^{13}CH_3]^+$, which is relatively insignificant as compared to $[^{12}CH_4]^+$. The relative intensity of each of these peaks is not as easily derived as those obtained in the mass spectra of elements since we have to consider the relative probability of the ions formed and include the isotopes too.

Example: Analyzing the mass spectrum of ethane

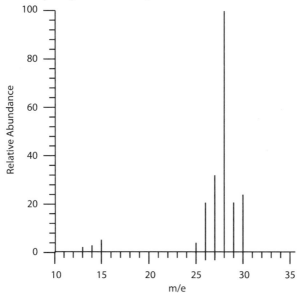

- The mass spectrum of ethane shows the molecular ion peak at $m/e = 30$. This value corresponds to the M_r of $^{12}\text{C}_2\text{H}_6$.
- The peaks at m/e values close to 30 are produced by fragment ions when a variable number of C–H bonds are cleaved:

m/e ratio	Examples of ions responsible for the peak observed
31	$[^{12}\text{CH}_3^{13}\text{CH}_3]^+$, $[^{13}\text{CH}_3^{13}\text{CH}_2]^+$
30	$[^{12}\text{CH}_3^{12}\text{CH}_3]^+$, $[^{12}\text{CH}_3^{13}\text{CH}_2]^+$, $[^{13}\text{CH}_2^{13}\text{CH}_2]^+$, etc.
29	$[^{12}\text{CH}_3^{12}\text{CH}_2]^+$, $[^{12}\text{CH}_3^{13}\text{CH}]^+$, $[^{12}\text{CH}_2^{13}\text{CH}_2]^+$, $[^{12}\text{CH}^{13}\text{CH}_3]^+$, etc.
28	$[^{12}\text{CH}_2^{12}\text{CH}_2]^+$, $[^{12}\text{CH}_3^{13}\text{C}]^+$, $[^{12}\text{CH}_2^{13}\text{CH}]^+$, $[^{12}\text{CH}^{13}\text{CH}_2]^+$, $[^{12}\text{C}^{13}\text{CH}_3]^+$, $[^{13}\text{C}^{13}\text{CH}_2]^+$, $[^{13}\text{CH}^{13}\text{CH}]^+$, etc.
27	$[^{12}\text{C}_2\text{H}_3]^+$, $[^{12}\text{CH}_2^{13}\text{C}]^+$, $[^{12}\text{C}^{13}\text{CH}_2]^+$, $[^{12}\text{CH}^{13}\text{CH}]^+$, $[^{13}\text{C}_2\text{H}]^+$, etc.
26	$[^{12}\text{C}_2\text{H}_2]^+$, $[^{12}\text{CH}^{13}\text{C}]^+$, $[^{13}\text{C}_2]^+$, etc.
25	$[^{12}\text{C}_2\text{H}]^+$, $[^{12}\text{C}^{13}\text{C}]^+$
24	$[^{12}\text{C}_2]^+$

• At lower m/e values are the fragment ions obtained from the cleavage of the C–C bond along with a variable number of C–H bonds:

m/e ratio	Examples of ions responsible for the peak observed
15	$[^{12}CH_3]^+$, $[^{13}CH_2]^+$
14	$[^{12}CH_2]^+$, $[^{13}CH]^+$
13	$[^{12}CH]^+$, $^{13}C^+$
12	$^{12}C^+$

Taking into account the carbon-13 isotope, you will have noticed the myriad ions that contribute to each peak. Do we need to analyze the spectrum in such explicit detail as to determine every possible fragment ion detected?

Because of the complex yet unique fragmentation pattern obtained from specific molecular structures, the chances that two compounds have identical mass spectra is small. This means that we can identify unknown samples by using "molecular fingerprinting," i.e. by comparing their mass spectra against reference mass spectra to find a match.

But the real advantage of the carbon-13 isotope lies in the (M + 1) peak it contributes. This particular peak can be used to determine the number of carbon atoms in an organic molecule when compared against the molecular ion (M) peak.

The ratio of ^{12}C to ^{13}C in naturally occurring carbon is approximately 98.9 to 1.1. Based on statistical probability, the relative abundance of the M ion to the (M + 1) ion can be calculated as follows:

	M		(M+1)
Molecular ion	$[^{12}CH_4]^+$		$[^{13}CH_4]^+$
Number of C isotope per 1000 methane molecules	989	:	11
Molecular ion	$[^{12}C_2H_6]^+$		$[^{12}CH_3^{13}CH_3]^+$
Number of C isotope per 1000 ethane molecules	$2 \times 989 = 1978$	:	$2 \times 11 = 22$
Molecular ion	$[^{12}C_3H_8]^+$		$[^{12}CH_3^{12}CH_2^{13}CH_3]^+$
Number of C isotope per 1000 propane molecules	$3 \times 989 = 2967$	:	$3 \times 11 = 33$

The proportion of molecules containing at least one ^{13}C atom increases with the number of carbon atoms (n). Mathematically, the trend suggests that the ratio of the relative abundances of the M and (M + 1) peaks should be 100 to $1.1n$. Hence, if given a mass spectrum, we can measure the relative abundances of both peaks (M and (M + 1)) and solve for n (rounded to the nearest integer) to find the total number of carbons:

$$n = \frac{100}{1.1} \times \frac{I_{(M+1)}}{I_M},$$

where n is the number of carbon atoms in the molecule, and I_{M+1} and I_M are the relative intensities of the (M + 1) and M peaks.

Exercise: Compound **X** produces a molecular ion peak at $m/e = 96$ with relative abundance of 25%. Another peak at $m/e = 97$ has relative abundance of 1.96%. How many carbon atoms are there in a molecule of **X**? [**Answer:** Since $n = 7$, there are 7 carbon atoms in the molecule.]

14.6 Analyzing the Mass Spectrum of Chloroalkanes

Example: Analyzing the mass spectrum of chloromethane and dichloromethane.

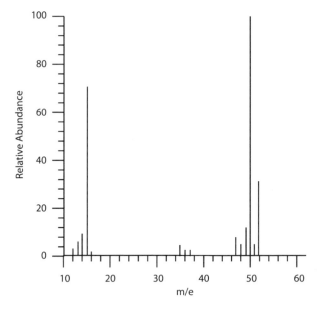

The ratio of ^{35}Cl and ^{37}Cl in naturally occurring chlorine is 3 : 1. Unlike the ^{13}C isotope, whose associated peaks have low intensity, the peaks of both chlorine isotopes have rather strong intensity due to their much higher relative abundances. For any chlorine-containing molecule, the distinct pair of two peaks at m/e of 2 units apart with relative intensity of 3 to 1 is observed. In the mass spectrum of chloromethane, these peaks are at m/e 50 and 52, attributed to the $[^{12}CH_3^{35}Cl]^+$ and $[^{12}CH_3^{37}Cl]^+$ ions, respectively. In the mass spectrum of dichloromethane, the characteristic pair of peaks is found at m/e 49 and 51, associated with the $[^{12}CH_2^{35}Cl]^+$ and $[^{12}CH_2^{37}Cl]^+$ respectively.

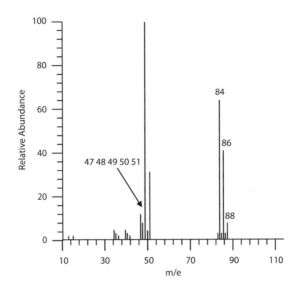

As seen in Section 14.4, the mass spectrum of chlorine has another set of distinct peaks, namely M, (M + 2) and (M + 4) in the ratio of 9 : 6 : 1. This set of trio peaks in the specific ratio can be observed for compounds containing two chlorine atoms just as shown in the mass spectrum of dichloromethane at the following m/e values:

m/e value	84 (M)		86 (M+2)		88 (M+4)
Molecular ion	$[^{12}CH_2^{35}Cl_2]^+$		$[^{12}CH_2^{35}Cl^{37}Cl]^+$		$[^{12}CH_2^{37}Cl_2]^+$
Relative abundance	$\frac{3}{4} \times \frac{3}{4} = \frac{9}{16}$	:	$2 \times \frac{3}{4} \times \frac{1}{4} = \frac{6}{16}$	:	$\frac{1}{4} \times \frac{1}{4} = \frac{1}{16}$
	9	:	6	:	1

The set of M, (M+2) and (M+4) peaks are not only useful in detecting two chlorine atoms but also two bromine atoms in a compound. A ratio of 9 : 6 : 1 denotes the presence of two chlorine atoms. If a compound contains two bromine atoms, the relative intensity of these three peaks will be in the ratio of 1 : 2 : 1 since both isotopes of bromine are in equal abundance (see Section 14.4). If only one bromine atom is present, there will be two peaks at m/e of 2 units apart with equal intensity.

Q: What is expected to be observed in the mass spectrum of a compound containing one chlorine and one bromine atom?

A: There will still be three distinct peaks attributed to the M, (M + 2) and (M + 4) ions, but of different relative intensities as shown in the calculations for bromochloromethane below:

m/e value	128 (M)	130 (M+2)	132 (M+4)
Molecular ion	$[^{12}CH_2^{35}Cl^{79}Br]^+$	$[^{12}CH_2^{35}Cl^{81}Br]^+,$ $[^{12}CH_2^{37}Cl^{79}Br]^+$	$[^{12}CH_2^{37}Cl^{81}Br]^+$
Relative abundance	$2 \times \frac{3}{4} \times \frac{1}{2} = \frac{6}{8}$:	$\left(2 \times \frac{3}{4} \times \frac{1}{2}\right) +$ $\left(2 \times \frac{1}{4} \times \frac{1}{2}\right) = \frac{8}{8}$:	$2 \times \frac{1}{4} \times \frac{1}{2} = \frac{2}{8}$
	3 :	4 :	1

Exercise: For trichloromethane ($CHCl_3$), identify the possible molecular ions and calculate the relative intensities of the corresponding peaks.

Answer:

m/e value	118 (M)	120 (M+2)	122 (M+4)	124 (M+6)
Molecular ion	$[^{12}CH^{35}Cl_3]^+$	$[^{12}CH^{35}Cl_2^{37}Cl]^+$	$[^{12}CH^{35}Cl^{37}Cl_2]^+$	$[^{12}CH^{37}Cl_3]^+$
Relative abundance	$\frac{3}{4} \times \frac{3}{4} \times \frac{3}{4}$: $= \frac{27}{64}$	$3 \times \frac{3}{4} \times \frac{3}{4} \times \frac{1}{4}$: $= \frac{27}{64}$	$3 \times \frac{3}{4} \times \frac{1}{4} \times \frac{1}{4}$: $= \frac{9}{64}$	$\frac{1}{4} \times \frac{1}{4} \times \frac{1}{4}$ $= \frac{1}{64}$
	27 :	27 :	9 :	1

The multiple of 3 input in the calculation of relative intensities of the (M + 2) and (M + 4) peaks corresponds to the three possible different arrangements of halogens in the molecule, i.e.

m/e value	120 (M+2)	122 (M+4)
Molecular ion	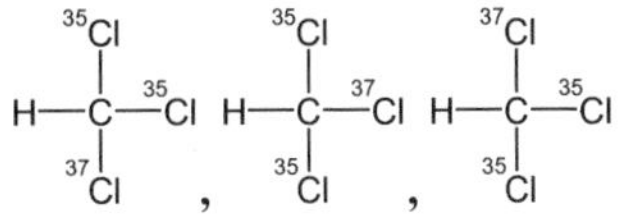	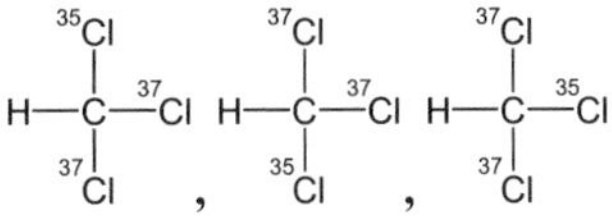

Q: Is it possible to differentiate 1,1-dichloroethane from 1,2-dichloroethane? Won't their mass spectra show the same M, (M + 2) and (M + 4) peaks with the same intensities?

A: Don't forget that based on molecular structures, these isomers have different bonding patterns which lead to different fragmentation. Apart from the molecular ion peaks, we can work on other peaks, i.e. in the mass spectrum of 1,1-dichloroethane, we will see peaks at m/e 83, 85 and 87 due to the $[^{12}CH^{35}Cl_2]^+$, $[^{12}CH^{35}Cl^{37}Cl]^+$ and $[^{12}CH^{37}Cl_2]^+$, respectively. These peaks are not observed if it is 1,2-dichloroethane. Can you figure out what are the relative intensities of these peaks?

Let us now take a look at the mass spectra of the isomers, propanal and propanone. We expect to find differences in the daughter ion peaks which highlight the use of fragmentation patterns to distinguish between isomers and hence elucidate the correct structure.

14.7 Analyzing the Mass Spectra of Isomers

Shown below are two unlabeled mass spectra which belong to either propanal or propanone. Our task is to match each spectrum to the correct isomer.

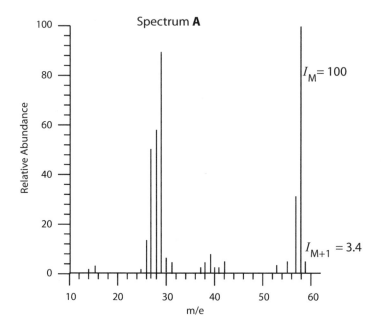

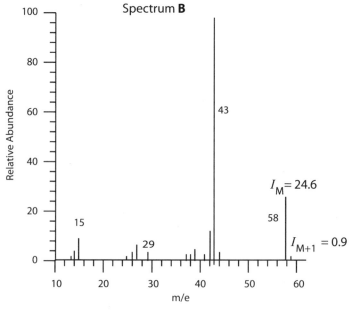

Using the ratio of the relative abundances of the M and (M + 1) peaks, we can calculate the number of carbon atoms.

For mass spectrum **A**, $\frac{100}{1.1} \times \frac{3.4}{100} = 3$.

For mass spectrum **B**, $\frac{100}{1.1} \times \frac{0.9}{24.6} = 3$.

Both calculations show that there are 3 carbon atoms in a molecule of each compound and at $m/e = 58$, we have the $[C_3H_6O]^+$ molecular ion responsible for it. The absence of M, $(M + 2)$ and $(M + 4)$ peaks is expected as there are no halogens in the compounds.

Let us find differences in the daughter ion peaks and identify possible ions that contribute to these peaks based on the structures of the two isomers.

- In mass spectrum **A**, there are distinct peaks either not found in mass spectrum **B** or of very weak intensity in the latter. These are at $m/e = 26$ to $m/e = 29$ and $m/e = 57$.
- In mass spectrum **B**, there are distinct peaks either not found in mass spectrum **A** or of very weak intensity in the latter. These are at $m/e = 15$ and $m/e = 43$.

We will now attempt to cleave the bonds in each structure to identify the possible fragment ions that contribute to the peaks at the distinct m/e values stated. We will ignore the higher isotopes of carbon and oxygen since these are in low abundance and focus only on the ions containing ^{12}C and ^{16}O.

Cleaving each bond in turn at the carbonyl group in propanal gives the following fragment ions:

$$\left[\begin{array}{c} \overset{H}{\underset{H}{\overset{|}{C}}} - \overset{H}{\underset{H}{\overset{|}{C}}} - \overset{O}{\overset{\|}{C}} - H \\ \end{array} \right]^{+\bullet} \longrightarrow [CH_3CH_2CO]^+ + H^\bullet \quad \text{or} \quad [CH_3CH_2CO]^\bullet + H^+$$

Propanal

$m/e = 57$ (for first) $\quad m/e = 1$ (for second)

$$\left[\begin{array}{c} \overset{H}{\underset{H}{\overset{|}{C}}} - \overset{H}{\underset{H}{\overset{|}{C}}} - \overset{O}{\overset{\|}{C}} - H \\ \end{array} \right]^{+\bullet} \longrightarrow [CH_3CH_2]^+ + [CHO]^\bullet \quad \text{or} \quad [CH_3CH_2]^\bullet + [CHO]^+$$

$m/e = 29 \quad\quad\quad m/e = 29$

It can be seen that the fragmentation pattern for propanal can account for the distinct peaks at $m/e = 29$ and 57 in mass spectrum **A**. Further cleavage of bonds in the $-CH_2CH_3$ group can explain the peaks shown at m/e 26 to 28. The fragment $[CO]^+$ can account for the peak at $m/e = 28$, but this ion can also be obtained from propanone, hence it does not offer any conclusion. At this point, it is highly probable that mass spectrum **A** belongs to propanal. To verify this, we will look into the fragmentation pattern of propanone.

Cleaving either one of the bonds at the carbonyl group in propanone gives the following fragment ions:

$$\left[\begin{array}{c} H \ O \ H \\ H-C\text{-}C\text{-}C-H \\ H \ \ \ H \end{array} \right]^{+\bullet} \longrightarrow [CH_3]^+ + [COCH_3]^\bullet \quad \text{or} \quad [CH_3]^\bullet + [COCH_3]^+$$

$$m/e = 15 \qquad\qquad\qquad m/e = 43$$

It seems that the fragmentation pattern for propanone accounts well for peaks at $m/e = 15$ and 43 in mass spectrum **B**. To affirm our conclusions above, we can seek to justify that it is not possible for propanone to give peaks at $m/e = 29$ and 57. Looking at the bonds that can be cleaved in a molecule of propanone, it is indeed not possible to get a peak at $m/e = 29$. But it is still possible to produce the fragment ion $[CH_3COCH_2]^+$ that gives a peak at $m/e = 57$.

Doing likewise with the structure of propanal, it is possible to obtain the $[CH_3]^+$ that accounts for the peak at $m/e = 15$ and the fragment $[CH_2CHO]^+$ can give a peak at $m/e = 43$.

At this point in time, the only lead we have is the peak at $m/e = 29$, which is not attainable from the fragmentation pattern of propanone. Based on this, we conclude that mass spectrum **A** belongs to propanal and mass spectrum **B** belongs to propanone.

Q: Since the relative abundances of the $[CH_2CHO]^+$ $(m/e = 43)$ and $[CH_3COCH_2]^+$ $(m/e = 57)$ species are high in the spectrometer, this shows that these two species are relatively more stable in the spectrometer. Is there any reason for it?

A: Yes indeed both species are relatively more stable in the spectrometer as both the $[CH_2CHO]^+$ and $[CH_3COCH_2]^+$ are resonance stabilized, as shown below:

The electron deficiency is delocalized to the more electron-rich oxygen atom, or more "spread out." In addition, if you look at the mass spectrum for propanone, you will notice that at $m/e = 43$, which corresponds to the $[CH_3CO]^+$ species, the relative intensity is also very high due to resonance stabilization:

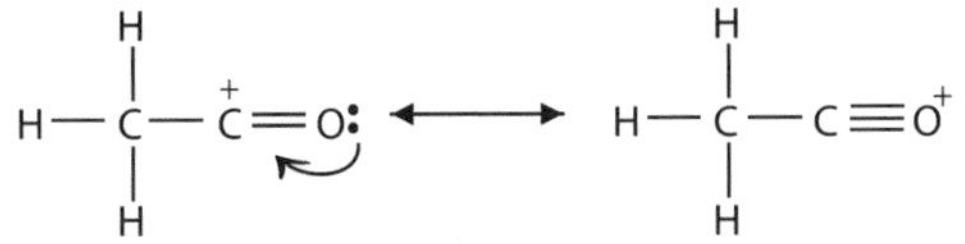

Q: What if we are stuck in a sticky situation of having to distinguish between the following isomers $(CH_3)_2CHCHO$ and CH_3CH_2 CH_2CHO, and it boils down to the following given information?

	Fragment ion from methylpropanal	Fragment ion from butanal
$m/e = 29$	$[CHO]^+$	$[CH_3CH_2]^+$

A: In this kind of situation, we have to rely on the spectrometer's capabilities. There are high resolution mass spectrometers that can measure m/e ratios to a whopping accuracy of five significant figures, or 1 part in 100,000! This means that we can use the accurate relative atomic mass to calculate the mass of the ions, and these values are reflected in the m/e values on the mass spectrum:

Element	Accurate A_r
H	1.0078
C	12.000
O	15.995

Mass of $[CHO]^+ = 12.000 + 1.0078 + 15.995 = 29.0028$.

Mass of $[CH_3CH_2]^+ = 2(12.000) + 5(1.0078) = 29.039$.

The difference in the masses is 36 parts in about 29,000, or about 0.1%, which is within the resolution capabilities of such spectrometers. Thus, if the mass spectrum indicates a peak at $m/e = 29.0390$, we know that the $[CH_3CH_2]^+$ ion is responsible for it and not $[CHO]^+$.

The use of high-resolution mass spectrometry hence allows the accurate determination of M_r and molecular formula.

14.8 Deducing the Structure of an Unknown Compound from Mass Spectrum Analysis

It is now time to bring together what we have learnt to determine the structure of an unknown organic compound based on its mass spectrum. Remember the following useful pointers when doing mass spectrum analysis:

- Measure the relative intensities of the M and (M + 1) peaks to calculate the number of carbon atoms in the molecule using the formula $n = \frac{100}{1.1} \times \frac{I_{(M+1)}}{I_M}$.
- Measure the relative intensities of the M, (M + 2) and (M + 4) peaks, if any, to determine the presence of chlorine and bromine.
- Identify the fragment ions that account for the peaks observed, and these collectively give the structure of the parent molecule.

Table 14.2 lists some m/e values commonly associated with the fragment ions stated. Likewise, the absence of peaks at some of these m/e

Table 14.2

Mass	Possible fragment ion	Possible fragment lost
15	$[CH_3]^+$	CH_3
17	$[OH]^+$	OH
18	$[H_2O]^+$	H_2O
28	$[C_2H_4]^+$, $[CO]^+$	C_2H_4, CO
29	$[C_2H_5]^+$, $[CHO]^+$	C_2H_5, CHO
31	$[CH_3O]^+$	CH_3O
42	$[C_3H_6]^+$, $[CH_2CO]^+$	C_3H_6, CH_2CO
43	$[C_3H_7]^+$, $[CH_3CO]^+$	C_3H_7, CH_3CO
44	$[C_3H_8]^+$, $[CONH_2]^+$,	C_3H_8, $CONH_2$
45	$[C_2H_5O]^+$, $[COOH]^+$	C_2H_5O, $COOH$
57	$[C_4H_9]^+$, $[CH_3CH_2CO]^+$	C_4H_9, CH_3CH_2CO
77	$[C_6H_5]^+$	C_6H_5
91	$[C_6H_5CH_2]^+$	$C_6H_5CH_2$
105	$[C_6H_5CO]^+$	C_6H_5CO

values indicates the common uncharged fragments that are lost from the parent (M) ion. For example, $(M - 18)$ indicates the loss of a H_2O moiety and $(M - 45)$, the loss of a –COOH group.

Example: Deduce the structure of Compound **A** given its molecular formula $C_nH_6O_2$.

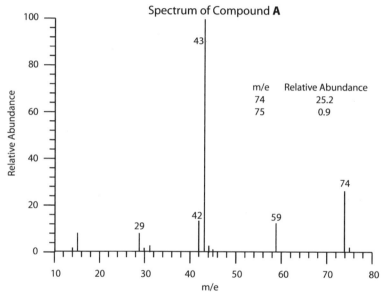

- The largest m/e ratio is at 74. This gives the M_r of Compound **A**.
- With the molecular ion peak at $m/e = 75$, $n = \frac{100}{1.1} \times \frac{0.9}{25.2} = 3$. There are a total of three carbon atoms in Compound **A**.
- With reference to Table 14.2, the list of possible fragment ions that account for the peaks with relatively strong signals is as follows:

	Possible fragment ion
$m/e = 15$	$[CH_3]^+$
$m/e = 29$	$[C_2H_5]^+$, $[CHO]^+$
$m/e = 42$	$[C_3H_6]^+$, $[CH_2CO]^+$
$m/e = 43$	$[C_3H_7]^+$, $[CH_3CO]^+$
$m/e = 59$	
$m/e = 74$	$[C_3H_6O_2]^+$

	Possible fragment lost
$(M - 59) = 15$	CH_3
$(M - 43) = 31$	CH_3O
$(M - 42) = 32$	
$(M - 29) = 45$	C_2H_5O, $COOH$
$(M - 15) = 59$	

Based on the oxygen-containing fragment ions and the fact that Compound **A** contains two oxygen atoms, the most probable fragments include the CH_3CO- group and the CH_3O- group. Starting with these fragments, we can obtain a possible structure, CH_3COOCH_3, that satisfies the number of atoms of each element.

To confirm if the structure is a feasible structure, we will try to cleave the bonds in it to account for the peaks seen.

$$\left[\begin{array}{c} H \; O \;\;\;\; H \\ H-C-C-O-C-H \\ H \;\;\;\;\;\;\;\; H \end{array} \right]^{+\bullet} \longrightarrow [CH_3]^+ + [COOCH_3]^\bullet \quad \text{or} \quad [CH_3]^\bullet + [COOCH_3]^+$$

$$m/e = 15 \hspace{4cm} m/e = 59$$

$$\left[\begin{array}{c} H \; O \;\;\;\; H \\ H-C-C-O-C-H \\ H \;\;\;\;\;\;\;\; H \end{array} \right]^{+\bullet} \longrightarrow [CH_3CO]^+ + [OCH_3]^\bullet \quad \text{or} \quad [CH_3CO]^\bullet + [OCH_3]^+$$

$$m/e = 43 \hspace{4cm} m/e = 31$$

The peak at $m/e = 42$ is due to the $[CH_2CO]^+$ ion produced as follows:

$$[CH_3CO]^+ \rightarrow [CH_2CO]^+ + H^\bullet.$$

As for the peak at $m/e = 29$, it is due to the $[O–C–H]^+$ fragment ion formed from the cleavage of two C–H bonds in the $-OCH_3$ group.

CHAPTER 15

Ultra-Violet and Visible Spectroscopy

15.1 Introduction

The covalent bond is the resultant of the net electrostatic attraction between the shared electrons and the two positively charged nuclei, after taking into consideration the presence of inter-electronic and inter-nuclei repulsion. According to valence bond (VB) theory, a single covalent bond is formed when two electrons with their spins paired are shared by the overlap of two valence atomic orbitals, one from each bonding atom.

As shown in Fig. 15.1, there are two ways that the orbitals can overlap with each other: The head-on overlap of orbitals forms the sigma bond; and the side-on overlap of orbitals forms the pi bond.

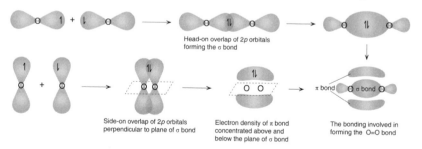

Head-on overlap of 2p orbitals
forming the σ bond

Side-on overlap of 2p orbitals
perpendicular to plane of σ bond

Electron density of π bond
concentrated above and
below the plane of σ bond

The bonding involved in
forming the O=O bond

Fig. 15.1. Covalent bond formation through orbitals overlap.

In Chapter 2, we saw that the idea of the overlapping of valence atomic orbitals is inadequate to account for experimental data such as bond lengths and bond angles. To mediate this, hybridization theory was formulated to explain covalent bond formation through the overlap of hybridized valence orbitals, instead of "normal" atomic orbitals. In this view, hybridization theory does not seek to rival VB theory, but actually complements the latter — the fundamental idea is still the overlap of valence orbital on each bonding atom.

Yet there are still shortcomings with these models. Take, for instance, the diatomic oxygen molecule. The O atoms are doubly bonded to each other and, accordingly, there is no unpaired electron on any of these O atoms, i.e. O_2 is seen as a diamagnetic compound unaffected by magnetic fields:

Q: What is a diamagnetic compound?
A: A diamagnetic compound refers to a compound that is repelled by the applied magnetic field when the compound is placed within the field. This would mean that when the compound is placed in an external magnetic field, there would be a magnetic field created within the compound, and this created magnetic field repels the external applied field. Diamagnetism is a characteristic of elements and compounds in which all the electrons are paired.

However, experimental data indicates that O_2 is paramagnetic, which means there must be at least one unpaired electron somewhere in the molecule. The only way to get an unpaired electron is to assume that there is only a single bond between the O atoms:

But this structure is still not a good fit to experimental data, which points out that it cannot be an O–O single bond. This then brings us to another modern theory of covalent bonding based on the principles of wave mechanics, i.e. molecular orbital (MO) theory.

Q: What is paramagnetism?
A: Paramagnetism refers to the phenomenon whereby a compound when placed in an external magnetic field is attracted by the applied magnetic field, i.e. with a magnetic field aligning with the

applied field. It is the opposite of diamagnetism. Paramagnetism usually arises because of the presence of unpaired electrons in the atom or molecule. When there is/are unpaired electron/s, the particle has a permanent magnetic moment, and all these magnetic moments are randomly arranged without the substance having a net magnetic force because of thermal motion. But when an external magnetic field is applied, these magnetic moments are arranged in such a way that the substance has a magnetic force that is aligned with the external applied magnetic field. An aligned magnetic force with the external magnetic field is more stable than an opposing one. If the electrons in an atom or molecule are all paired up, then there would be no net magnetic moment as the spins of a pair of electrons are opposite.

Q: Is paramagnetism same as ferromagnetism?

A: No. A paramagnetic substance does not retain its magnetism after the removal of the external applied magnetic field, but a ferromagnetic substance does.

Q: What is wave mechanics?

A: In wave mechanics, subatomic particles, such as electrons, are perceived to have wave characteristics. This does not mean that the electron moves in a "wavy" manner. In fact, it is the probability of locating the electron at a specific point in space that follows the wave mathematical function, which is sinusoidal in nature. That is, we can use a sine function to describe the probabilistic distribution of the electron in space with respect to time. Thus, because of the probabilistic nature of the distribution plus the high speed of the electron, what we would perceive externally is an electron cloud and not a stationary electron. An orbital is a region of this electron cloud where the total probability of locating an electron is ∼95%. And it is also because of this specific wave function for a particular electron in space that we have different shapes for different types of orbital. For example, an *s*-orbital is spherical, a *p*-orbital is dumb-bell shaped, etc.

Q: What is so good about knowing wave mechanics?

A: By perceiving an electron as having wave probability, we can use the various mathematical tools associated with the manipulation of wave functions to look at covalent bond formation as a result of the interaction of these wave functions.

15.2   Molecular Orbital Theory (MO Theory)

In VB theory, the overlap of two atomic orbitals gives the covalent bond. MO theory takes it further — when the two atomic orbitals combine, two molecular orbitals are formed, and the manner in which bonding electrons are distributed over the molecular orbitals explains covalent bonding. Hence, we do not talk just about the overlapping of atomic orbitals, but the linear combination of these atomic orbitals which gives us the molecular orbitals. Like an atomic orbital, each molecular orbital can only contain two electrons with the opposite spin, i.e. $+1/2$ and $-1/2$. Importantly, the molecular orbital that "houses" the two electrons spreads throughout the whole molecule, but with concentration of electron density within the inter-nuclei region, i.e. the probability of locating an electron within the inter-nuclei region is $\sim 95\%$.

The linear combination of atomic orbitals (LCAO) is essentially a mathematical method that employs wave mechanics to describe the molecular orbitals. In considering the wave nature of electrons, there is both constructive and destructive interference when these waves combine. In the former case, the intensities of the waves are added, while in the latter, the intensities are subtracted or cancelled. Thus, we can simply perceive a molecular orbital as a mathematical wave function describing the wave-like behavior of an electron in a molecule, whereas an atomic orbital describes that of an electron in an atom. Of course, the wave-like behavior is none other than the probability function that we are talking about.

Constructive Interference   Destructive Interference

Take, for example, the simplest molecule, H_2. When the two $1s$ atomic orbitals overlap, one molecular orbital is formed from the constructive interference of the electron waves, and it is called the bonding MO. The build-up of electron density within the inter-nuclei region draws the two atoms closer together — a feature of strong bonding. Thus, electrons found in a bonding MO stabilize the molecule, resulting in a lower energy level for the MO that is formed from the atomic orbitals. The bonding MO is denoted by the symbol σ_{1s}, in which the subscript represents the atomic orbitals from which the MO is generated.

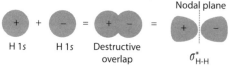

The destructive interference of electron waves reduces the electron density between the two nuclei. This is shown by the lack of electron density at the nodal plane. The MO formed in this way is termed the antibonding MO. It is denoted as σ_{1s}^{*}. Since there is a lower probability of finding electrons in the inter-nuclei region, there is greater repulsion between the positively charged nuclei, which causes them to be further apart. This weakens the bond, and thus electrons found in antibonding MOs tend to destablize the molecule. As a result of the increased inter-nuclei repulsion, the energy level of an antibonding MO is higher than the atomic orbitals that formed the antibonding MO.

Q: What is a nodal plane?

A: If you perceive an electron as having a wave function, then when two waves interfere destructively, the region with zero amplitude, which translates to zero probability to locate the electrons, is known as a node. So two-dimensionally, it is called a nodal plane.

Nodal plane

H 1s   H 1s   Destructive overlap $\sigma_{H\text{-}H}^{*}$

Q: If there is a nodal plane in between the two lobes of the molecular orbital, how can an electron move from one lobe to the other? In the process of moving through the plane, won't the probability of finding the electron be non-zero?

A: The electron moves from one lobe to another through this phenomenon known as quantum mechanical tunneling. In classical physics, when a particle moves, the position and velocity of the particle are both known at the same time. But for sub-atomic particles, it is not possible to know the position and velocity of the particle at the same time, i.e. either you know the position, but with the velocity unknown, or vice versa. This is the basis of the famous theory known as the Heisenberg uncertainty principle. Based on this theory, we can visualize that if we try to see whether the electron is at a particular point in space, i.e. the x-value along the x-axis, then the unknown velocity would create a region of "fuzziness," such that there is no way to know the exact position, i.e. what we would get would be Δx and not an exact value for x. Thus, at the so-called nodal plane, we actually do not know specifically whether the electron is there or not when we try to observe it. In fact, this is the case for every point in space.

Q: Other than the bonding and anti-bonding molecular orbitals, is there any other type of molecular orbital?

A: There is another type, which is known as the non-bonding molecular orbital. The pair of electrons in this non-bonding molecular orbital is equivalent to the lone pair of electrons in the Lewis structure. Molecular orbitals are formed from the linear combination of atomic orbitals and the non-bonding molecular orbitals arise from atomic orbitals that have no other atomic orbitals to combine with it. As a result, the electrons in the non-bonding molecular orbitals mostly reside on the atom which provides the atomic orbitals. This would also mean that the energy level of the non-bonding molecular orbital is the same as that of the atomic orbital from which it has originated. Thus, electrons in the non-bonding molecular orbital do not affect the strength of the covalent bond that forms.

Based on the relative stabilizing effect, the energy level of the bonding MO is lower than that of the antibonding MO, as shown in Fig. 15.2. Similarly to the atomic orbitals in an atom, the energy levels of the MOs are also quantized, and these are filled with electrons in the same way as atomic orbitals are filled. Electrons are first placed in the lowest energy MO (Aufbau principle) and each orbital can contain a maximum of two electrons of opposite spins (Pauli exclusion principle). When filling a set of degenerate orbitals, an electron goes to an empty orbital first, before completely filling up any of these similar energy orbitals (Hund's rule of multiplicity).

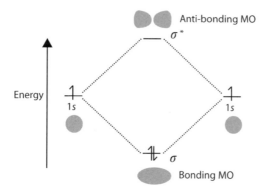

Fig. 15.2.   Energy levels of different molecular orbitals.

For the oxygen molecule, we have to consider not only the σ molecular orbitals but also the π molecular orbitals. Focusing on the valence shell, the overlap of two $2s$ orbital contributes to two MOs, namely the σ_{2s} and σ_{2s}^* MOs. These MOs have the same shape as the σ_{1s} and σ_{1s}^* MOs respectively. The head-on overlap of $2p$ orbitals also produces two MOs, i.e. considering the head-on overlap of the $2p_x$ orbitals, we have the σ_{2p_x} and $\sigma_{2p_x}^*$ MOs.

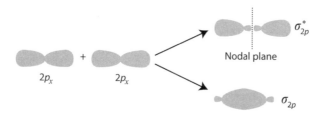

Q: Is the $2p_x$ orbital always the one involved in σ bond formation? Can it be the $2p_z$ or $2p_y$ orbitals instead?

A: No, it is not necessarily the $2p_x$ orbital. In fact, we do not know which of the p-orbitals is involved in σ bond formation. We have arbitrarily used the $2p_x$ orbitals here. It does not matter which of the p-orbitals you use to form σ bond; the outcome is still the same.

The side-on overlaps of the remaining pairs of $2p_y$ and $2p_z$ orbitals give the π molecular orbitals; one pair of bonding and antibonding

MOs for each set of atomic orbital overlap. These π MOs are perpendicular to the x-axis:

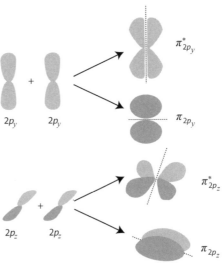

Q: Why are we not considering a MO that is formed between the $2s$ orbital of one oxygen atom with, let's say, the $2p_x$ of the other oxygen atom?

A: You can try considering a MO that is formed between the $2s$ orbital of one oxygen atom with the $2p_x$ of the other oxygen atom. But you would find that this MO does not exist in a real oxygen molecule. Why? This is because when you place two oxygen atoms side by side, the only way is to formulate a model where by the two $2s$ orbitals overlap due to their similarities in energy and symmetry. Take note that a $2s$ orbital has lower energy than a $2p$ orbital. Thus, the model of explanation that you formulate must be able to account for the experimental observation that you have made. Take, for instance, if you try to invoke the sp^3 or sp hybridization model to explain the shape of the carbon atom in ethene — you would find that it does not fit what you have observed experimentally. Only by using sp^2 hybridization would it work.

Q: Is it possible to have a MO formed from the overlap of different types of atomic orbitals such as the $2p_x$ and $2p_y$ orbitals?

A: No! You can't. This is because the distributions of the electron density for these two atomic orbitals are perpendicular to each other. Mathematically, we say that these two wave functions are *orthogonal* to each other. Or, physically, there is no way for these two orbitals to overlap as the electron densities are perpendicularly distributed in space.

Summing up, the MO diagram for the oxygen molecule looks like this:

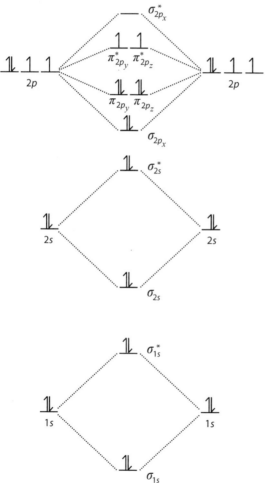

Actually, the amount of energy decreased during the formation of a bonding MO is slightly greater than the amount of energy increased for the formation of the anti-bonding MO. This is because each MO that forms interacts with other MOs to a different extent. In addition, the attractive force that the nuclei have on each of the MOs is different, thus resulting in different changes in energy. Such a phenomenon is similar to the "compression" of energies of the atomic orbitals in an atom as the atomic number increases. For example, the $1s$ orbital of a beryllium atom has a higher energy than the $1s$ orbital of magnesium.

The σ_{2p_x} MO is lower in energy than the π_{2p_y} and π_{2p_z} MOs. This is expected since the head-on overlap of atomic orbitals is more effective in accumulating a higher amount of electron density in the internuclei region than the side-on overlap of orbitals. As a result of the greater stabilization of the σ_{2p_x} MO than the π_{2p_y} and π_{2p_z} MOs, there is also greater destabilization of the $\sigma^*_{2p_x}$ MO than the $\pi^*_{2p_y}$ and $\pi^*_{2p_z}$ MOs (from the law of conservation of energy, the amount of energy of stabilization is equivalent to the amount of energy of destabilization). Hence, the $\sigma^*_{2p_x}$ MO is of a higher energy level than the $\pi^*_{2p_y}$ and $\pi^*_{2p_z}$ MOs.

When we apply the rules in filling the MOs with the 16 electrons contributed by the two O atoms, we end up with two π^* orbitals each containing a single electron — this feature helps to account for the paramagnetic behavior of the oxygen molecule. In accordance with Hund's rule of multiplicity, the most stable form of the oxygen molecule is paramagnetic (which agrees with experimental data) and not diamagnetic, as suggested by VB theory. In addition, it is good to know that the highest molecular orbital that contains electrons is known as a HOMO (Highest Occupied Molecular Orbital), whereas LUMO stands for Lowest Unoccupied Molecular Orbital. In the case of oxygen, the HOMOs are the $\pi^*_{2p_y}$ and $\pi^*_{2p_z}$ MOs whereas the LUMO is the $\sigma^*_{2p_x}$ MO. It is important to know these two terms for electronic transition, as an electron would usually transit from a HOMO to a LUMO which has the smallest energy difference. We will discuss this in more detail in Section 15.3.

The other experimental data we have to deal with is the O=O bond in the oxygen molecule. Under MO theory, the bonding pattern is explained using the concept of bond order, which is defined as the number of electron pairs (or chemical bonds) between two atoms. Its formula is given as:

Bond order
$$= \frac{\text{number of bonding electrons} - \text{number of anti bonding electrons}}{2}$$

A bond order of one translates to a single bond between the bonding atoms. A bond order of two indicates a double bond, a bond order of three indicates a triple bond, and so on. Referring to the MO diagram of the oxygen molecule, the bond order of O_2 is two, which corresponds to an O=O double bond:

$$\text{Bond order of } O_2 = \frac{10 - 6}{2} = 2$$

A greater bond order corresponds to a stronger bond, since there is greater accumulation of electron density within the inter-nuclei region. A bond order of zero means no bond is present between the atoms, and such a molecule does not exist. Take Ne for instance; if two Ne atoms were bonded with each other, the MO diagram would look like this:

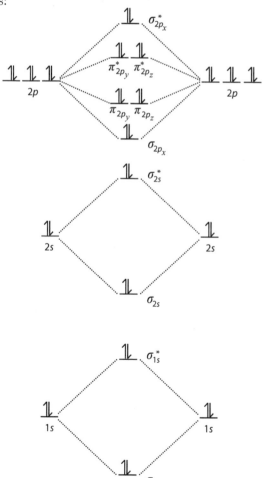

Since there is an equal number of bonding and anti-bonding electrons, the bond order of Ne_2 is zero — Ne_2 does not exist, which is consistent with experimental data. You may say that the use of Lewis structures based on VB theory can give us the same conclusions on the bonding pattern for simple molecules such as H_2 and N_2 without much fuss in calculations. Indeed, even though VB theory has flaws that seem to be taken care of by MO theory, it is still widely used because of its simplicity of usage. Both VB and MO theories complement each other; what is lacking in one theory is made up for by the use of the other.

With that said, one serious flaw of VB theory is in its assumption that the covalent bond is a localized bond. This problem arises when attempting to explain the bonding pattern of certain species and trying to match that with the experimental data. Take the example of benzene — VB theory addresses the bonding involved by picturing it to consist of two equivalent resonance structures:

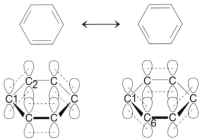

But in such a model suggested by VB theory, the idea still centers on localized bonds. For instance, a π bond is formed between C1 and C2 in one structure, and it can also be formed between C1 and C6. But the latter results in another Lewis structure.

MO theory takes a different view by stating that bonding electrons belong to the molecule as a whole, and they are not just restricted to residing in an atom or between bonding atoms. This view, when applied to benzene, elegantly explains why there is only one observed type of carbon–carbon bond length. In addition, the bond order is 1.5, that is, between that of a single and a double bond. The benzene molecule is perceived by using a single depiction of its electronic structure — the simultaneous overlap of six p-orbitals in creating an extensive π molecular orbital, which has the six electrons delocalized over the six C nuclei:

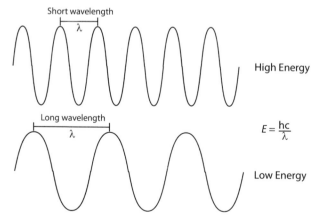

15.3 Basic Principles of Spectroscopy

Spectroscopy is a subject area that deals with the interaction between electromagnetic radiation and matter. Electromagnetic (EM) radiation is made up of packets of energy called photons that have neither mass nor charge. It exhibits both the properties of waves and particles, but here in spectroscopy, we shall focus on its wave-like nature. There are many forms of EM radiation, and these can be described by the frequency (f), wavelength (λ) or energy (E) of the photons. The relationships among these quantities are shown by Planck's equation:

$$E = hf = \frac{hc}{\lambda},$$

where h = Planck constant = 6.63×10^{-34} J s, f = frequency (s^{-1}) and c = speed of light = 3.00×10^8 m s^{-1}.

In order of increasing frequency, we have radio waves, microwaves, infrared radiation, visible light, ultra-violet radiation, X-rays and gamma rays (see Fig. 15.3). Visible light is but a small portion of the entire EM spectrum that we can see, and it is composed of the colors, in order of increasing wavelength, violet, blue, green, yellow, orange and red (or decreasing frequency). Notice that there are overlapping frequencies that correspond to more than one type of EM radiation.

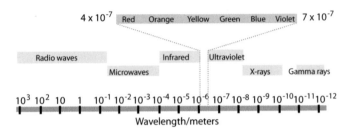

Fig. 15.3.   The electromagnetic spectrum.

How does matter interact with EM radiation? First, we must make a distinction between two general kinds of spectra — the continuous spectrum and the discrete spectrum. Visible light is an example of a continuous spectrum in which radiation is distributed over a continuous range of frequencies and not just over a few defined frequency ranges. Solids, liquids and dense gases emit a continuous spectrum, and this includes the ordinary incandescent light bulb and flame from a burning candle.

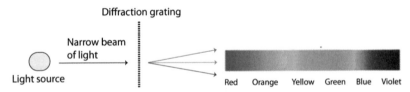

The basic experimental set-up includes focusing the radiation that is emitted from the sample into a narrow beam which is then split into its component wavelengths by a prism or diffraction grating. Subsequently, the emitted radiation is captured on a photographic plate or detector for analysis.

If an electrical discharge is passed through a low-density hot gas, we do not get a continuous spectrum but a line spectrum:

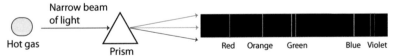

Such a spectrum with specific lines or bands that correspond to a certain narrow range of frequency is called a discrete spectrum. Those lines shown in the emission spectrum correspond to frequencies of radiation that are given out by the gas. Notice that the wavelengths of the emission lines converge at the shorter wavelength or higher frequency range? It is through this observation that we can conclude that the energy levels in an atom converge too. The consequence of this is that the rate of increase in atomic size is not as great for higher period number than for lower period.

What causes the formation of a discrete spectrum? When electrical or thermal energy is supplied to the gaseous sample, electrons in each atom take in some of this energy, and they either move faster or they overcome the attractive force of the nucleus and move to a higher energy level, which is further away from nucleus. Since electrons are found in orbitals of discrete energy (or we say that the energy level of the orbital is quantized), radiation corresponding to the difference in the energy levels (ΔE) is needed for the promotion of an electron to a higher energy state (a process known as excitation). This energy difference is of a fixed value, or is quantized!

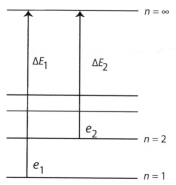

The excited state is not a stable state. In time, the electrons in the excited state will revert back to the stable ground state as they fall back to the lower energy level. When the latter happens, radiation corresponding to ΔE is given out and this appears in the emission spectrum as discrete lines of various frequencies.

Take, for example, the element hydrogen. Its atom has only one electron, which occupies the lowest energy $1s$ orbital in the $n = 1$ principal quantum shell. Excitation of this electron allows it to move up to one of the higher energy levels. For a sample of hydrogen gas, the emission spectrum displays lines associated with the energy given out when an electron falls back to a lower energy level but not necessary back to the $n = 1$ principal quantum shell. Hence, the lines observed in the spectrum are grouped into spectral series based on the lowest energy level that the electron falls back to. One of these spectral series is called the Balmer series, named after its discoverer, that involves electronic transitions from $n = \infty$ to the $n = 2$ principal quantum shell. Figure 15.4 shows the spectral lines in the Balmer series that correspond to frequencies within the visible region of the EM spectrum. Notice that transitions associated with a greater ΔE emit radiation of shorter wavelength or higher frequency.

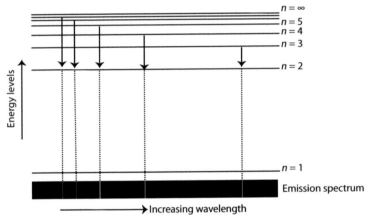

Fig. 15.4.   Electronic transition for the Balmer series.

Q: If the discrete emission spectrum that is obtained corresponds to the release of energy when an electron falls from a higher energy level to a lower one, does it mean that if I pass a beam of radiation through the hydrogen gas which includes the wavelengths corresponding to the emission lines, the very same wavelengths of radiation would be absorbed by the hydrogen gas?

A: Yes, indeed. The hydrogen gas would absorb the wavelengths of radiation that correspond to the promotion of electron from the lower energy level, which is discrete, to the higher one. Such a spectrum is known as an absorption spectrum. Both the absorption and emission spectra are unique to the element per se, and these spectra are very informative for the elucidation of the presence of certain elements, for example in the sun.

Q: When an electron absorbs a photon of radiation and gets promoted from the $n = 1$ to $n = 3$ principal quantum shell, can the electron "land" itself in any of the subshells in the $n = 3$ principal quantum shell?

A: No, it can't. In quantum mechanics, a photon of EM radiation has an angular momentum value. In electronic transition, the total angular momentum must be conserved after transition. Thus, $s \rightarrow s$ or $p \rightarrow p$ or $d \rightarrow d$ types of transition are forbidden due to the lack of change of angular momentum (both s-orbitals have same angular momentum value, etc.) after the electron has absorbed the photon of energy which itself carries an angular momentum. Similarly, for emission transition, the electron must transit from one orbital to another such that there is a change of angular momentum.

Apart from studying emission spectra, we can also look into the absorption spectrum which shows the frequencies of radiation that are absorbed when electrons are excited. The absorption spectrum of hydrogen (see Fig. 15.5) displays black lines that correspond to the radiation absorbed. Notice that the absorption lines have the same frequencies as the emission lines which indicate the transition of electrons between specific energy levels.

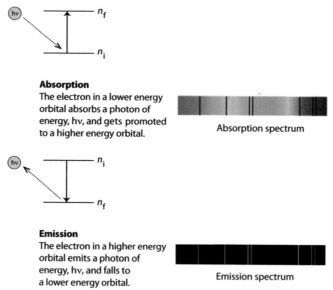

Absorption
The electron in a lower energy orbital absorbs a photon of energy, hv, and gets promoted to a higher energy orbital.

Absorption spectrum

Emission
The electron in a higher energy orbital emits a photon of energy, hv, and falls to a lower energy orbital.

Emission spectrum

Fig. 15.5. Emission and absorption spectra for hydrogen gas.

Different elements and molecules produce unique absorption and emission spectra, which means that we can use this reference data in "fingerprint matching" to identify the chemical composition of unknown substances.

Q: Why do different elements produce different emission spectra?

A: This is because the energy levels of the electronic shells of two different elements are not the same due to the differences in the number of electrons and the nuclear charges. For instance, the energy level of the $2s$ orbital of carbon atom is lower than that for boron because of the higher nuclear charge.

Spectroscopy is therefore used to investigate the structure of compounds by examining either the absorption or emission spectra obtained when EM radiation is applied to the sample. Take note that the emission spectrum is a complement to the absorption spectrum. The type of interaction that was discussed previously pertains to the transition of electrons when they are excited. But depending on the type of EM radiation, different atomic or molecular behavior can be induced in the same atom or molecule other than electronic transitions, which we will be discussing it below.

15.4 Types of Spectroscopic Methods

The energy-induced behavior of a molecule that has absorbed a photon of EM radiation can be one of the following:

- translation — the three-dimensional movement of the molecule as a whole.
- rotation — the rotation of the molecule as a whole about an axis.
- vibration — the periodic motion of atoms in a molecule that arises from the bending or stretching of bonds.
- electronic transition — the movement of electrons between energy levels.
- nuclei transition — the rotation or spinning of the nuclei.

All these forms of transition require energy that is quantized, and each is associated with a particular region of the EM spectrum, as shown in Fig. 15.6.

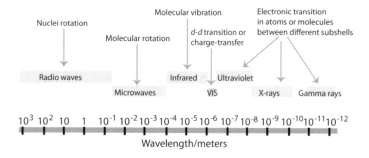

Fig. 15.6. Different types of electronic and molecular transitions.

In this book, we will discuss the following spectroscopy techniques in the order given:

- Ultra-violet (UV)/Visible spectroscopy
- Infra-red (IR) spectroscopy
- Nuclear magnetic resonance (NMR) spectroscopy

Each of the techniques analyzes a specific behavior of molecules. UV spectroscopy deals with electronic transitions, IR spectroscopy is all about the molecular vibrations and rotations, and NMR spectroscopy studies the transition of nuclei.

15.4.1 *Ultraviolet–Visible (UV-Vis) Spectroscopy*

In Section 15.3, we discussed electronic transitions, which are the focal point of UV-Vis spectroscopy. It measures the intensity and wavelength of light in the near-UV and visible region that is being absorbed by a sample. The amount of energy in this region of the EM spectrum is sufficient to excite outer valence electrons to higher energy levels but not sufficient to promote inner core electrons (X-rays correspond to transition of inner core electrons). However, note that it is not the whole molecule that will absorb EM radiation in this region. The part of the molecule that is responsible for the absorption of the EM radiation is called a chromophore.

Q: What is a chromophore?

A: A chromophore is the part of a molecule that absorbs visible EM waves because the energy difference between two different molecular orbitals falls within the visible range of the EM spectrum.

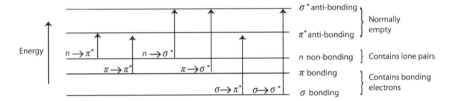

Fig. 15.7. Electron transition between different types of MOs.

Figure 15.7 shows the possible electronic transitions that can occur between the three different types of MOs — the bonding MOs, the anti-bonding MOs and the non-bonding MOs, that contain lone pair of electrons. Of these six electronic transitions, only the $n \rightarrow \pi^*$ and $\pi \rightarrow \pi^*$ transitions can occur with the absorption of light in the near-UV and visible region of the spectrum (200–800 nm). Other types of electronic transitions would need more energetic EM radiation, which would fall out of the UV-Vis range. This means that the chromophore would normally consist of an unsaturated group that contains π-electrons or has atoms with lone pairs of electrons. The excitation of the electron is from the highest occupied molecular orbital (HOMO) to the lowest unoccupied molecular orbital (LUMO) since the energy gap between these MOs is the smallest (Table 15.1).

Table 15.1

Chromophore	Example	Electronic Transition	$\Lambda_{\max}$/nm
C=C	Ethene	$n \rightarrow \pi^*$	171
C≡C	1-Hexyne	$\pi \rightarrow \pi^*$	180
C=O	Ethanal	$n \rightarrow \pi^*$	290
		$\pi \rightarrow \pi^*$	180
N=O	Nitromethane	$n \rightarrow \pi^*$	275
		$\pi \rightarrow \pi^*$	200
C–X (X=Br, I)	Bromomethane	$n \rightarrow \sigma^*$	205
	Iodomethane	$n \rightarrow \sigma^*$	255

Chromophores with a conjugated pi system that consists of a series of alternating single and double bonds show absorbance at longer wavelengths, i.e. at the lower energy end of the spectrum. The extended pi network of orbitals reduces the energy gap between the π and the π^* MOs. The more conjugated a system is, the lower the amount of energy that is required to excite an electron, and absorption would thus be seen at longer wavelengths.

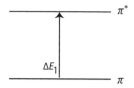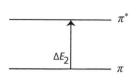

Increases in pi conjugation
decreases the energy gap
between the $\pi \rightarrow \pi^*$ transition

Q: Why does an increase in pi conjugation lead to a decrease in the energy difference between two MOs?

A: Two atomic orbitals linearly combine to produce two MOs, three atomic orbitals produce three MOs, and n atomic orbitals produce n MOs. As the number of atomic orbitals increases, the number of MOs formed also increases. This would result in a smaller energy gap between the various MOs. For example:

	$\pi \rightarrow \pi^*$ energy gap/kJ mol^{-1}	λ/nm
R-CH=CH-R	690	190
RCH=CH-CH=CH-R	520	230
R-CH=CH-CH=CH-CH=CH-R	235	275

Physically, the greater the number of atomic orbitals that make an MO, the greater the dispersal of the negatively charged electrons over a greater surface area when it occupies the MO, and hence the lower the energy of the electron due to a decrease in inter-electronic repulsion. Remember the Golden Rule: Accumulation of charge leads to an increase in energy while dispersal of charge decreases in energy.

In addition, if an atom with a lone pair of electrons is incorporated into the molecule as compared to one that does not have any lone pair of electrons, we would observe a shift of absorbance to the longer wavelengths of absorption as the $n \rightarrow \pi^*$ energy gap is smaller than the $\pi \rightarrow \pi^*$ gap. For example:

CH=CH Absorption at 296 nm, appears colorless as absorption is out of the Vis region.

N̈=N̈ Absorption at 400 nm, i.e. blue-violet, appears orange in color. Color observed is complementary to the color absorbed.

Although the energy of absorption is discrete (this is because the energy level of each MO is also discrete), we will never get an absorption spectrum that corresponds to only a few specific lines of EM radiation being absorbed. Instead, a complex spectrum would be obtained. This is because other than the pure electronic transition, there is also the superposition of rotational and vibrational transitions onto the electronic transitions, and these give rise to a combination of overlapping spectra lines. These vibration and rotation transitions are themselves discrete too.

Thus, the actual electronic transition spectrum appears as a continuous absorption band rather than discrete lines. Not to mention that sometimes the solvent that is being used to dissolve the compound would also affect the absorption spectrum obtained. A typical UV spectrum corresponding to that for ethenzamide, a common analgesic, is shown below:

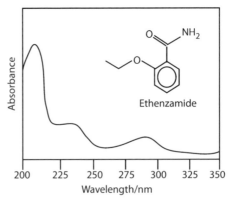

There are three absorption peaks due to three different electronic transitions ($n \rightarrow \pi^*$ or $\pi \rightarrow \pi^*$ or $n \rightarrow \sigma^*$), but it is not possible to allocate a specific transition to each wavelength due to the broadness of the peaks. Thus, if one wanted to use the position of the absorption peaks to determine the identity of a compound, it could be fruitless. And this is being made worse because many different chromophores absorb at nearly about the same wavelength. The main use of UV/Vis spectroscopy is thus more to analyze quantitatively how much of a particular compound is present in a solution. This makes use of Beer's law:

$$A = \log_{10}(I_o/I) = \varepsilon cl$$

where
A = absorbance
I_o = intensity of radiation into the sample
I = intensity of radiation out from the sample
ε = the molar extinction coefficient $(\mathrm{dm}^3\,\mathrm{mol}^{-1}\,\mathrm{cm}^{-1})$
c = concentration of compound (in $\mathrm{mol\,dm}^{-3}$)
l = path length of the absorbing solution (in cm)

Q: What is the significance of ε?
A: ε is an indication of the amount of radiation absorbed per unit concentration and path length when a particular amount of radiation passes through the substance. The higher the molar extinction coefficient, the lower the amount of compound needed to absorb a specific amount of radiation. Compounds with high molar extinction coefficients are excellent materials for blocking out undesired radiation, like in sunglasses.

The concentration of a particular chromophore in solution is proportional to the intensity of the radiation absorbed (A). The molar extinction coefficient reflects how strongly the chromophore is able to absorb the radiation per se, and it is an intrinsic property of the chromophore. To quantitatively determine the concentration of a compound in question, just follow the following procedure:

1. Determine the UV/Vis spectrum of the compound in question, and of other compounds that may be present in the mixture by making a scan through the various wavelengths.
2. Select a significant absorption peak which corresponds only to the compound of interest.
3. Next, prepare a series of standard solutions of the compound in question, to cover the range of concentrations expected.
4. Determine the absorption values for each of these solutions to obtain the calibration plot, which basically is a plot of absorbance versus concentration.

5. Place the sample of unknown concentration in the spectrophotometer and measure the absorbance for this unknown.
6. Determine the concentration of the compound in question from the calibration curve.

If the molar extinction coefficient, ε is known, the concentration can be calculated directly from the absorbance using the Beer's law equation.

CHAPTER 16

Infrared Spectroscopy

16.1 Introduction

Infrared (IR) spectroscopy covers the infrared radiation of the EM spectrum that has a longer wavelength and lower energy than visible light. The energy available is insufficient to cause electronic transitions, but it is enough to cause molecular vibrations and rotations.

Molecules are not motionless entities with the rigid bond lengths and angles that we are used to seeing in books. The atoms in a molecule actually display periodic motions known as vibrations, which arise from the bending or stretching of bonds.

A H_2O molecule can "shake" in three fundamental ways, and these are normally termed vibrational modes (see Fig. 16.1). Just as with the execution of dance moves, "performing" each of these stretching and bending modes requires a different amount of effort, in terms of energy, which IR radiation could supplement.

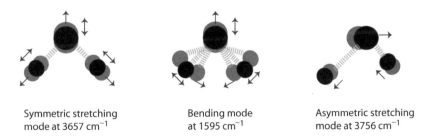

Symmetric stretching
mode at 3657 cm^{-1}

Bending mode
at 1595 cm^{-1}

Asymmetric stretching
mode at 3756 cm^{-1}

Fig. 16.1. Vibrational modes of the H_2O molecule.

Figure 16.1 shows the energy required for each vibrational mode of the H_2O molecule, depicted as a wavenumber (e.g. $3657\,cm^{-1}$). Wavenumber is the reciprocal of wavelength (i.e. $1/\lambda$), and it has the unit cm^{-1}. It is proportional to frequency but not equivalent to it, as both variables have different units. A larger wavenumber corresponds to a shorter wavelength and thus higher energy. Hence, the two stretching modes of H_2O require more energy to be executed than the bending mode.

Q: If a molecule is in constant vibration, how can we have fixed bond lengths and angles?

A: The bond length and bond angle actually indicate average separation of atoms; these are the measurements that we would obtain when we try to determine it. The instrument that we use to determine the bond length would not be sensitive enough to give us the instantaneous bond length at a specific point of time.

Q: Why are stretching modes associated with higher energies than the bending mode?

A: In the bending mode, at least two bonds are being compressed towards each other or are being pulled apart from each other, without changing the bond length. Thus, we have only effectively changed the bond angle during bending mode. With a decrease of the bond angle, inter-electronic repulsion between the pairs of bonding electrons increases. But with an increase in bond angle, although the inter-electronic repulsion between the bonding electrons decreases, the inter-electronic repulsion between the bonding and lone pair electrons (if any) increases. Thus, for bending mode to happen, energy is required. Now, during the stretching mode, two atoms are being pulled apart or pushed towards each other. When we pull two atoms apart, essentially we are trying to break the covalent bond. When we push two atoms towards each other, we increase the inter-electronic repulsion as the bonding electrons now occupy a smaller region of space. At the same time, inter-nuclei repulsion would also increase. Hence, energy needs to be absorbed during the stretching mode. Generally, more energy is associated with the stretching mode than the bending mode because it is more energetically demanding to break a bond.

Q: Does it mean that it is more energetically demanding to bend two double bonds than two single bonds?

A: Certainly. There would be greater inter-electronic repulsion between eight bonding electrons when two double bonds are being bent, compared to the inter-electronic repulsion between four bonding electrons when we try to bend two single bonds.

Q: Does it also mean that it is more energetically demanding to stretch a double bond as compared to a single bond?

A: Yes, you are right. A double bond is stronger than a single bond because the net attractive force the two nuclei have on four bonding electrons is stronger than that on only two bonding electrons. Thus, more energy would be required to separate the two nuclei from each other to cause it to vibrate at the same frequency for a double bond as compared to a single bond. This would be reflected as a higher wavenumber or higher frequency or shorter wavelength recorded.

As for the linear CO_2 molecule, besides the two different modes of stretching, it can either bend along the plane of the paper or perpendicular to the plane, making a total of four possible types of vibrational modes (see Fig. 16.2). On closer inspection, the two bending modes are actually similar motions that only differ in the plane in which they occur, hence they have the same energy, i.e. they are degenerate. Take note that in reality, each atom in a molecule does not have the concept of three-dimensionality, thus to see whether there is a difference in the mode of vibration or not, the key point is to focus on whether there is any difference in the change of their relative position with respect to each other.

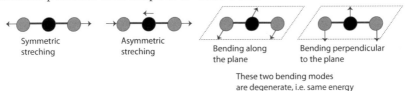

Symmetric streching

Asymmetric streching

Bending along the plane

Bending perpendicular to the plane

These two bending modes are degenerate, i.e. same energy

Fig. 16.2. Vibrational modes of the CO_2 molecule.

Q: Since both H_2O and CO_2 are triatomic, is it possible for H_2O to have two bending modes like CO_2?

A: Unlike CO_2, H_2O is a non-linear molecule. When it lies on the plane of the paper, we can imagine the two O—H bonds being bent towards each other on the plane of the paper. This bending motion results in a decrease in the bond angle. Now, if one were to try bending the molecule by pushing the bonds in the direction perpendicular to the plane of the paper, the resultant motion would not be that of bending, but rather the flipping of the entire molecule. There is no resultant change in the bond angle between the two O—H bonds; it is similar to just swinging a V-shape about the vertex. Hence, there is only one bending mode possible for H_2O.

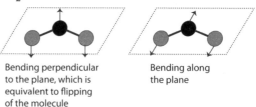

Bending perpendicular
to the plane, which is
equivalent to flipping
of the molecule

Bending along
the plane

The energy levels for the fundamental vibrations of H_2O and CO_2 are shown in Fig. 16.3. Note that there is only one energy level corresponding to the bending mode of CO_2. And although both H_2O and CO_2 molecules have the same types of vibrational modes, these are associated with different energies because they are affected by both the masses of the bonding atoms and bond strengths.

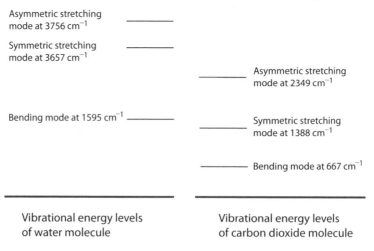

Asymmetric stretching
mode at 3756 cm^{-1}

Symmetric stretching
mode at 3657 cm^{-1}

Asymmetric stretching
mode at 2349 cm^{-1}

Bending mode at 1595 cm^{-1}

Symmetric stretching
mode at 1388 cm^{-1}

Bending mode at 667 cm^{-1}

Vibrational energy levels
of water molecule

Vibrational energy levels
of carbon dioxide molecule

Fig. 16.3.   Vibrational energy levels for H_2O and CO_2.

Q: CO_2 has a higher relative molecular mass than H_2O. The C=O bond energy $(740\,kJ\,mol^{-1})$ is larger than that of O–H $(460\,kJ\,mol^{-1})$. How can we use this information to account for the difference in energy level of the stretching modes for both molecules?

Shouldn't the stronger C=O double bond cause the stretching modes of CO_2 to be more energetically demanding than for H_2O?

A: The energy required for vibration is dependent on the strength of the bond as well as on the masses of the atoms. And yes, if one simply compares the bond strength, no doubt the stretching mode for CO_2 is more energetically demanding than H_2O, for reasons we have discussed previously. But if we look at the molecular structures of both CO_2 and H_2O, we would see that the peripheral atoms in CO_2, which are two oxygen atoms, are heavier than the two hydrogen atoms in H_2O. Imagine you have two balls of different masses, each hanging at the end of an elastic spring of the same force constant. More energy is needed to bring about the same amount of displacement for the lighter ball (imagine this is the hydrogen atom) than for the heavier ball (imagine this is the heavier oxygen atom). And when the force is released, the lighter ball would oscillate at a higher frequency than the heavier ball. Thus, if we were to bring the lighter hydrogen atom to an infinite amount of displacement (this implies that the bond has broken) as to be the same as the oxygen atom, then much more energy is required for the lighter hydrogen atom. In the case of the stretching mode for a H_2O molecule, it may be equivalent to the breaking of the O–H bond. Whereas for the stretching mode of CO_2, it would simply be displacement, without the breaking of the C=O bond. Hence, what we see here is that both the bond strength and the mass of the atoms act concertedly to influence the amount of energy required for stretching to happen in two different molecules. And in this comparison, the mass of the atom plays a more deterministic role than bond strength.

Q: Why is more energy required to displace a lower mass than a greater one?

A: Imagine there is no mass attached to a spring — can there be a displacement effected? The answer is no. Hence, as one moves asymptotically towards zero mass, a greater amount of force, or more energy, would be required to bring about the same amount of displacement for a lower mass particle.

Q: How do you account for why the energy for the bending mode of CO_2 is lower than that for H_2O molecule? Isn't it more difficult to bend two double bonds towards each other?

A: Yes, it is indeed more difficult to bend two double bonds towards each other, as there would be greater inter-electronic repulsion among the eight bonding electrons in CO_2 as compared to the four bonding electrons in the bending mode of H_2O. So the fact that experimental value indicates a less energetic bending mode for CO_2 than H_2O must mean that the inter-electronic repulsion between greater numbers of bonding electrons is not the deterministic explanatory factor here. Why is it so? If you look at the CO_2 molecule, the two double bonds are 180° apart, as compared to the approximate bond angle of 109.5° for H_2O. Thus, it would be more difficult to bend the bonds in H_2O than CO_2 in the first place, as the bonding electrons are already closer together in H_2O. Next, there are two lone pairs of electrons in H_2O, which means that when we increase the bond angle in H_2O during bending, we would at the same time increase inter-electronic repulsion between the bonding electrons and the lone pair of electrons, not to mention that the inter-electronic repulsion between the two lone pairs of electrons would be increased too. Thus, what we have seen here is that it becomes complicated to compare the energy needed for bending mode of molecules which have different molecular shape and types of electrons residing on the central atom.

In general, we can calculate the number of fundamental vibrations that a molecule exhibits in terms of the vibrational degree of freedom, as follows:

- for a non-linear molecule, its vibrational degree of freedom $= 3n - 6$;
- for a linear molecule, its vibrational degree of freedom $= 3n - 5$,

where n is the number of atoms in the molecule and must be greater or equal to 3.

If we apply the above formulae to H_2O and CO_2, it is calculated that the number of fundamental vibrations is 3 and 4, respectively, just as we have discussed.

16.2 What is the Concept of Degree of Freedom?

A molecule can move freely about in one of the three ways — translation, rotation and vibration. Each of these motions involves a change in the position and orientation of the atoms in the molecule in space. To describe a motion, we make use of the Cartesian coordinate axes (x, y and z). A single atom only has translational motion, and it can move along any of the three axes. Hence, we say that it has three degrees of freedom. It can rotate on its axis, but this spinning motion is not counted towards the degree of freedom since there is no displacement in position. A molecule containing n atoms has a total of $3n$ degrees of freedom when we treat each of the n atoms to be moving independently from each other. Regardless of the value of n, any molecule will have three degrees of freedom attributed to the translational motion along the three Cartesian axes. Shown below are the translational motions of both the H_2O and CO_2 molecules:

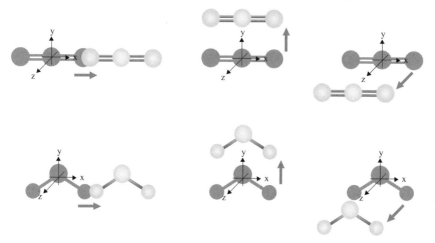

For non-linear molecules, the other 3 degrees of freedom correspond to rotation about each axis. For the H_2O molecule, its rotation is depicted as:

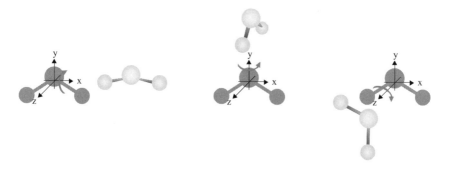

But for a linear molecule, such as CO_2, there are only two degrees of freedom pertaining to rotation about the axes. As shown below, rotating the molecule along the x-axis, in which the atoms lie, does not displace any atom from its position and hence does not count towards one degree of freedom:

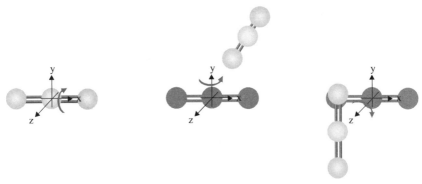

Subtracting the translational and rotational degrees of freedom (six for a non-linear molecule and five for a linear molecule) from the total of $3n$ degrees of freedom will give us the vibrational degrees of freedom, i.e.

- vibrational degree of freedom for a non-linear molecule $= 3n - 6$;
- vibrational degree of freedom for a linear molecule $= 3n - 5$.

For a given molecule, the transition from one vibrational mode to another requires EM radiation that happens to fall within the IR region of the electromagnetic spectrum. Since the energy of each vibrational mode is quantized, only photons of IR radiation with energy that fit the transition from one vibrational level to the other will be absorbed, and the absorption will be reflected as an absorption peak on the IR spectrum. Knowing the number of fundamental vibrations allows us to predict the number of absorption peaks in an IR spectrum. But the number of peaks actually observed may be fewer than expected. This is because in order for a particular vibrational mode to be IR active, the vibrational mode must involve a change in the net dipole moment of the molecule during vibration. In addition, the energy levels of some vibration modes may be too close to each other such that the different peaks become convoluted.

16.3 How is an IR Spectrum for a Molecule Determined?

A monochromatic beam of IR radiation is passed through the sample, and the wavelength of the radiation is varied over time. As the radiation passes through the sample, those wavelengths that resonate with the vibration modes are absorbed. But how does one know whether a particular wavelength has been absorbed? Before the IR radiation is passed through the sample, the original intensity is recorded and "remembered" by the instrument. As the radiation passes through the sample, the outgoing radiation is then compared against the "memorized" intensity. Any differences in intensity would be recorded as the amount of radiation being absorbed.

Nowadays, modern IR spectrometers are mostly of the double-beam type (Fig. 16.4). Basically, a monochromatic beam is split into two different beams of the same intensity. One of the beams acts as the reference beam while the other is passed through the sample. The outgoing beam is then compared against the reference beam to detect absorption signals.

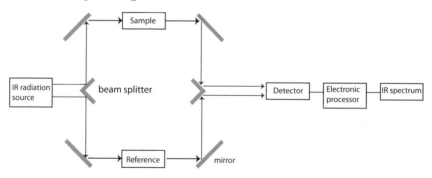

Fig. 16.4. A simplified schematic of a double-beam IR spectrometer.

Modern IR spectrometers allow samples in various physical states to be used. But take note that if the same sample undergoes different types of sample preparation, the spectra obtained would be slightly different due to the differences in the sample's physical states. If the sample to be analyzed is in the gaseous form, the sample is being contained in a gas cell with a long path-length of about 5 to 10 cm in length. The long path-length is required to make up for the diluted concentration of the gaseous sample. If the sample is in liquid form, usually a thin layer of the liquid sample is smeared between two solid plates made up of NaCl or KBr. These two substances are transparent to the transmission of IR radiation. Lastly, solid samples are usually ground together with solid KBr using a mortar and pestle made with marble or agate. The mixed powder is then transformed into a thin translucent disc using a high-pressure mechanical press.

16.4 Analyzing the IR Spectrum of Simple Molecules

The study of IR spectroscopy commonly deals with absorption spectra which are a plot of percentage absorbance or transmittance against wavenumbers that span from 4000 to $400\,\text{cm}^{-1}$.

The absorption spectrum for CO_2 is shown in Fig. 16.5. Based on calculations, it is predicted that CO_2 molecule has four fundamental vibrations, but its absorption spectrum only shows two bands. The strong band at $2349\,\text{cm}^{-1}$ is attributed to the asymmetric stretch mode, and the other band at $667\,\text{cm}^{-1}$ is attributed to both the bending modes of CO_2, which are actually equivalent and thus degenerate. The band attributed to the symmetric stretch mode seems to be "missing."

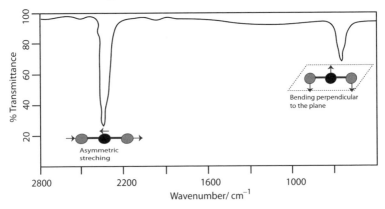

Fig. 16.5. IR absorption spectrum for CO_2.

More appropriately speaking, the symmetric stretch mode of CO_2 is IR inactive (does not absorb in the IR spectrum), and the reason behind this lies in the fact that this vibrational mode does not produce a change in the net dipole moment of the molecule.

Q: Why must there be a change in net dipole moment in order for a vibrational mode to be IR active?

A: If there is a change in the net dipole moment while the molecule vibrates, there will be a changing electric field, right? This changing electric field will then be able to interact with the IR radiation, which is none other than a "photon of changing electric field." Do not forget that IR radiation is a kind of electromagnetic radiation.

That said, symmetrical diatomic molecules such as H_2 and N_2 are IR inactive. Each of these molecules, with just one bond, only exhibits one stretching mode. And since this vibrational mode is not associated with a change in net dipole moment, these molecules are IR inactive. However, unsymmetrical diatomic molecules such as CO are IR active.

Q: How can I tell if a vibration causes a change in net dipole moment?

A: First, the bond must be a polar bond, which means that the two atoms that form the bond must be dissimilar. Now, in order to see whether there is a change in the net dipole moment, one needs to consider each bond as a vector quantity and then apply the vector resolution method on the various polar bonds.

Q: Can a molecule be non-polar in nature yet IR active?

A: Yes, it can. Take, for, instance methane, CH_4; it is a non-polar molecule although all the four C—H bonds are polar. But when the molecule vibrates, there are certain vibrational modes that have a net dipole moment, resulting in the molecule being IR active. Thus, because CH_4 is IR active, it actually contributes to the greenhouse effect if too much methane is released into the atmosphere.

Q: If the number of bands actually observed is fewer than expected, can I attribute those "missing" bands to be associated with IR inactive vibrational modes?

A: No, you cannot. There are other reasons that account for "missing bands," and these include absorption out of the range of 4000 to $400\,\text{cm}^{-1}$, absorption that produces a very weak band and absorption bands that are so close together that the instrument cannot resolve them.

16.5   Analyzing the IR Spectrum of Organic Molecules

Capitalizing on the fact that both masses of the atoms and bond strength affect the energy levels of vibrational modes, IR spectroscopy proves to be a valuable technique in determining the functional groups present in samples since they give absorption bands at characteristic frequencies and intensities attributed to their structure and chemical environment (refer to Fig. 16.6 and Table 16.1).

The IR spectrum of a molecule is a fingerprint for the identification of the molecule. But as the number of atoms in a molecule increases, the IR spectrum becomes more complicated. How can one determine which absorption band corresponds to which vibration? In a practical sense, it is not crucial to assign every single absorption peak to a particular vibrational mode. One needs only to be able to identify whether certain functional groups, such as –OH, –NH$_2$, –CO–, are present in the molecule. Such an application is known as functional group analysis, and it works because we can approximate certain vibrational modes as being only due to the atoms that make up the functional group, thus ignoring the other vibrational effects of the rest of the atoms in the molecule.

In addition, one can also simply compare the IR spectrum of an unknown compound against a library of spectra for qualitative analysis purposes. This application is useful for monitoring the progress of a synthetic organic reaction.

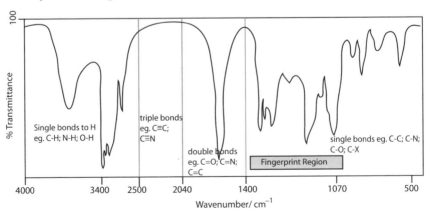

Fig. 16.6.   Spectrum showing characteristic absorption of some common bonds.

Table 16.1 **Table of Characteristic Absorption Due to Stretching Vibrations.**

Functional Group	Bond	Wavenumber (cm^{-1})	Band Description
alcohols, phenols	O–H, "free"	3640–3610	strong, sharp
alcohols, phenols	O–H, hydrogen-bonded	3550–3200	strong, broad
amides, amines*	N–H	3500–3200	medium
alkynes	C–H	3330–3270	strong, narrow
carboxylic acids	O–H	3300–2500	medium, broad
arenes	C–H	3100–3000	strong
alkenes	C–H	3100–3000	medium
alkanes	C–H	2950–2850	medium
aldehydes	C–H	2830–2695	medium
nitriles	C≡N	2260–2210	medium
alkynes	C≡C	2260–2100	weak
carbonyls, carboxylic acids, esters	C=O	1780–1680	strong
amides	C=O	1690–1630	strong
alkenes	C=C	1680–1610	medium
alcohols, carboxylic acids, esters, ethers	C–O	1320–1000	strong
chloroalkanes	C–Cl	800–700	medium
bromoalkanes	C–Br	690–515	medium

*For both primary and secondary amides or amines.

Before we start, it is useful to take note that the IR spectrum is recorded in such a way that higher wavenumber is on the left whereas smaller wavenumbers are on the right-hand side. In analyzing the IR spectrum, we can first divide it into two regions — the left-hand region that lies above $2000\,cm^{-1}$ and the right-hand region that is below this value. The left-hand region contains relatively fewer bands but nonetheless provides valuable indicative information on the types of functional groups present.

Some features to look out for in the left-hand region ($>2000\,cm^{-1}$) include:

(i) C–H stretching bands around $3000\,cm^{-1}$
  - Bands above $3000\,cm^{-1}$ indicate unsaturated carbon–carbon bonds;
  - Bands below this value indicate saturated carbon chains.

(ii) A broad O–H stretching band around 3550–3200 cm^{-1}

- Such a band indicates the presence of an alcohol or phenol.
- If a carboxylic acid is present, the band is seen at lower values around 3300–2500 cm^{-1}, and bands that correspond to C=O and C–O stretching can also be observed.

(iii) N–H stretching bands around 3500–3100 cm^{-1}

- If two bands are observed, it corresponds to a primary amide.
- If one band is observed, it indicates a secondary amide.
- Tertiary amides do not produce any N–H absorption bands.
- To distinguish between amines and amides, we can look out for a C=O stretching band, the presence of which indicates the latter.

(iv) N–H stretching bands around 3500–3300 cm^{-1}

- The presence of one or two bands indicates a primary amine or secondary amine respectively.
- Tertiary amines do not produce any N−H absorption bands.

(v) C≡N and C≡C stretching bands around 2260–2100 cm^{-1}.

Q: Why does the C–H bond of both the ≡C–H and =C–H groups vibrate at higher frequencies than the C–H bond of a saturated carbon atom?

A: In the ≡C–H group, the C—H bond is formed via the overlapping of a *sp* hybridized orbital with the 1*s* orbital of a hydrogen atom. In the =C–H group, the C—H bond is formed via the overlapping of a *sp*^{2} hybridized orbital with the 1*s* orbital, and finally, for a saturated carbon atom, the C–H bond involves using a *sp*^{3} hybridized orbital. The percentage of *s* character is the highest in a *sp* hybridized orbital and lowest in a *sp*^{3} hybridized orbital. Since a *s*-orbital has higher penetrating power than a *p*-orbital, i.e. on the average a *s*-orbital is closer to the nucleus than *p*-orbital, the greater percentage of *s* character in a hybridized orbital would means that the electrons that occupy it would be subjected to stronger attractive force by the nucleus. Hence, the resultant bond forms would be stronger.

Q: Why are the stretching frequencies for O–H or N–H bonds higher than that for a C–H single bond?

A: Both the O–H and N–H bonds are polar bonds, which are inherently stronger than the essentially non-polar C–H due to the additional attractive force between the dipole that is created within the polar bond. Or, from another perspective, the effective nuclear charge of O and N atom is greater than that of a C atom. As a result, the shared covalent electrons are being more strongly attracted in the O–H and N–H bonds than in the C–H bond.

Q: Why is the O–H stretching band usually broad for alcohol?

A: The broad O–H stretching band obtains because of hydrogen bonding occurring between the alcohol molecules. A sample of alcohol is made up of a huge number of molecules. The extent of hydrogen bonding may differ from one molecule to another since hydrogen bonds form and break freely at room temperature. As a result, the chemical environment (refer to Chapter 17 on NMR) around the various O–H bonds of different alcohol molecules varies. Each different O–H bond with a slightly different chemical environment absorbs at a slightly different frequency. Hence, we do not get a sharp band for the O–H stretching mode but rather a broad band that encompasses the various absorption frequencies. But, we can minimize, if not prevent hydrogen bonding from occurring so that the O–H bonds in the molecules experience a similar environment that translates into a sharper absorption band. To do so, we can work with a very dilute sample or a gaseous sample to ensure the O–H bonds are "free" from hydrogen bonding since there is low probability of molecules coming in close contact under such conditions.

Q: If we dilute the alcohol sample, won't we be introducing a lot of water molecules that are capable of hydrogen bonding?

A: By dilution, it does not necessary mean that we have to add water. Instead, we can make use of non-polar organic solvents such as CCl_4 or CS_2 which do not interact much with our sample molecules. Another way would be to grind the sample together with solid KBr and then press it into a thin piece of pellet disc which is transparent to the IR radiation.

Q: Does that also mean that hydrogen bonding would also cause the absorption peak for an amine to be very broad?

A: No. As a matter of fact, the absorption peak for amines is much sharper than that for alcohols. This is because the N–H stretching in an amine is not as sensitive to the effect of hydrogen bonding as for alcohol. This arises because the hydrogen bonding between amine molecules is much weaker than that for alcohol; as a result, the electron density surrounding the N–H bond is not "disturbed" very much when the hydrogen bond forms and breaks freely at room temperature.

Q: Why is the absorption peak of the O–H bond for carboxylic acid found at a smaller wavenumber than that of an alcohol?

A: In carboxylic acid, the O−H group is bonded to an electron-withdrawing carbonyl functional group. As such, the electron density within the O–H bond of the –COOH functional group is more "diluted" than that present in an alcohol functional group. A weaker bond results in a smaller wavenumber of IR radiation being absorbed. In addition, the hydrogen bonding between the carboxylic acid molecules further dilutes the electron density between the O–H bond. How does it happen? When the lone pair of electrons approaches the electron-deficient H atom, the lone pair of electrons repels the bonding electrons within the O–H bond. Note that all these explanations apply for the observation that the N–H bond for an amide has a smaller wavenumber than that of an amine.

Q: But won't the effect of hydrogen bonding also present for the alcohol?

A: Yes, indeed it is also present for the alcohol. But for carboxylic acid, due to the stronger hydrogen bonding between the acid molecules (refer to Chapter 10), the effect of hydrogen bonding on the electron density of the O−H bond is greater for the carboxylic acid than for the alcohol molecule.

Q: Why is the absorption frequency of C≡N and C≡C so much smaller than C–H? Shouldn't a triple bond be stronger than a double bond?

A: Yes, we should expect the stretching frequency of C≡N and C≡C to be higher than that for a single bond based on the strength of the triple bond. Unfortunately, the case is complicated by the fact that the masses of the atoms that are being bonded next to the carbon atom of the C≡N, C≡C and C–H do make a difference to the stretching frequency.

The right-hand region usually contains many bands of varying intensities that include stretching modes of various single bonds and the lower-energy bending modes. Coupled with the different chemical environment these bonds are in, we have a complex but unique spectrum that serves as the "fingerprint" region, and we can match the spectrum of a sample against the database of spectra. Still, there are some strong bands to look out for in the right-hand region below $2000\,\text{cm}^{-1}$:

(i) C=O stretching bands around $1700\,\text{cm}^{-1}$

- If such a band is observed, we can look out for other characteristic bands to determine whether an aldehyde, ester, carboxylic acid or amide is present. To distinguish between the isomeric aldehyde and ketone, we have to check on the band at 2830–$2695\,\text{cm}^{-1}$ which corresponds to the C–H stretching mode of the aldehyde functional group.

(ii) C–O stretching bands around 1320–$1000\,\text{cm}^{-1}$

- Such a band indicates the presence of alcohols, carboxylic acids, esters and ethers. As before, to differentiate one functional group from the other, we look towards finding other characteristic bands. The absence of any O–H or C=O stretching bands indicates the presence of an ether.

Example 16.1: The following spectrum is obtained for compound **W**, C_2H_6O. Identify the bonds that are responsible for the bands seen at 1055, 2981 and $3391\,\text{cm}^{-1}$. Hence, deduce the structure of compound **W**.

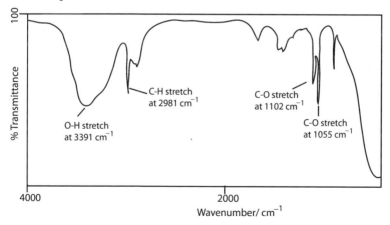

Solution:

- At $3391\,\text{cm}^{-1}$, the strong and broad band observed indicates the stretching of the O–H bond. Thus, an alcohol is present.

- At 2981 cm^{-1}, the band is attributed to the stretching of the C–H bond of a saturated carbon chain. Thus, compound **W** is a saturated alcohol.
- At 1055 cm^{-1}, the band is attributed to the stretching of the C–O bond, and this confirms that compound **W** contains an alcohol functional group.

Compound **W** is ethanol, i.e.

$$
\begin{array}{ccc}
\text{H} & \text{OH} & \\
| & | & \\
\text{H}-\text{C}-\text{C}-\text{H} \\
| & | & \\
\text{H} & \text{H} &
\end{array}
$$

Example 16.2: The following spectra are obtained for the isomers butan-2-one and butanal. Identify which spectrum belongs to which isomer.

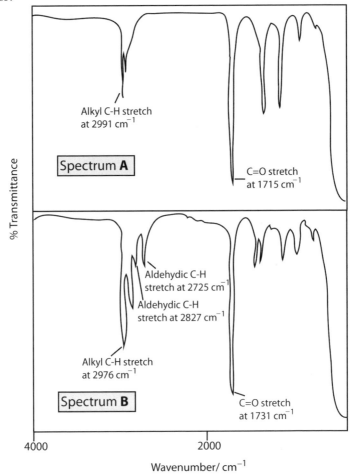

Solution: The characteristic C=O stretching band at around $1700\,\mathrm{cm}^{-1}$ confirms both are carbonyl compounds. The difference between the aldehyde and the ketone lies with the aldehydic C–H bond. The stretching band of this particular bond is around 2830–$2695\,\mathrm{cm}^{-1}$, and it is observed in spectrum **B**. Hence, spectrum **B** belongs to butanal and spectrum **A** belongs to butan-2-one.

Example 16.3: The following spectrum is obtained for compound **X**, $C_9H_{10}O_2$. Use the spectrum to identify the functional groups present in compound **X**.

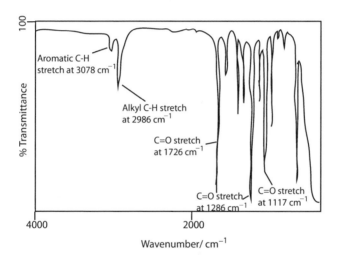

Solution:

- At $3078\,\mathrm{cm}^{-1}$, the band observed indicates the stretching of the C–H bond in which the C atom is sp^2 hybridized. This would mean that either an arene or alkene is present.
- At $2986\,\mathrm{cm}^{-1}$, the band is attributed to the stretching of the C–H bond of a saturated carbon chain, indicating the presence of an alkyl chain.
- At $1726\,\mathrm{cm}^{-1}$, the band is attributed to the stretching of the C=O bond, which means a carbonyl group is present.
- At 1286 and $1117\,\mathrm{cm}^{-1}$, the bands are attributed to the stretching of the C–O bond, which means either a carboxylic acid or ester is present.

Since no O–H stretching band is seen at around 3300–2500 cm^{-1}, the carboxylic acid functional group is absent. Hence, compound **X** is an ester. The high C:H ratio indicates the presence of a benzene ring.

Thus, compound **X** contains both the ester and arene functional groups.

Based on the deductions, it seems impossible to deduce the structure of compound **X**. It can either be ethyl benzoate, phenyl propanoate or even methyl 2-methylbenzoate, just to name a few. As we have it, IR spectroscopy is very useful in deducing the functional groups present, but to elucidate the structure of an unknown sample, we have to triangulate with data obtained from other analytical methods.

Q: Can't we just rely on matching the obtained spectrum against known ones in the database?

A: Yes, you can. But it is good to have at least some idea of what functional groups are present before you do the matching. There are countless of spectra in the database, and it is not very efficient to do the comparison straight away.

CHAPTER 17

Nuclear Magnetic Resonance Spectroscopy

17.1 Introduction

Previously, in the chapter on IR spectroscopy, we talked about the three ways to describe the movement of a molecule in space, namely translational, rotational and vibrational motion. The rotational motion about an axis is analogous to the spinning motion of a top. Other than for molecule, such spinning motion (or spin, for short) is also an intrinsic property of particles such as electrons, protons and neutrons. Similar to any linear translation, we can ascribe a magnitude and direction to the spin of a particle. This would mean that we can describe a particle's spin in terms of how fast it rotates about a particular axis. The direction of the rotation about this axis can be crudely perceived as being clockwise or anti-clockwise.

You may have come across the following orbital-as-box diagram, which depicts orbitals as boxes. These are occupied by electrons, with the spins denoted by half-headed arrows as follows:

$$\boxed{\uparrow\downarrow} \quad \boxed{\uparrow\downarrow} \quad \boxed{\uparrow|\uparrow|\uparrow}$$
$$\text{1s} \qquad \text{2s} \qquad \text{2p}$$

The up and down directions of the arrows denote the two non-degenerate spin states of an electron. Metaphorically, it spins either in a clockwise direction or in an anticlockwise direction. An electron is said to be a spin-1/2 particle. Other spin-1/2 particles include the proton.

491

Both the electron and proton are charged particles that generate a magnetic field when they spin about their own axis. If both the electron and proton spin in the same direction, the magnetic field generated would be in opposite directions.

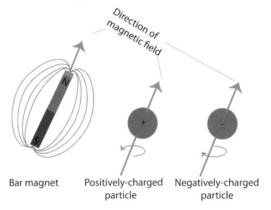

| Bar magnet | Positively-charged particle | Negatively-charged particle |

Q: What is a magnetic field?

A: A magnetic field is a force field that is generated when a charge particle moves, either in a translational or rotational manner. When a charge particle is stationery, it has a static electric field surrounding it. As it moves along a path, there is a "track" of the changing electric field. This changing electric field results in a force which is termed the "magnetic force." The nature of this magnetic force is dissimilar to that of the electrostatic force between electric charges. One can see that without a changing electric field, there is no magnetic field.

The nucleus of the hydrogen atom contains only one proton and, according to quantum mechanical calculation, it was found to have two possible spin states which have different energies. Due to the spinning nature of the proton, two opposite magnetic moments are created. In the absence of an external magnetic field, a bunch of hydrogen nuclei will be statistically in either of the two spin states, and their magnetic moments are oriented in random directions, cancelling out each other's effect. Thus, one would not actually know whether the nucleus is actually spinning, or in which direction it spins. The number of nuclei for each spin state is dependent on the temperature of the surroundings. The concept of spin and the evidence for its presence is only visible when the sample is placed in an external magnetic field (B_o). When this happens, these nuclei will have to re-orientate themselves such that the magnetic moment either aligns with the external magnetic field or against it. It is through this re-orientation that we have two clearly different energy levels being created, thus reinforcing the concept of a spinning nucleus.

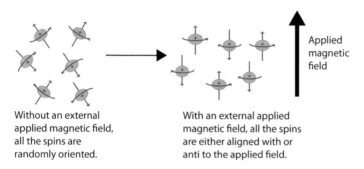

Without an external applied magnetic field, all the spins are randomly oriented.

With an external applied magnetic field, all the spins are either aligned with or anti to the applied field.

Applied magnetic field

When the magnetic moment of a nucleus opposes the external magnetic field, it is in the higher-energy level known as the excited state, just as when you are trying to shove your way through a crowd that's moving in the opposite direction. This higher-energy state is known as the $-1/2$ spin state. The energetically more favorable lower energy state is known as the $+1/2$ spin state.

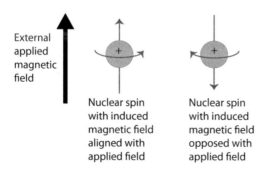

External applied magnetic field

Nuclear spin with induced magnetic field aligned with applied field

Nuclear spin with induced magnetic field opposed with applied field

Q: If the alignment of the magnetic moments with the external applied magnetic field results in a lower energy state, then why not *all* the magnetic moments be aligned with the external applied magnetic field?

A: They can't. This is because the number of nuclei with a particular spin state is statistically determined by the temperature of the surrounding. At any time, from Maxwell–Boltzmann's distribution of energy concept, there would be some nuclei that can "afford" to spin in a manner such that their magnetic moments oppose the applied magnetic field. These nuclei would be those that have higher energy content and thus can do work against the applied magnetic field.

Q: How can I determine the direction of the magnetic field created by a particular direction of spin?

A: You can use the right hand grip rule, or the so-called corkscrew rule to determine the direction of the magnetic field generated.

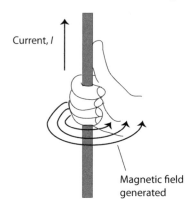

All you need to do is to use the thumb of your right hand to indicate the direction of an electric current or the motion of a positively charged particle such as a proton; the direction where the fingers curl would then indicate the direction of the magnetic field generated. If the particle is an electron or other negatively charged particle, let's say moving from left to right, one just needs to consider the direction of motion as from right to left, which is opposite to the actual motion of the negatively charged particle. The rest would be the same as when we try to determine the direction of the magnetic field generated by a moving positively charged particle. This would mean that the direction of the magnetic field generated by a proton spinning clockwise is the same as for an electron spinning anti-clockwise.

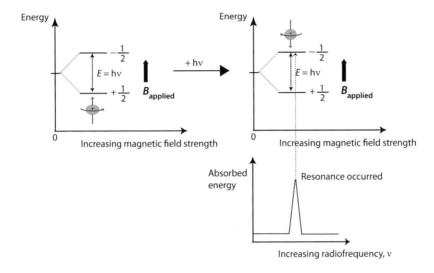

Usually, for a sample of hydrogen nuclei, the proportion of nuclei occupying the lower energy state, i.e. $+1/2$ spin state, is only slightly greater than that of those at the higher energy state. The transition from the lower energy state to the higher energy state can be achieved when a nucleus absorbs electromagnetic radiation corresponding to the energy difference between the two spin states $(\Delta E = hf)$.

This energy difference, ΔE, is dependent on the strength of the external magnetic field as follows:

$$\Delta E = \gamma(h/2\pi)B_{o},$$

where h = Planck constant = 6.63×10^{-34} J s, B_{o} = strength of the external magnetic field in tesla (T) and γ = gyromagnetic ratio for the nucleus under probe $(s^{-1}\ T^{-1})$. The gyromagnetic ratio for a particular nucleus is the ratio of its magnetic dipole moment to its angular momentum. For ^{1}H, the value of γ is $2.675 \times 10^8\ s^{-1}\ T^{-1}$.

Since ΔE is directly proportional to the frequency of electromagnetic radiation in $s^{-1}(f)$ as shown by the Planck's equation,

$$\Delta E = hf,$$

the frequency of radiation absorbed is also dependent on the strength of the external applied magnetic field. The larger the applied magnetic field strength, B_o, the larger the ΔE and the higher the frequency of radiation needed. For example,

$$\text{when} \quad B_o = 1.5\,\text{T}, \Delta E = 4.234 \times 10^{-26}\,\text{J},$$

$$B_o = 3.0\,\text{T}, \ \Delta E = 8.468 \times 10^{-26}\,\text{J}.$$

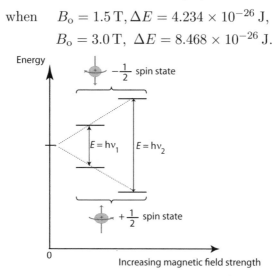

ΔE is a very small value, and it corresponds to electromagnetic radiation within the radio frequency spectrum (at frequencies of about 60–100 MHz). One needs to appreciate that in order to create a measureable energy gap, the magnitude of the applied magnetic field is relatively large if it is compared against the earth's magnetic field, which is only about 5.7×10^{-5} T.

This thus forms the basis of nuclear magnetic resonance (NMR) spectroscopy, which studies the absorption of radio frequency radiation to effect the transition between nuclear spin states, i.e. when spin-flipping occurs, under the influence of an external magnetic field.

Q: Since the ^{1}H atom is NMR active, what are the criteria for a nucleus to be NMR active?

A: Other than the proton, a neutron also possesses spin. However, only nuclei with an odd number of protons or/and an odd number of neutrons have a nuclear spin that we can detect using the NMR spectrometer. Nuclei such as ^{1}H, ^{2}H and ^{13}C are NMR-active whereas ^{12}C and ^{16}O are not. Why is this so? A proton is positively charged. Thus, when a nucleus contains an odd number of protons, there will always be a net magnetic moment. However, if there is an even number of protons, the magnetic moment of the even number of protons would be paired up, hence resulting in zero net spin.

Q: If a nucleus contains an odd number of neutrons but an even number of protons, the nucleus would still be NMR active. But neutrons are not charged particles. How is it possible that they generate a magnetic field when they spin?

A: Well, the main reason is that a neutron is not a fundamental sub-atomic particle. It is in fact made up of even smaller sub-atomic particles known as quarks, which are charged particles. Thus, although a neutron does not have a net charge, the uneven distribution of these quark particles in the spherical neutron creates a net electric dipole moment. Hence, when a neutron spins, the changing electric field creates a magnetic field. To have a better idea on this topic, you can refer to books on particles physics for more details.

We shall now focus our discussion on ^{1}H NMR spectroscopy, which studies the simplest nucleus ^{1}H, but yet is the foremost technique used in the structural elucidation of organic compounds.

17.2 Principles of ^{1}H NMR Spectroscopy

Q: Since a spinning charge particle creates a magnetic field and an electron is a charged particle, wouldn't the magnetic field from the electron spin affect the interaction between the nuclear spin and the external applied magnetic field?

A: Indeed it does. Under the influence of an external magnetic field (B_o), the electron would be forced to move in such a way that it generates a secondary magnetic field $(B_{induced})$, which opposes the external applied field. As a result, the effective magnetic field that is actually "felt" by the spinning proton would be diminished. Hence, we say that the electrons surrounding a proton, to a certain extent, shield it from the full strength of the applied field, i.e.

$$B_{net} = B_o - B_{induced}.$$

Q: Wait a minute, why would the moving electron under the influence of an external applied magnetic field "be forced" to move in such a way as to create a magnetic field that will oppose the applied field?

A: First, let us consider the movement of a positively charged particle in an external applied magnetic field. When a positively charged particle moves in a magnetic field, there would be a force acting on it according to Fleming's left-hand rule:

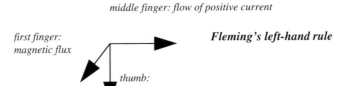

Thus, when the positively charged particle moves in the direction as indicated by the middle finger, its motion creates an induced magnetic field that would reinforce the external applied magnetic field. You can use the right hand grip rule or the corkscrew rule to reaffirm the direction of the induced magnetic field being created. Now, since the direction for the movement of electrons is considered to be opposite to that of the positively charged particles, the induced magnetic field generated by the moving electrons would be opposing the external applied magnetic field.

Now, knowing that an applied external magnetic field strength causes two different spin states with a difference in energy levels to be created, and the induced magnetic field from the surrounding electrons also affects the net magnetic field strength experienced by a proton, we can proceed to understand how we can make use of ^{1}H NMR spectroscopy in the structural identification of an organic molecule.

A ^{1}H atom in, let's say, CH_4, is different from a ^{1}H atom in ethane, CH_3CH_3. This is because the distribution of electron densities (this includes both the electron densities from other ^{1}H atoms bonded to the same carbon atom and those electron densities that come from the various carbon atoms that are bonded to it) are different. Each unique distribution of electron density is termed the *chemical environment*, and it shields the proton from the applied external magnetic field to a particular extent. The greater the electron density, the more the proton will be shielded. This is because a stronger opposing induced magnetic field would be generated. The net magnetic field felt by this proton would therefore be weaker. Thus, the energy gap between the two different spin states for this proton would be smaller.

Q: So, is it always true that the electrons near the proton would always generate a magnetic field in opposition to the external applied magnetic field?

A: No! This is not always true. Later on in the chapter, you will learn about the "magnetic anisotropic effect," and from this you will learn that the magnetic field that is generated by the moving electrons can indeed reinforce the external applied magnetic field.

17.3 How Is a ^{1}H NMR Spectrum for a Molecule Determined?

If we now apply an external magnetic field with a particular field strength, we are going to obtain a range of energy gaps that arises due to the different types of chemical environment present. Each energy gap would correspond to a particular magnitude within the radio frequency spectrum. If we now do a radio frequency sweep, we can record which particular radio frequency is being absorbed. The greater the energy gap, which of course results from a less shielded/more deshielded chemical environment, the higher would be the frequency absorbed by the protons.

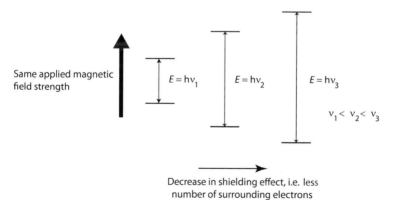

Same applied magnetic field strength

$E = h\nu_1$ $E = h\nu_2$ $E = h\nu_3$

$\nu_1 < \nu_2 < \nu_3$

Decrease in shielding effect, i.e. less
number of surrounding electrons

Other than the above, we can also fix a particular radio frequency magnitude, which means that the energy gap must be of this particular magnitude in order for protons to be able to absorb this frequency of radiation. But this particular energy gap would only correspond to a particular chemical environment. Since there may be more than one dissimilar chemical environment, how do we then make the protons in these different chemical environments absorb only this particular radio frequency? All we need to do is to vary the strength of the external applied magnetic field. Remember that different magnetic field strength would cause a different energy gap to be created? As the magnetic field strength is increased from low field to high field strength, the protons that are less shielded/more deshielded will have the energy gap that matches the magnitude of the radio frequency. Thus, these protons would be able to absorb the applied radio frequency. At the end of the high magnetic field strength, the more shielded/less deshielded protons will then have the necessary energy gap that matches the applied radio frequency.

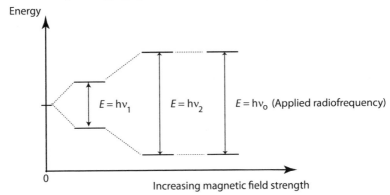

The two simplest methods that have been discussed so far in obtaining the spectrum are known as the continuous wave (CW) method. In the CW method, either a magnetic field sweep or a radio frequency sweep is employed (Fig. 17.1).

In doing a magnetic field sweep, the sample is subjected to a constant frequency of radio frequency radiation while the external magnetic field strength is varied. Take, for instance, bromoethane. Its molecule contains two groups of chemically equivalent protons — a group of two H_a protons and the other comprising the three H_b protons. In a group of chemically equivalent protons, each member proton has the same chemical environment, i.e. they are bonded to exactly the same type of neighbors at the same proximity. Although the two-dimensional diagram of bromoethane seems to portray one H_b atom

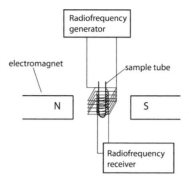

Fig. 17.1. Schematic of an NMR spectrometer.

to be further away from the neighboring H_a and Br atoms than the other two H_b atoms, it is definitely not the case. This is because the carbon atom which the three H_b protons are bonded to is tetrahedral in shape, and there is also free rotation about the C–C single bond.

$$Br-\overset{\overset{\displaystyle H_a}{|}}{\underset{\underset{\displaystyle H_a}{|}}{C}}-\overset{\overset{\displaystyle H_b}{|}}{\underset{\underset{\displaystyle H_b}{|}}{C}}-H_b$$

Two signals will appear in the ^1H NMR spectrum of bromoethane, each signal corresponding to one set of chemically equivalent protons. The H_b atoms are more shielded/less deshielded (experience weaker net magnetic field strength, have smaller energy gap, hence need stronger external applied magnetic field strength) than the H_a atoms. When a field sweep is carried out at increasing external magnetic field strength, the H_a atoms will resonate (spin-flip) first at lower field strength followed by the H_b atoms at a higher field strength.

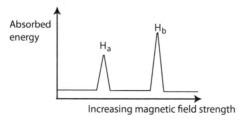

If a radio frequency sweep is conducted, the sample is placed under a constant external magnetic field strength while the magnitude of the radio frequency is varied. The H_a atoms, being less shielded/more deshielded (experience greater net magnetic field strength, have greater energy gap, hence require greater magnitude of radio frequency) than the H_b atoms, will resonate and give an absorption peak at a higher frequency.

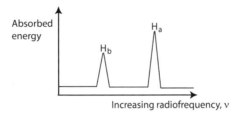

17.3.1   *Chemical Shifts*

In practice, the NMR absorption spectrum is neither plotted against magnetic field strength nor radiation frequency as the differences between absorption peaks are too small. This makes it hard to distinguish one peak from the other and hence the type of protons responsible for them. Furthermore, with the shielding or deshielding effect coming into play, it is better to have some sort of reference point from which useful data can be studied and compiled into a database. A common reference or internal standard used is tetramethylsilane (TMS), $(CH_3)_4Si$, and a small amount of it is added to the sample prior to placing it into the spectrometer.

Q: Why is TMS suitable for use as an internal standard?

A: TMS is a suitable choice based on the following reasons: (1) it is chemically unreactive and hence it does not interfere with the original composition of the sample, (2) it is soluble in most organic solvents just as what is necessary in dealing with organic samples, (3) with just a small concentration, it produces a very strong sharp signal, which is attributed to the high number of 12 chemically equivalent protons, (4) its protons produce a peak at a much higher magnetic field strength than those in many organic compounds since they are highly shielded due to the relatively smaller electronegativity of Si as compared to carbon atoms, (5) since it is non-polar, it is volatile, which makes it easy to be removed from, in particular, valuable samples that can be recovered.

The single absorption peak attributed to the protons in TMS is arbitrarily assigned the value of 0 ppm on a scale expressed in terms of **chemical shift** (δ). All the peaks produced by the protons in the original sample are measured relative to this reference peak — by the "distance they are shifted" from the reference point.

Q: What is "ppm?"

A: It stands for *parts per million*. This notation is basically a ratio used in dealing with very small quantities. 1 ppm stands for 1 part per million, which in terms of mass equates to 1 milligram per kilogram.

Protons in TMS are brought into resonance at a certain external magnetic field strength (say B_{TMS}) just as would be the protons in the sample, but at different magnitudes. With reference to bromoethane, we have B_a and B_b, which are the magnetic field strengths that are needed to bring the H_a and H_b protons into resonance. The difference (ΔB) between each of these values and that of B_{TMS} is measured.

However, since no two magnets are exactly the same, and each of the spectrometers may be operating at different radio frequency frequencies such as 60 MHz and 100 MHz, the absolute frequency at which a particular proton resonates (and hence ΔB) is different from one machine to another. Therefore, to make chemical shift data universally applicable and independent of the operating frequency of the spectrometer used, the chemical shift value for a particular proton is expressed as a ratio of ΔB to B_o. The B_o corresponds to the external applied magnetic field strength that is required to cause the absorption of the operating radio frequency — for example, it would be 60 MHz or 100 MHz. And since ΔB is much smaller than B_o, in the range of about 10^6 times, the ratio is then multiplied by a factor of 10^6, giving us chemical shift values in terms of ppm. Such values in ppm are much more manageable than those smaller values. For instance, the chemical shift for a H_a proton in bromoethane is calculated as follows:

$$\delta \; of \, H_a = \left(\frac{\Delta B}{B_o} \times 10^6 \right) \text{ppm} \quad \text{where} \quad \Delta B = B_{TMS} - B_a.$$

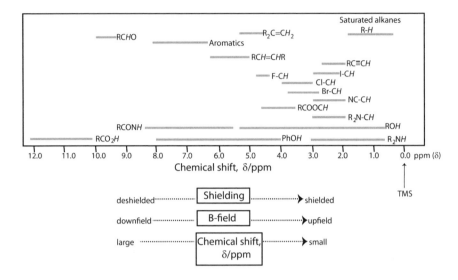

Fig. 17.2.   Chemical shift values of some organic functional groups.

The larger the chemical shift value, the more deshielded/less shielded a particular proton is as compared to those in TMS, hence the more down field (towards a smaller applied magnetic field strength) the signal would be found. Thus, we expect the chemical shift value for H_a to be greater than that for H_b in bromoethane. Figure 17.2 shows the chemical shift values of protons in various chemical environments as found in common organic molecules.

Chemical shift values are found to the left of the TMS reference point, and the δ values increase from right to left. Protons with greater δ values are more deshielded/less shielded, and they resonate at lower magnetic field strength.

Q: Why does a particular type of proton have a range of δ values and not a precise one?

A: It all boils down to the chemical environment the particular type of proton is in. For example, the electron density for the CH_3 — group in ethane is not going to be the same as for those in 2,2-dimethylpropane, right?

Q: Is it possible to obtain signal peaks at δ values in the negative range, i.e. less than zero ppm?

A: Certainly. The δ value for the methyl carbanion ion in $LiCH_3$ falls in the negative value range. This is because the protons in the CH_3^- ion are more shielded/less deshielded than the CH_3- group in TMS, as it is negatively charged. Hence, the electron density surrounding the protons in CH_3^- ion is higher than in the CH_3–group for TMS.

Q: Why are the protons in TMS highly shielded than the rest in Fig. 17.2? Why are the H_a protons in bromoethane more deshielded than the H_b protons?

A: The protons in TMS are highly shielded, even more so than those in saturated alkanes. This goes to show that it has something to do with the Si atom. More specifically, it is due to the lower electronegativity of Si than C. In fact, the same electronegativity factor also accounts for why H_a protons in bromoethane are more deshielded/less shielded than the H_b protons — the H_a protons are nearer to the electronegative Br atom, which draws electron density away from these protons causing them to be deshielded. The impact of this electron-withdrawing effect on the H_b protons is smaller because of the further distance away from the highly electronegative Br atom.

17.3.2 *Factors that Account for the Relative Chemical Shifts*

The chemical shift values can be accounted for by the chemical environment that a particular proton is found in. Essentially, it has got to do with the electron density around the proton.

A higher electron density implies that:

- the proton is more shielded/less deshielded and thus
- it resonates at a higher external magnetic field strength (upfield), that results in
- its smaller δ value.

The opposite holds true. A decrease in electron density that leads to the deshielding of a proton can be caused by the following factors:

- presence of an electronegative atom (e.g. N, O and the halogens); and
- presence of unsaturated bonds — the magnetic anisotropic effect.

17.3.2.1 *Effect of electronegative atom on chemical shift values*

Electronegative atoms such as N, O and the halogens draw electron density away from nearby H atoms through the sigma bonds (in what is termed the inductive effect). As mentioned, with a reduced electron density, the proton is more deshielded/less shielded and it has a larger δ value and hence would appear more down field.

A more electronegative atom has the ability to withdraw electron density to a greater extent and in turn causes the proton to have a larger δ value. Take, for instance, the halogenoalkanes, in which the electronegativity of the halogen decreases from F to Cl to Br to I. Consequently, there is a decrease in δ values from CH_3F to CH_3Cl to CH_3Br to CH_3I:

Type of proton	CH_3F	CH_3Cl	CH_3Br	CH_3I
δ (ppm)	4.3	3.0	2.7	2.2

Some other examples include the following: (i) as C is more electronegative than Si, the protons of $(CH_3)_4C$ have a larger δ value than those in $(CH_3)_4Si$; (ii) as O is more electronegative than S, the proton in an ether functional group (RO–CH) has a larger δ value than the one in a sulfide group (RS–CH).

The deshielding effect, and accordingly the δ value, increases with an increase in the number of electronegative atoms. An example would be the bromoalkanes, whose δ values are as follows:

Type of proton	$CHBr_3$	CH_2Br_2	CH_3Br
δ (ppm)	7.0	4.9	2.7

Apart from electronegative atoms, there are also electron-withdrawing groups to consider, such as the carbonyl, cyano and nitro groups, which withdraw electron density via both the inductive and resonance effects (see Chapter 6).

Note that the electron-withdrawing effect is influenced by proximity — it decreases with increasing distance. Comparing the chemical shifts of the protons in butane and 1-bromobutane, we see that the electron-withdrawing effect of the Br atom not only causes the δ value of the protons in the latter to be larger than those in butane, it also affects, albeit in decreasing influence, the δ value of the H_w, H_x, H_y and H_z atoms:

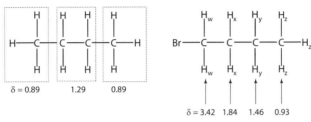

Another example is that the proton in an aldehyde group (RCO–H) is more deshielded than a proton in a ketone group (RCO–CH–) since the proton is directly bonded to the carbon atom of carbonyl functional group.

Q: Why do the chemical shift values of the following protons show an increase with an increase in the number of electron-donating alkyl groups? Shouldn't the proton be more shielded with a higher number of electron-donating alkyl groups?

Type of proton	δ (ppm)
$R–CH_3$	0.9
$R_2–CH_2$	1.3
$R_3–CH$	1.5

A: There is a misconception here! If you consider the two CH_3–groups in propane, why should the CH_3–group donate electron to the $–CH_2$–group? The electron-donating effect works ONLY if there is a demand for it, for example, in a carbocation or radical. The carbocation or radical is relatively more electron-deficient than the surrounding alkyl groups; as a result, there is a tendency for electron density to flow from the surrounding alkyl groups into it. From the perspective of the alkyl group, we say that the alkyl group is electron-donating. And this sometime creates confusion as we would think that the electron-donating effect is inherent in the alkyl group itself.

So, how do we account for the differences in the chemical shift value then? First, we need to know that C atom is still more electronegative than a H atom. A carbon atom that is bonded to three H atoms in $R-CH_3$ has more electron density to "pull" from than the one in R_3-CH. In R_3-CH, the carbon atom is bonded to a H atom and three other carbon atoms. Since all carbon atoms have the same electronegativity, we have the carbon atom attracting electron density from only a H atom in R_3-CH. Thus, the H atom in R_3-CH is more deshielded/less shielded, which accounts for the greater chemical shift value.

17.3.2.2   *The magnetic anisotropic effect*

So far, when we discuss the effect of the magnetic field that is generated from the surrounding electrons, we are assuming that the magnetic flux is homogeneously distributed round the proton of consideration. This is only partially true for electrons in sigma bonds as they are less polarizable. For organic molecules that contain the more polarizable pi electron cloud, the magnetic field that is generated by the pi electrons is not the same for all protons that are near to the influence from this induced magnetic field. As a result, some protons may experience a shielding effect while others a deshielding effect. We term this phenomenon the magnetic anisotropic effect, whereby the term "anisotropic" is Greek for "different in different directions." The table below indicates the approximate chemical shift values for protons which have experienced the magnetic anisotropic effect. These are very useful values as the chemical shift value is very unique to the specific type of proton being considered, hence it can be used to identify the specific kind of functional group that is present in the molecule.

Q: Why is a sigma electron cloud less polarizable than a pi electron cloud?

A: A sigma electron cloud accumulates within the inter-nuclei region whereas the pi electron cloud is out of the inter-nuclei region. As a result, the sigma electron cloud is more strongly attracted by the nuclei as compared to the pi electron cloud.

Structure	Aldehydic proton	Aromatic proton	Vinylic proton	Acetylenic proton
δ / ppm	9.0 – 10.0	6.0 – 8.5	4.5 – 6.5	2.5 – 3.0

Take, for instance, the aromatic proton in a benzene molecule. According to the right hand rule (Section 17.2), the magnetic field that is generated by the circulating pi electron cloud, which consists of two concentric rings above and below the six-membered carbon ring, would be as shown below.

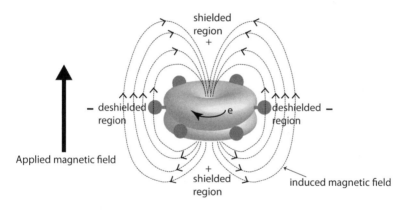

Do you notice that the magnetic field is reinforcing (indicated by a "−" sign) the external applied magnetic field only outside the ring? Whereas it is opposing (indicated by a "+" sign) the external applied field within the ring. Thus, for protons experiencing the reinforcing magnetic field, they would be deshielded. This deshielding effect is the opposite of the shielding effect caused by the sigma electrons that we have been discussing.

Q: If the induced magnetic field is reinforcing the external applied magnetic field, why is the "−" sign use to symbolize it? We should use the "+" sign instead, shouldn't we?

A: Well, you should know that the energy gap between the two types of spinning protons is dependent on the magnetic field strength experienced by the protons. Consider two types of protons, one experiencing a reinforced magnetic field while the other is not. The one that is experiencing the reinforced magnetic field would need a weaker external applied magnetic field to have an energy gap that is large enough to absorb the radio frequency that is being applied to the proton. Thus, the "−" sign simply indicates that it is no longer necessary to apply such a strong magnetic field to bring about nuclear magnetic resonance if the induced field reinforces the external applied magnetic field.

There are other similar magnetic anisotropic effects observed for other pi electron cloud systems, as shown below.

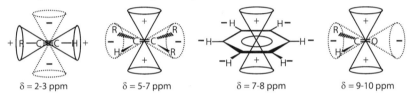

δ = 2-3 ppm δ = 5-7 ppm δ = 7-8 ppm δ = 9-10 ppm

17.4   Important Features on the ^{1}H NMR Spectrum of Organic Molecules

After learning about the principles and the factors that affect the spinning of a nucleus, let us learn how to extract as much information as we can from an NMR spectrum, so as to help us in the elucidation of the constitutional/structural formula of an organic molecule. Let us use bromoethane as an example. Figure 17.3 shows a simplified ^{1}H NMR spectrum of bromoethane:

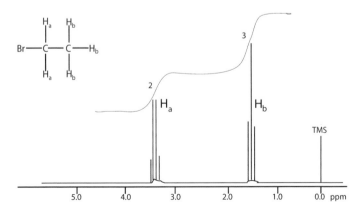

Fig. 17.3.   NMR spectrum of bromoethane.

In analyzing ^{1}H NMR spectra, there are four features to look out for, and summing these, we not only can get information about functional groups present but also the bonding pattern.

The four features are:

(1) the number of signals
(2) the δ value of each signal
(3) the relative areas under the peaks
(4) the splitting patterns of the signals

It is to be noted that the spectra that we have used in this chapter are simplified ones, made so for the purpose of clearer illustration. The actual NMR spectrum is not as "clean." Nonetheless, they retain the essential features that are needed for the introduction of concepts in this chapter.

17.4.1 *The Number of Signals*

The number of signals informs us of the number of groups containing chemically equivalent protons. In Fig. 17.3, there are two signals that originate from bromoethane — the two H_a protons produce one signal and the three H_b protons produce another. The signal at 0 ppm is due to the internal standard TMS. In short, the number of signals indicates the number of different chemical environments that are present in the molecule.

17.4.2 *The δ Value of Each Signal*

Counting the signals alone does not give us useful information. The next thing to do is to analyze the δ values, which when matched against those in the proton chemical shift table allow us to determine the type of chemical environment each different group of protons is in.

In Fig. 17.3, the signal at about $\delta = 3.4$ ppm informs us of a more deshielded/less shielded set of protons than those at $\delta = 1.7$ ppm. The former would be the H_a atoms which are closer to the electron-withdrawing Br atom than the H_b atoms are. In addition, for protons that experience magnetic anisotropic effect, the δ values for such protons would be characteristics of the functional groups that are present in the molecule.

17.4.3   The Relative Areas under the Peaks

Up to this point, we know the types of protons and the chemical environment they are in. Wouldn't it be good to know the number of protons that are present in a particular chemical environment? It would definitely be helpful in deciphering the structure of the molecule. The solution lies in looking at the intensity of the signal, which is proportional to the quantity of the protons responsible for it. The more intense the signal is as compared to others in the same spectrum, the greater the number of a distinct type of proton present in that chemical environment.

However, it is not possible to know the absolute number of each type of protons. This is because when getting the NMR spectrum for a compound, the concentration of the compound is not of concern. Thus, there is no way to quantify the number of protons present in each chemical environment. As a result, the high intensity of a signal could be due to a large number of protons present in a particular chemical environment in the molecule that contribute to the signal, or it could just be due to a high concentration of sample molecules being used in the analysis.

So, what we can do is to compare the intensities of all the signals that come from the sample under study. From this, we can obtain the relative proportion of each set of chemically equivalent protons. Thus, what can be done is an integration of the relative areas under the peaks, and this is computed automatically by the instrument. One of the ways that the integrated data is presented is in the form of an integrator trace, which is a continuous line that overlays with the spectrum (see Fig. 17.3). Along this line are steps that correlate with the area under the peaks. The height of the step, or simply the step height, is proportional to the number of protons responsible for that particular signal. The step heights are then measured to arrive at a ratio, which gives the relative number of each type of proton present in each chemical environment. Looking back at Fig. 17.3, the ratio of the intensities of the signals due to H_a and H_b is 2 : 3. Alternatively, some NMR spectrometers actually print out the relative ratio for the user, and these numbers are usually printed below the x-axis of the spectrum.

Q: If the NMR spectrum only gives us the ratio of the distinct sets of protons, how do we use this information to conclude the total number of protons in the molecule and hence its structure?

A: In practice, elemental analysis is carried out first to determine what elements are present in the sample. Then the percentage quantity of each of the elements after the mass spectrum has been obtained (Remember that the mass spectrum would give us the relative molecular mass of the molecule? Refer to Chapter 14). If the ratio of peak heights is 2 : 3 and through elemental analysis and the mass spectrum, the molecular formula indicates 15 H atoms in total, then the absolute number of protons responsible for each signal is 6 and 9 respectively.

Q: There should only be two signals indicating two different chemical environments for bromoethane right? If we look back at Fig. 17.3, there seem to be two signals, with each comprising a series of peaks that are close together. How can we account for this observation?

A: In a low-resolution spectrum, we may well observe two signals each consisting of a single peak. But when we analyze the high-resolution spectrum of the same sample, we will indeed see two separate sets of peaks. And now if we observe more closely, we would be able to identify a particular splitting pattern for each of the signals. The splitting pattern at each δ value is in fact a critical piece of information that the low-resolution spectrum cannot provide — it indicates the number of neighboring protons on adjacent carbon atoms which are chemically equivalent. This information is useful for structural elucidation. In the next section, we will look into the concept behind splitting patterns and learn how to interpret them.

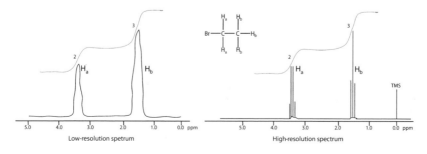

Low-resolution spetrum   High-resolution spectrum

17.4.4 *The Splitting Patterns of the Signals*

Splitting patterns arise from the phenomenon known as spin-spin coupling interaction that occurs between neighboring protons that are chemically non-equivalent. If we translate it literally, it basically means that the effect of a spinning proton (referring to the magnetic field that it generates) would affect the spinning effect of another proton (the overall magnetic field experienced by this proton) when these two different spinning effects interact.

Take, for instance, the signal produced by the H_b protons in bromoethane. The splitting pattern is known as a triplet, identified by a set of three peaks whose relative intensities are in the specific ratio of $1 : 2 : 1$, as shown in the following diagram:

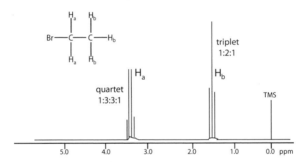

We have seen how the secondary magnetic field generated by the movement of electrons surrounding a proton affects the strength of the magnetic field at which the proton resonates. Recall that a proton can either have its magnetic moment aligned with the external applied magnetic field or against it. The H_b protons have two neighboring H_a protons on the adjacent carbon atom. Considering the fact that each of them can exist in either one of the two spin states (i.e. ↑ or ↓), we have the following configurations of spin states that the two H_a protons can adopt: ↑ ↑ or ↓ ↓ or ↑ ↓ (↑ ↓ is the same as ↓ ↑).

These local magnetic fields generated by the H_a protons exert a different influence on the field that is needed to bring H_b protons into resonance (assuming a fixed radio frequency is being applied):

- when both spins of the H_a protons are aligned with the external magnetic field ($\uparrow\uparrow$), a weaker external applied magnetic field needs to be used as the rest of the magnetic field that is required to bring about resonance can come from the magnetic field that is generated by the spins of the H_a protons. Thus, the H_b protons would resonate at a slightly lower field strength (down field), which translates into a greater δ value;
- when both spins of the H_a protons are against the external magnetic field ($\downarrow\downarrow$), there is a need to apply a stronger external magnetic field because part of this external magnetic field would be "used" to counteract the opposing magnetic field that is generated by the spins of the H_a protons. Thus, the H_b protons would resonate at a slightly higher field strength (upfield), which translates into a smaller δ value;
- when the spins of the H_a protons are opposite each other ($\uparrow\downarrow$), their influence on the resonance frequency of the H_b protons cancel each other out, causing the H_b protons to resonate at the "expected" field strength.

These points are illustrated in the following diagram:

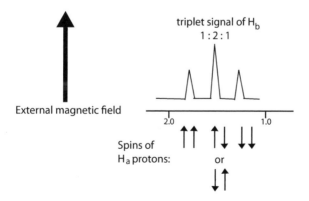

Q: The orientations of the spins of both H_a protons account for the three peaks in the signal of the H_b protons. But how do we account for the differences in the relative intensities of these three peaks?

A: When the spins of the two H_a protons are in opposite directions, either of the two protons can have its spin align with the external magnetic field while the other has its spin against the applied field. Thus, the probability for bromoethane molecules to have both spins of the H_a protons in the $\uparrow \downarrow$ permutation is twice that for both spins to be oriented in the same direction (either $\uparrow \uparrow$ or $\downarrow \downarrow$), and this translates to the peak signal with twice the intensity.

A similar explanation accounts for the signal due to the H_a protons. The coupling interaction between these protons and their three neighboring H_b protons produces a quartet defined by a set of four peaks whose relative intensities are in the specific ratio of 1:3:3:1.

The three H_b protons can adopt either one of the two spin states, giving us a total of eight possible permutations, and these cause differences in the strength of the external applied magnetic field that is needed to bring the H_a protons into resonance:

- when the spins of the three H_b protons are aligned with the external magnetic field ($\uparrow \uparrow \uparrow$), a weaker external applied magnetic field needs to be used as the rest of the magnetic field that is required to bring about resonance can come from the magnetic field that is generated by the spins of the H_b protons. Thus, the H_a protons would resonate at a slightly lower field strength (down field), which translates into a greater δ value;

- when one, two or all the spins of these H_b protons are against the external magnetic field, there is a need to apply a stronger external magnetic field because part of this external magnetic field would be "used" to counteract the opposing magnetic field that is generated by the spins of the H_b protons. Thus, the H_a protons would resonate at a slightly higher field strength (upfield), which translates into a smaller δ value.

- the probability of the molecules having a set of spins where by two out of the three H_b protons are oriented in opposite directions is thrice that for a set of spins that are oriented in the same direction. This translates to the two peaks that are in the middle of the signal, having thrice the intensity of the other two peaks at the peripheral.

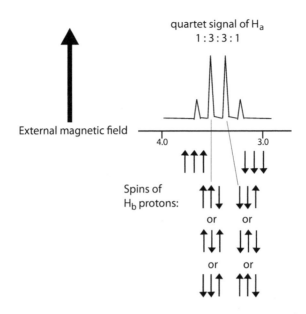

Q: Do we need to draw out all the possible permutations in order to arrive at the splitting pattern and the relative intensities of the peaks?

A: What we need to do is to identify the number of neighboring protons and apply the $(n + 1)$ rule in deciding the multiplicity of the signal. If there is only one such chemically equivalent neighboring proton, the signal contains $1 + 1 = 2$ peaks, which is known as a doublet. If there are two or three such chemically equivalent neighboring protons, the signal is a triplet $(2 + 1 = 3)$ and quartet $(3 + 1 = 4)$ respectively, and so forth. To determine the relative intensities of the peaks that make up the signal, we apply the number pattern known as Pascal's triangle. Refer to Fig. 17.4 for illustration of the use of these concepts. Note that the information that is obtainable from Fig. 17.4 can only be applied to the coupling effect provided the neighboring protons are chemically equivalent among themselves.

Q: How do we define the kind of "neighboring protons" that can be involved in a coupling interaction?

A: First, these protons must not be chemically equivalent to the proton(s) whose signal is being studied. For example, the two

n^*	Number of peaks $(n+1)$	Multiplicity of signal	Relative intensities of peaks (using the Pascal's Triangle)
0	1	singlet	1
1	2	doublet	$1:1$
2	3	triplet	$1:2:1$
3	4	quartet	$1:3:3:1$
4	5	quintet	$1:4:6:4:1$
5	6	sextet	$1:5:10:10:5:1$

n^* stands for the number of neighboring protons that are chemically equivalent among themselves.

Fig. 17.4.   Pascal'striangle.

CH_3–groups in ethane are chemically equivalent, thus there would not be any splitting of the signal for the CH_3–group, which means the NMR spectrum for ethane is a singlet. Second, significant coupling interactions occur with neighbors that are no more than three covalent bonds away from the proton being studied — these are the couplings that we shall focus on in this book.

Q: Can coupling occur for protons that are less than three covalent bonds apart?

A: Yes. At this proximity, we are looking at protons bonded to the same carbon atom (known as geminal protons). Coupling between these protons is possible as long as they are chemically non-equivalent. This scenario is observed for mono- and di-substituted alkenes. An example is bromoethene:

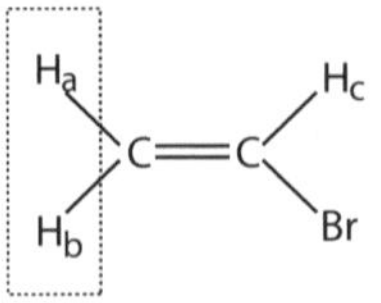

Both the H_a and H_b protons are bonded to the same carbon atom and in turn have the same connectivity to the $=CHBr$ group. Just by looking at its structure, it seems that the H_a and H_b protons are chemically equivalent. Based on this, we expect

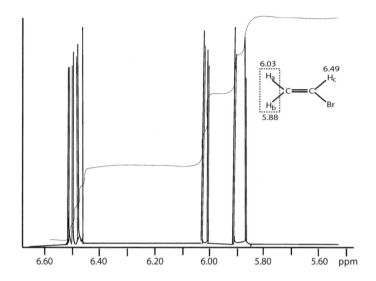

Fig. 17.5. NMR spectrum of bromoethene.

to observe two signals in its ^{1}H NMR spectrum — one due to the H_c proton and the other due to both the H_a and H_b protons. However, this is not the case in the actual ^{1}H NMR spectrum of bromoethene (refer to simplified spectrum in Fig. 17.5). The spectrum shows three signals instead. This means that H_a and H_b are in fact chemically non-equivalent.

Q: Why are the H_a and H_b protons chemically non-equivalent?

A: On closer inspection, the H_b proton is on the same side of the Br atom, i.e. *cis* to the Br atom, whereas the H_a is *trans* to it. Thus, we expect the electron cloud due to the non-bonding electrons of the Br atom to have more impact on the H_b proton than the H_a proton.

With reference to Fig. 17.5, the signal furthest down field (greater δ value) is due to the most deshielded proton in the molecule, i.e. H_c is the most deshielded as it is closest to the electronegative Br atom. The next strongly deshielded proton is H_a, and the least deshielded is H_b.

Q: Why is H_a slightly more deshielded than H_b?

A: Since H_b is closer to the Br atom, the magnetic field generated by the electron cloud of the non-bonding electrons of the Br atom provides greater shielding effect for the H_b proton than the H_a.

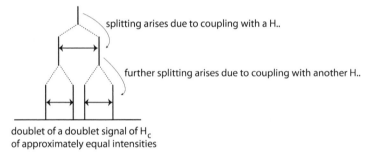

The signal produced by H_c is a multiplet known as a doublet of doublets. The splitting pattern arises from coupling with its two neighboring protons. But since these neighboring H_a and H_b protons are chemically non-equivalent, they influence to a different extent the magnetic field that is needed to bring H_c into resonance — the proton that exerts a greater coupling effect splits the signal into a doublet and each of the peaks in the doublet is further split into another pair of doublets by the other proton (see Fig. 17.6).

splitting arises due to coupling with a H..

further splitting arises due to coupling with another H..

doublet of a doublet signal of H_c
of approximately equal intensities

Fig. 17.6. Stick diagram showing the splitting pattern of a doublet of doublets.

Note that a doublet of a doublet is not the same as a quartet. Although both signals contain four peaks, the ratios of peak intensities is different.

Q: Is it right to say that the H_a proton causes the first splitting, followed by the H_b proton?

A: No! You would not be able to know that. Which ever proton causes the splitting first, the outcome is the same. So, it doesn't matter.

Q: Shouldn't the signal produced by H_a, and likewise for H_b, be a doublet of a doublet too, since it has two neighbors which are also chemically non-equivalent?

A: Yes, the splitting pattern should be that of a doublet of a doublet. But sometimes this pattern is not clearly observed due to the relatively low resolution of the spectrometer. Peaks with chemical shift values too close to each other may overlap with each other and not be distinctly portrayed.

17.5 Analyzing the ^{1}H NMR Spectrum of Simple Organic Molecules

Example 17.1: The following spectrum is obtained for methyl propanoate. Account for the main features in its ^{1}H NMR spectrum.

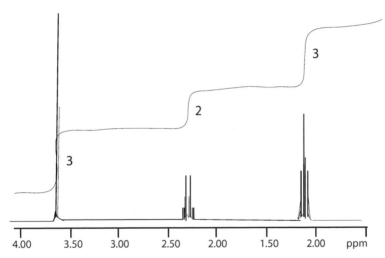

Solution:

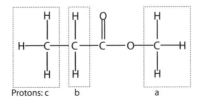

- There are three signals indicating three sets of protons in three different chemical environments.
- Based on step height measurements, the ratio of these three sets of protons is 3 : 2 : 3 in decreasing order of δ values. Since the

molecular formula is $C_4H_8O_2$, the ratio actually represents the absolute number of each type of protons.

- The signal at around δ = 3.6 ppm is attributed to the H_a protons in the $-CH_3$ group, which are the most deshielded as they have their electron density withdrawn by the nearby electronegative O atom, which is only two bonds away. This signal is a singlet since there are no chemically non-equivalent protons nearby which can exert a coupling effect (the closest protons are five covalent bonds away).
- The signal at δ = 2.3 ppm is attributed to the H_b protons in the $-CH_2-$ group. These protons are more deshielded than the H_c protons as they are closer to the electron-withdrawing $-C=O$ group, where the O atom is three bonds away from the H_b protons. The signal is a quartet due to coupling with the three chemically equivalent H_c protons in the adjacent $-CH_3$ group.
- The triplet at around δ = 1.1 ppm is attributed to the highly shielded H_c protons in the $-CH_3$ group that couple with the two H_b protons in the adjacent $-CH_2-$ group.

Example 17.2: Determine the structure of compound **W**, $C_5H_{10}O$, from its ^{1}H NMR spectrum.

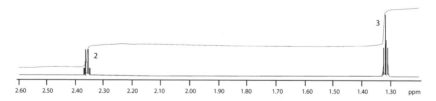

Solution:

- There are two signals indicating two sets of protons in two different chemical environments.
- The quartet at around δ = 2.4 ppm and the triplet at δ = 1.3 ppm collectively suggest the presence of a $-CH_2$ group and an adjacent $-CH_3$ group whose protons couple with one another. (Take note that this is an important feature indicating the presence of an ethyl group!)
- Based on chemical shift correlations, the quartet is attributed to the protons in the $-COCH_2$ group, which are deshielded by the electron-withdrawing $-C=O$ group.

- The triplet is attributed to the protons in the $-CH_3$ group, which are further from the carbonyl group and hence less deshielded.
- Based on step height measurements, the ratio of the quartet to the triplet is 2 : 3, and this is in line with the deductions that a $-CH_2CH_3$ group is present.
- Since the molecular formula indicates a total of 10 protons, the absolute number of protons responsible for the quartet and triplet is 4 and 6 respectively. It can be concluded that there are two $-CH_2CH_3$ groups that share the same chemical environment as a whole, i.e. there is a plane of symmetry in the molecule.
- Hence, compound **W** is identified as $CH_3CH_2COCH_2CH_3$.

Example 17.3: The following spectrum is obtained for compound **X**, $C_9H_{10}O_2$. Use it to deduce its structure.

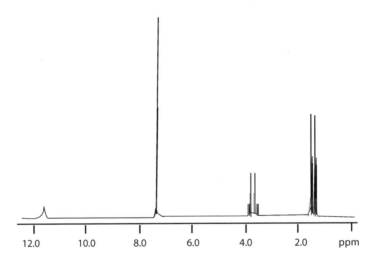

Solution:

- There are four signals indicating four sets of protons in four different chemical environments.
- Based on step height measurements, the ratio of these four sets of protons is 1 : 5 : 1 : 3 in decreasing order of δ values which, based on the molecular formula, are the actual number of protons present in the molecule.
- The singlet at $\delta = 11.7$ ppm indicates the presence of a carboxylic acid functional group.

- The singlet at $\delta = 7.3$ ppm is attributed to the presence of a benzene ring, and its relative intensity suggests that it is mono-substituted.
- The quartet at $\delta = 3.6$ ppm indicates a –CH– structural unit in which the proton has three neighboring protons on an adjacent C atom.
- The doublet at $\delta = 1.5$ ppm indicates a –CH_3 group, and it supports the idea of three protons that couple with an adjacent proton, i.e. the presence of the –$CH(CH_3)$– structural unit. Since the signal is most upfield, these methyl protons are further from the electron-withdrawing –COOH group than their adjacent proton, which concludes one segment of the molecule as having the –$CH(CH_3)COOH$ group.
- Thus, based on the above deductions, the compound is

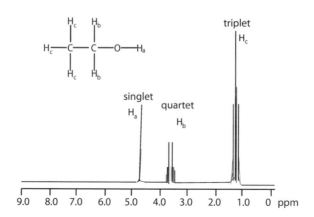

17.6 Labile Protons and Proton Exchange Reactions

Based on Fig. 17.7, there are three signals indicating the three types of protons in ethanol. The quartet at around $\delta = 3.6$ ppm and the

Fig. 17.7. ^{1}H NMR spectrum of ethanol.

triplet at $\delta = 1.1$ ppm collectively point towards the presence of a $-CH_2CH_3$ group. But why isn't there any further splitting of the quartet signal, which arises from the methylene ($-CH_2-$) protons, by the adjacent hydroxyl proton? Why is the signal for the hydroxyl proton a singlet and not a triplet?

The answer is attributed to the fact that the hydroxyl proton is a labile proton. It undergoes a rapid proton exchange reaction with a hydroxyl proton of another ethanol molecule:

$$R_1-O-H_1 + R_2-O-H_2 \rightleftharpoons R_1-O-H_2 + R_2-O-H_1.$$

This reaction is catalyzed by trace amounts of acids or bases, and it occurs so fast that the hydroxyl protons do not reside long enough in a particular molecule to couple with other chemically non-equivalent protons. From another perspective, the protons do not "feel" the presence of the magnetic field from the hydroxyl proton, and vice versa. Other labile protons include those in the $-COOH$, $-NH_2$ and the phenolic $-OH$ functional groups.

Q: How does an acid or base catalyze the proton transfer?

A: The acid can protonate the hydroxyl group, "thinned" out the electron density of the O—H bond that is already present, thus making it more susceptible to breaking, whereas the base would facilitate the transfer by extracting the proton. The phenomenon of the labile proton is very interesting as it demonstrates that at the molecular level, covalent bonds are not as permanent as what most of us have thought!

Proton exchange also occurs with the trace of water molecules (Remember that water is a stronger acid than alcohol?) that are present:

$$R-O-H_1 + H-O-H_2 \rightleftharpoons R-O-H_2 + H-O-H_1.$$

Because the switch is so rapid between the two types of labile protons, the spectrometer can only identifies the average of the two chemical environments and records a singlet at a δ value in between what is expected for the alcohol and the water $-OH$ groups.

Since the signal of a labile proton occurs over a relatively large range of chemical shift values, the way to identify the presence of a labile proton is to find its signal missing or diminishing after the addition of D_2O to the sample prior to insertion into the spectrometer. In the case of ethanol, when D_2O is added, the following proton exchange occurs:

$$R-O-H + D-O-D \rightleftharpoons R-O-D + D-O-H.$$

When this happens, the signal due to the hydroxyl proton disappears as the deuterium nucleus resonates at a much higher external field strength beyond the range seen for the ^{1}H nucleus. If water is present in the sample, there is also proton exchange with D_2O:

$$H-O-H + D-O-D \rightleftharpoons 2\ D-O-H.$$

In all, a new signal is detected at around $\delta = 4.7$ ppm due to the proton in DOH.

CHAPTER 18

Chromatography and Electrophoresis

18.1 Introduction

In a typical organic synthesis process, which can be more than one step, it is very important to trace the quantity or concentration of each of the intermediates in the whole synthetic pathway. In addition, each of the intermediates needs to be isolated and purified, so as to "feed" into the next synthetic process. Not to mention that at the end of the synthesis, the desired product needs also to be isolated and purified from other side products. Hence, the knowledge of effective and efficient separation techniques is important. An effective separation method is one that allows maximum recovery of the substance being isolated whereas efficiency would relate to the amount of time and resources that are needed to effect the separation.

There are numerous types of separation techniques, which employ various physical principles. For our current discussion, we are going to focus on separation techniques which make use of physical properties such as solubility and adsorption. You will later notice that the common aspect of the separation techniques being discussed is the requirement of two different phases. Our desired organic components to be isolated will distribute themselves differentially between these two phases due to different physical interactions with each of the two phases. Obviously, the greater the differences in interaction of the desired component with each of the two phases, the more efficient the separation technique would be.

Q: What is adsorption? Is it the same as absorption?

A: Adsorption refers to the phenomenon where particles "cling" onto the surfaces of a substance through electrostatic forces of interaction. So, you can actually say that the particles are "bonded" to the surfaces of the substance. Adsorption is different from absorption as, in the case of the latter, the particles have actually gone below the surface, i.e. they have been subsumed by the substance under consideration, like a sponge absorbs water. Adsorption, on the other hand, is a surface phenomenon.

Q: Let's say I need to extract a particular desired product from a plethora of other side products. Is there a way to check which two phases are optimal for the isolation of a particular component of interest?

A: Yes. Basically, the differential distribution of a particular substance **X** over two different phases is characterized by a unique equilibrium constant known as the partition coefficient, $P_{1/2}$, which is defined as follows:

$$P_{1/2} = \frac{\text{concentration of substance X in phase 1}}{\text{concentration of substance X in phase 2}}$$

A large $P_{1/2}$ indicates that the interaction of substance **X** with the particles of phase 1 is more favorable than those of substance **X** with the particles of phase 2. Like all other equilibrium constants, partition coefficient is temperature dependent. Thus, what one needs to do is to compare our desired product with something which has already been documented and try using the documented desired phases. But, if one really cannot find something that is similar, then one probably has to resort using the trial-and-error method if one knows the physical property, such as the polarity, of the substance of interest.

18.2 Paper Chromatography — An Introduction to Chromatography Technique

You would probably have come across chromatography as a technique used in separating component dyes in black ink. The basic procedure (see Fig. 18.1) first involves using a pencil to draw a baseline on a rectangular strip of chromatography paper. Sample black ink is spotted onto the baseline and the entire strip is then inserted upright into a beaker containing a suitable solvent, making sure the baseline is above the solvent level. Immediately, the solvent creeps up the paper by capillary action, and following its trail are colored spots moving at different speeds. Once the solvent is near the edge of the paper, the paper is removed and the solvent front is marked out. The developed paper with the spots in their final positions is called the chromatogram.

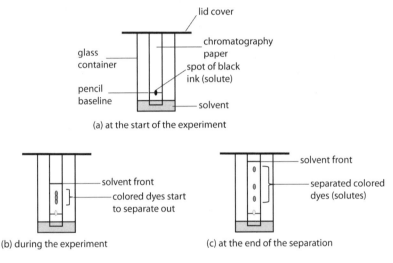

Fig. 18.1. The basic features and procedure for paper chromatography.

Q: What is capillary action?

A: Capillary action refers to the phenomenon where a liquid level rises through a thin narrow tube of small diameter or through porous materials such as cloth or paper. If one looks closely at a piece of cloth or paper, one would notice that it is made up of multiple small little capillaries or "channels." The rising of the liquid level is against gravity, and it is possible because of strong interactive forces between the liquid molecules and the surfaces of the substance. It is an interfacial adsorption phenomenon!

Q: Why does it happen only for capillaries? What happens if the diameter of the tube is too big?

A: If the diameter of the tube is too big, then as the molecules cling onto the wall at the side, the weight of the molecules at the center would pull it down. In a capillary, because it is too narrow, there aren't many molecules at the center of the tube, hence the weight that is pulling down at the molecules at the side is not too great. Therefore, the molecules can "climb" up the wall.

Q: Why is a lid placed over the container?

A: The container is covered with a lid so as to ensure that a constant stable vapor atmosphere encloses the chromatogram. If not, the evaporation of the solvent would create a draft, and this would interfere with the movement of the solvent front and the separation of the different types of solute. In addition, the solvent molecules at the solvent front would also evaporate at a different rate. This would cause a non-uniform solvent front being obtained, hence posing problems in the calculation of the retardation factor.

Q: Why must the pencil baseline be above the solvent level?

A: If the baseline is below the solvent level, the sample would dissolve in the solvent.

Not only do we know that black ink is a mixture of a few colored dyes, we can also identify what those colors are — this goes to show that both separation and identification can go hand in hand.

To identify an unknown sample, we first have to separate it into its individual components. The principle behind the technique lies in the different extent of adsorption of the solutes in each of the two phases needed to carry out the experiment. The solutes are the colored dyes and the solvent in the beaker is the mobile phase in which the dyes can dissolve. That is why these spots are observed to "travel" across the paper. But the blue dye travels the least and the red dye the furthest. Why do these spots move across different distances? This is where we bring in the idea of the second phase.

The second phase is the stationary phase across which the mobile phase moves. The two phases differ in their nature of polarity. So what we have here is a solute that has higher affinity for one phase than the other. If a solute dissolves better in the mobile phase, it will be "carried" further across the chromatography paper. If it has higher affinity for the stationary phase, it will reside longer at a particular spot, opposing the "tidal force" of the mobile phase as the solvent moves forward.

Q: What is the stationary phase in paper chromatography?

A: In paper chromatography, the stationary phase consists of the water molecules that are embedded within the porous structure of the cellulose fibers. Hence, this stationary phase is polar in nature as the cellulose is actually made-up of poly(glucose). This would mean that solute particles that are polar in nature would have greater affinity for this polar water/cellulose stationary phase, hence being adsorbed more strongly onto the stationary phase. Therefore, they would be "retained" further behind the solvent front. In contrast, if the solute is less polar and thus interacted much better with the non-polar mobile phase instead, then we would expect this less polar solute to move more in line with the solvent front. Thus, such differential interaction with the mobile and stationary phases allows different solutes to be separated.

Celluslose - A Poly(glucose)

Q: So, if the stationary phase is polar, does it mean that we have to use an absolutely non-polar mobile phase?

A: Not really. If the stationary phase is polar, we can still employ a polar mobile phase. Importantly, there must be a substantial difference between the polarities of the two different phases to allow a significant differential distribution of the solute between these two phases. For example, we can use the polar ethanol as the mobile phase, since ethanol is relatively less polar than the water/cellulose stationary phase.

Despite all the above, there is still a chance that solutes may not be well separated from each other. One solution to this problem is to perform a two-dimensional (2D) paper chromatography, which is just an extension of the normal procedure. The sample is spotted onto the corner of a square piece of chromatography paper instead of a rectangular one. Once a chromatogram is obtained, it is dried and rotated 90°. In this new orientation, the chromatogram is placed in a beaker containing a solvent of different polarity from the first. The same procedure is performed again, with the solutes once again distributed between two phases and the experiment is stopped when the solvent front is near the edge of the chromatography paper (see Fig. 18.2). This technique is useful for the separation of a mixture of components of similar physical properties, such as amino acids.

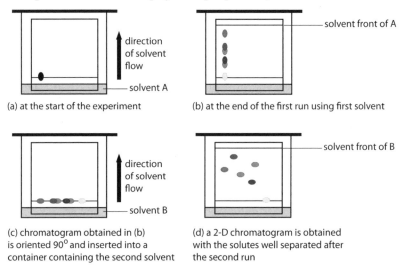

Fig. 18.2. A two-dimensional chromatographic separation.

Q: Dyes are colored substances. But how can paper chromatography be used on organic substances such as amino acids, which are colorless?

A: One way to show that the various spots contain different amino acids is to place the chromatogram in a beaker containing some iodine crystals. The iodine would vaporize and the molecules would adsorb onto the less polar spots, making the spots visible. Alternatively, we can spray ninhydrin onto the chromatogram. The ninhydrin would form a violet-colored product with the amino acids (refer to Chapter 12 on Amino Acids).

Once separation is done, it is time to identify each of the components in the mixture by calculating the **retardation factor** (R_f) represented in the following formula:

$$R_f = \frac{\text{distant travelled by solute}}{\text{distant travelled by solvent}}$$

The retardation factor is a ratio of the distance travelled by the solute to the distance travelled by the solvent (see Fig. 18.3).

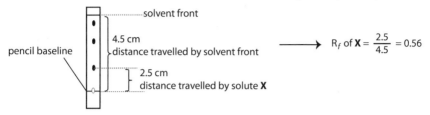

Fig. 18.3. Calculation of retardation factor, R_f.

This is why it is important to mark the solvent front immediately at the end of the experiment, before it evaporates from the paper. Each substance has a different R_f value. The unknown substance can be identify by matching its R_f value obtained against a database of R_f values of some known compounds.

Since the R_f values are obtained experimentally, there are a few factors that we need to look into that would affect the data obtained. These factors include the type of paper used, the solvent and its concentration, the amount of solutes being spotted and the temperature:

- Different types of paper contain different amounts of water adsorbed onto the cellulose polymer, which itself varies across paper types. Thus, the overall "polarity" of the stationary phase would be different, resulting in differential strengths of interaction between the solutes and the stationary phase.
- Different solvents have intermolecular forces of different strengths, which result from differences in the polarities of the solvent molecules. Thus, the strengths of the interactive forces between the solute and different types of solvent would differ.
- If the solvent that is being used consists of a mixture of components, such as ethanol mixed with water, then the overall polarity of this mixture is different from the pure components. In addition, the composition of each of the components present would also affect the overall polarity of the mixture.
- If a large amount of solutes is being spotted, more time would be required to separate each of the individual components from one another. As a result, each of the components would be stretched over a greater surface area. Hence, determination of the centric

point for the calculation of the retardation factor would pose some difficulty. In addition, some of the spots may be too smudged such that they overlap with each other.

- Temperature affects the solubility of a particular solute in a solvent and its adsorptive strength with the stationary phase. At a higher temperature, solubility is likely to increase while adsorption on the stationary phase would decrease.

As it is difficult to ensure constancy for all the above variables, sometimes several known reference components are applied together with the mixture of unknown compounds. Comparison of the retardation factor of the unknown component against the known reference components would indicate to us the identity of the unknown component.

To improve on the limitation of paper chromatography and increase the reproducibility and efficiency of the chromatographic separation techniques, thin layer chromatography was developed. We shall be discussing the features and advantages of this separation technique in the next section.

18.3   Thin Layer Chromatography (TLC)

As the name thin layer chromatography implies, there is a thin layer of the stationary phase, which usually is a solid such as alumina or silica being mounted onto an inert base such as a piece of glass or plastic. Similar to paper chromatography, this thin layer of alumina or silica is the polar stationary phase which allows the less polar mobile phase to travel over it. Separation of different solutes is effected due to the differential in the adsorptive strength with the polar stationary phase and the differential in solubility in the mobile phase. In addition, if the stationary phase consists of silica, which is acidic in nature, then solute components that are basic in nature would have greater affinity for the stationary phase. But if alumina, which is slightly basic in nature, is used, then an acidic solute would be more strongly attracted to the alumina. This would cause the solute to lag much further behind the solvent front.

The detection of solute components that are colorless is quite similar to the method used for paper chromatography. But in addition, if the solute components possess chromophores (refer to Chapter 15), such as a highly conjugated pi system of electron clouds, that are capable of absorbing UV radiation and reemitting it as visible white light, then each solute component maybe identified under a UV lamp enclosed in a dark box.

Q: Shouldn't a particle reemit the frequency of radiation that it has absorbed? That is, if the particle absorbs UV radiation, then shouldn't it reemit UV radiation? How can the particle emit visible light after absorbing UV radiation?

A: Recall that in Chapter 15, we mentioned that there are vibrational and rotational energy levels embedded within a particular electronic energy level. Let's say that after the particle has absorbed an UV photon and got promoted into a higher energy level, instead of releasing this exact amount of radiation and falling back to the original energy level. What happens is that the particle can lose some energy within this excited electronic level via the rotational or vibrational energy levels, to the extent that the final transition from a higher energy state to a lower one may involve the release of a photon in the visible light region. This way of decreasing in energy within an excited electronic level is known as electronic relaxation. The light that is emitted by a firefly is a result of one such relaxation mechanism known as chemiluminescence.

In addition to the use of TLC to check for purity of samples and to monitor the progress of a reaction, TLC is also very useful for the isolation of a desired component. What is usually done after all the components have been well separated, is to scrap away the solid phase with our desired component adsorbed on it. The desired component is then desorbed from the solid phase through the use of a solvent which the desired component has greater affinity for than for the solid phase. Another way would be to add a reagent which would react with the desired component, resulting in an intermediate that has a lower affinity for the solid phase, which is then desorbed away from the solid phase.

Generally, the applications of TLC center on the purpose of qualitative analysis. To quantitatively separate the components in a sample more effectively, instead of using a short plate, a long column embedded with the stationary phase is usually used, and the name for this technique is called column chromatography.

18.4 Column Chromatography (CC)

The principle behind CC is the same as that for TLC — separation of components is based on their differential in affinity for the mobile and stationary phases. But instead of mounting a layer of silica or alumina onto an inert base, the stationary phase is packed into a column. Hence, if more time is required to separate the two components, one just needs to use a longer column, which is much more versatile than TLC. The components are separated by "washing" them down the column through a reservoir of solvent, which trickles slowly into the column. This process is known as elution, and the solvent is known as the eluent, which travels through the column with the help of gravity. Each individual component, or elutant, is then collected as the solvent drips from the bottom of the column.

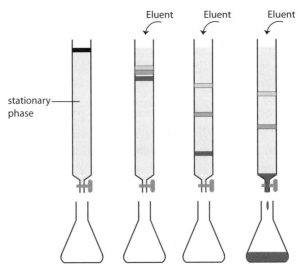

Q: It is easy to collect individual pure components when they are colored. How do we collect components that are colorless?

A: Well, hopefully your desired compound is UV sensitive so that you can use UV light to keep track of its progress of elution.

If it is positive, you would observe the different components as moving bands!

Q: If a solute adsorbs strongly to the stationary phase and moves down the column at a very slow rate, how long do we have to wait for it to be eluted?

A: The objective in performing the column chromatography is to separate and collect the individual components in a sample, right? Once the individual components have been separated well enough, we can just use another eluent that will flush out the solute at a faster rate. The eluent selected should be one that the solute has a higher affinity for. Alternatively, one can also apply pressure onto the eluent to "push" it through the column at a faster rate. This is the working principle behind high-performance liquid chromatography.

18.5 High-Performance Liquid Chromatography (HPLC)

The stationary phase that is normally used in column chromatography is usually composed of very fine particles of high surface area, which provides large surface area for adsorption. As a result of the uniform packing of these fine particles, a high resistance for the eluent to pass through the column is being created. To circumvent the problem, a high-pressure pump, which operates up to about 400 atm, is used to "force" the eluent through the column. This separation technique is known as high-performance liquid chromatography (HPLC).

Unlike column chromatography, where the stationary phase is polar in nature, in HPLC, the stationary phase can be non-polar in nature. The non-polar stationary phase consists of silica coated with a layer of non-polar hydrocarbon chains. The eluent used is polar in nature. This method is known as reversed phase HPLC, which is in contrast to the normal phase HPLC, utilizing a polar stationary and non-polar mobile phases.

In addition, if the solid stationary phase is coated with a single enantiomer of a chiral compound, then this column can be used to separate a mixture of enantiomers, where each of this pair of enantiomers would have different affinity for the chiral column. This process is known as chiral column chromatography, and it is very useful for the purification of enantiomers.

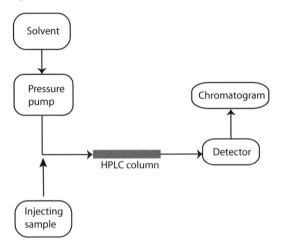

One additional obvious advantage of HPLC is the versatility of the type of detector used. For the high-end HPLC instrument, a mass spectrometer can be connected to monitor the appearance of different components being eluted out from the column.

18.6   Gas–Liquid Chromatography (GLC)

Gas–liquid chromatography, commonly known as gas chromatography, implies that the mobile phase is a gaseous component. The carrier gas used is usually helium, argon or nitrogen, with nitrogen as the most commonly used due to its lower cost. The stationary phase consists of a high boiling point liquid, which is polymeric in nature, adsorbed onto the surfaces of a porous solid. The rate at which each of the components emerges through the column would depend on the time the component spent moving with the mobile phase as opposed to being dissolved in the absorbed liquid. The higher the rate the component passes through the column, the lower the retention time within the column. To decrease the retention time, one can program the oven in such a way that it heats the column at different temperatures for different time intervals. The more volatile components would be the first to emerge, leaving the high boiling components lagging behind.

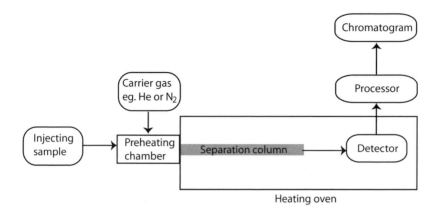

Heating oven

18.7 Electrophoresis

If a sample of components consists of particles of a varying degree of electrical charges, then it is possible to separate these charged particles via the application of an electric field. This separation technique is known as electrophoresis. This method is very useful for separating amino acids, peptides, proteins and DNA. Thus, other than being dependent on the amount of charge that a particle possesses, the rate at which the particle moves through the applied electric field is also dependent on the mass of the particle. A differential amount of charge on each particle can be created by the adjustment of the pH of the sample components using buffer solutions of different pH values. Overall, the rate of diffusion through the electric field follows the following equation:

$$\text{rate of diffusion} \propto \frac{\text{charge}}{\text{mass}}$$

There are numerous versions of electrophoresis, with gel electrophoresis being the most commonly discussed. The term gel here refers to a porous polymeric substance used as the matrix. Gels with different pore sizes are used for the separation of different particulate sizes, with larger particles requiring pores of larger sizes. The sample components are contained in small little "wells" before their separation. When an electric field is applied across the gel, the molecules move at a differential rate based on their charge over mass ratio. The negatively charged particles move towards the positive terminal (anode) and vice versa. After the various components have been separated, different components appear as distinct bands on the gel. Each of the components may subsequently be isolated from the gel matrix for other purposes.

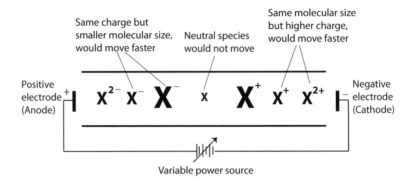

Index